Diversity of Marine Fungi as a Source of Bioactive Natural Products

Diversity of Marine Fungi as a Source of Bioactive Natural Products

Editor

Dehai Li

MDPI

Basel • Beijing • Wuhan • Barcelona • Belgrade • Novi Sad • Cluj • Manchester

Editor
Dehai Li
Ocean University of China
Qingdao
China

Editorial Office
MDPI
St. Alban-Anlage 66
4052 Basel, Switzerland

This is a reprint of articles from the Special Issue published online in the open access journal *Marine Drugs* (ISSN 1660-3397) (available at: https://www.mdpi.com/journal/marinedrugs/special_issues/6B6Z19CQ23).

For citation purposes, cite each article independently as indicated on the article page online and as indicated below:

Lastname, A.A.; Lastname, B.B. Article Title. *Journal Name* **Year**, *Volume Number*, Page Range.

ISBN 978-3-0365-9166-7 (Hbk)
ISBN 978-3-0365-9167-4 (PDF)
doi.org/10.3390/books978-3-0365-9167-4

Contents

About the Editor vii

Preface ix

Tao Chen, Wencong Yang, Taobo Li, Yihao Yin, Yufeng Liu, Bo Wang and Zhigang She
Hemiacetalmeroterpenoids A–C and Astellolide Q with Antimicrobial Activity from the Marine-Derived Fungus *Penicillium* sp. N-5
Reprinted from: *Mar. Drugs* **2022**, *20*, 514, doi:10.3390/md20080514 1

Chen Chen, Geting Ye, Jing Tang, Jialin Li, Wenbin Liu, Li Wu and Yuhua Long
New Polyketides from Mangrove Endophytic Fungus *Penicillium* sp. BJR-P2 and Their Anti-Inflammatory Activity
Reprinted from: *Mar. Drugs* **2022**, *20*, 583, doi:10.3390/md20090583 11

Xuewen Hou, Changlong Li, Runfang Zhang, Yinping Li, Huadong Li, Yundong Zhang, et al.
Unusual Tetrahydropyridoindole-Containing Tetrapeptides with Human Nicotinic Acetylcholine Receptors Targeting Activity Discovered from Antarctica-Derived Psychrophilic *Pseudogymnoascus* sp. HDN17-933
Reprinted from: *Mar. Drugs* **2022**, *20*, 593, doi:10.3390/md20100593 25

Yaodong Ning, Yao Xu, Binghua Jiao and Xiaoling Lu
Application of Gene Knockout and Heterologous Expression Strategy in Fungal Secondary Metabolites Biosynthesis
Reprinted from: *Mar. Drugs* **2022**, *20*, 705, doi:10.3390/md20110705 41

Maan T. Khayat, Khadijah A. Mohammad, Abdelsattar M. Omar, Gamal A. Mohamed and Sabrin R. M. Ibrahim
Fungal Bergamotane Sesquiterpenoids—Potential Metabolites: Sources, Bioactivities, and Biosynthesis
Reprinted from: *Mar. Drugs* **2022**, *20*, 771, doi:10.3390/md20120771 55

Reham Khaled Abuhijjleh, Dalia Yousef Al Saeedy, Naglaa S. Ashmawy, Ahmed E. Gouda, Sameh S. Elhady and Ahmed Mohamed Al-Abd
Chemomodulatory Effect of the Marine-Derived Metabolite "Terrein" on the Anticancer Properties of Gemcitabine in Colorectal Cancer Cells
Reprinted from: *Mar. Drugs* **2023**, *21*, 271, doi:10.3390/ md21050271 87

Honghua Li, Yanqi Fu and Fuhang Song
Marine *Aspergillus*: A Treasure Trove of Antimicrobial Compounds
Reprinted from: *Mar. Drugs* **2023**, *21*, 277, doi:10.3390/md21050277 115

Hao-Bing Yu, Zhe Ning, Bo Hu, Yu-Ping Zhu, Xiao-Ling Lu, Ying He, et al.
Cytosporin Derivatives from Arctic-Derived Fungus *Eutypella* sp. D-1 via the OSMAC Approach
Reprinted from: *Mar. Drugs* **2023**, *21*, 382, doi:10.3390/md21070382 133

Angela A. Salim, Waleed M. Hussein, Pradeep Dewapriya, Huy N. Hoang, Yahao Zhou, Kaumadi Samarasekera, et al.
Talarolides Revisited: Cyclic Heptapeptides from an Australian Marine Tunicate-Associated Fungus, *Talaromyces* sp. CMB-TU011
Reprinted from: *Mar. Drugs* **2023**, *21*, 487, doi:10.3390/md21090487 147

About the Editor

Dehai Li

Dehai Li is a professor at the School of Medicine and Pharmacy, Ocean University of China. He received his Bachelor's degree in Pharmacy from Shandong University in 2002 and obtained his Doctor of Medicine in 2007 from Ocean University of China. After that, he completed his postdoctoral research with Professor Raymond J Andersen at the University of British Columbia from 2007 to 2009. From 2014 to 2015, he moved to Professor Yi Tang's lab at the University of California, Los Angeles, as a visiting scientist. Currently, his research is focused on marine natural products, including: (1) isolation and identification of fungi and actinomycete strains from various marine samples; (2) searching for novel bioactive secondary metabolites from marine-derived microorganisms; (3) chemical synthesis and biosynthesis of natural products. In the past 15 years, his group has isolated over 5000 microorganism strains from marine environments and discovered more than 1000 new structures. He has published over 160 papers and has had 10 patents authorized.

Preface

Marine fungi are an ecologically diverse group of microbes which play critical roles in the matter cycle and energy flow of marine ecosystems and are important sources of bioactive natural products. The most investigated fungi by natural product chemists are laboratory-cultivated ascomycetes, such as Aspergillus, Penicillium and Fusarium. In the past decades, great efforts have been focused on the isolation and characterization of novel bioactive natural products, especially those with anti-infectious and anticancer properties. However, because of the limitations of traditional natural product mining methods, the discovery of new secondary metabolites has become increasingly difficult. In this aspect, researchers continue to make progress towards uncovering marine fungal diversity by exploring new habitats and leveraging modern techniques. The metabolite producing potential of a group of uncommon genera, including Basidiomycota, Blastocladiomycota, Chytridiomycota and Mucoromycota, has been uncovered.

In the meantime, a package of tools, including OSMAC, epigenetic modulation, genome mining and even combinatorial biosynthesis tools, has been developed and applied to explore the metabolic capacity of target strains, which has greatly improved the efficiency of discovering natural products. Moreover, collaborations between marine fungal ecologists, evolutionary biologists and natural product chemists have proved effective strategies for all endeavors involved in marine fungi studies. The development of techniques such as genomic and metabolomic approaches will help acquire more insightful knowledge on the physiological and biological properties of marine-derived fungi and those concepts are essential to target broader scientific topics related to environmental maintenance and human health.

Dehai Li

Editor

marine drugs

MDPI

Article

Hemiacetalmeroterpenoids A–C and Astellolide Q with Antimicrobial Activity from the Marine-Derived Fungus *Penicillium* sp. N-5

Tao Chen, Wencong Yang, Taobo Li, Yihao Yin, Yufeng Liu, Bo Wang * and Zhigang She *

School of Chemistry, Sun Yat-sen University, Guangzhou 510275, China
* Correspondence: ceswb@mail.sysu.edu.cn (B.W.); cesshzhg@mail.sysu.edu.cn (Z.S.)

Abstract: Four new compounds including three andrastin-type meroterpenoids hemiacetalmeroterpenoids A-C (**1–3**), and a drimane sesquiterpenoid astellolide Q (**15**), together with eleven known compounds (**4–14**) were isolated from the cultures of the marine-derived fungus *Penicillium* sp. N-5, while compound **14** was first isolated from a natural source. The structures of the new compounds were determined by analysis of detailed spectroscopic data, and the absolute configurations were further decided by a comparison of the experimental and calculated ECD spectra. Hemiacetalmeroterpenoid A (**1**) possesses a unique and highly congested 6,6,6,6,5,5-hexa-cyclic skeleton. Moreover, the absolute configuration of compound **14** was also reported for the first time. Compounds **1**, **5** and **10** exhibited significant antimicrobial activities against *Penicillium italicum* and *Colletrichum gloeosporioides* with MIC values ranging from 1.56 to 6.25 μg/mL.

Keywords: andrastin-type meroterpenoids; drimane sesquiterpenoid; marine-derived fungus; antimicrobials activities

Citation: Chen, T.; Yang, W.; Li, T.; Yin, Y.; Liu, Y.; Wang, B.; She, Z. Hemiacetalmeroterpenoids A–C and Astellolide Q with Antimicrobial Activity from the Marine-Derived Fungus *Penicillium* sp. N-5. *Mar. Drugs* **2022**, *20*, 514. https://doi.org/10.3390/md20080514

Academic Editor: Dehai Li

Received: 1 August 2022
Accepted: 12 August 2022
Published: 13 August 2022

1. Introduction

Andrastins are meroterpenoids characterized by a 6,6,6,5-tetra-carbocyclic skeleton. They are biogenetically derived from 3,5-dimethylorsellinic acid (DMOA) and farnesyl diphosphate (FPP), synthesized via a mixed polyketide-terpenoid pathway, and usually possess a keto-enol tautomerism at the cyclopentane ring [1–4]. To date, over 40 andrastins have been reported with multiple potential biological activities, including cytotoxic [5], anti-inflammatory [6], antiproliferative [7] and antimicrobial activity [4]. The complex structures and potential biological activities of andrastins have attracted much attention in recent years [8–10].

Marine fungus is known to be a natural source of structurally diverse and biologically active metabolites for drug discovery [11–16]. Recently, a series of novel bioactive natural products from marine fungi were reported by our group [17–22]. In our ongoing search for new bioactive secondary metabolites from marine fungi, the fungus *Penicillium* sp. N-5, isolated from the rhizosphere soil of mangrove plant *Avicennia marina*, led to the isolation of four new compounds, hemiacetalmeroterpenoids A–C (**1–3**) and astellolide Q (**15**). Especially, hemiacetalmeroterpenoid A (**1**) was a new andrastin-type meroterpenoid containing a unique 6,6,6,6,5,5-hexa-cyclic skeleton. Meanwhile, eleven known compounds, including 3-deacetyl-citreohybridonol (**4**) [23] citreohybridone A (**5**) [24], 3,5-dimethylorsellinic acid-based meroterpenoid 2 (**6**) [5], andrastins A–C (**7, 10, 13**) [25], andrastone C (**8**) [26], penimeroterpenoid A (**9**) [4], 23-deoxocitreohybridonol (**11**) [1], 6α-hydroxyandrastin B (**12**) [1], and compound V (**14**) [27] were also obtained from the fungus N-5 (Figure 1). All the isolated compounds were investigated for their antimicrobial activity against two phytopathogenic fungi and four bacterial strains. Herein, we report the isolation, structural characterization and antibacterial activity of these compounds.

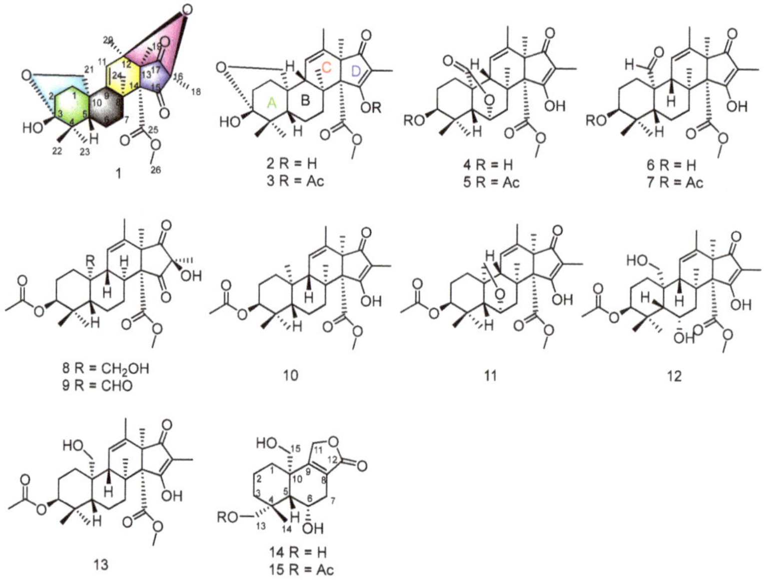

Figure 1. Structure of compounds **1–15**.

2. Results

2.1. Structure Identification

Hemiacetalmeroterpenoid A (**1**) was obtained as a white powder. It molecular formula was assigned as $C_{26}H_{34}O_7$ according to HRESIMS analysis at *m*/*z* 459.23709 $[M+H]^+$ (calcd. 459.23773), indicating ten degrees of unsaturation. In the ^{1}H NMR spectrum, the signal for one olefinic proton (δ_H 5.63), one methoxyl (δ_H 3.60), one methine (δ_H 1.47), four methylenes (δ_H 1.42, 1.62, 1.85, 1.91, 1.94, 2.11, 2.33 and 2.61) and six methyls (δ_H 1.03, 1.07, 1.19, 1.23, 1.33 and 1.49). The ^{13}C NMR data exhibited 26 carbon resonances, including two olefinic carbons for one double bond (δ_C 127.2, 150.1), three carbonyl carbons for two ketone (δ_C 203.7 and 203.8), and one ester carbonyl (δ_C 169.3), one methine (δ_C 49.1), five methylenes (one oxygenated), seven methyls (one oxygenated), eight quaternary carbons (one highly oxygenated: δ_C 99.8) (Table 1). The NMR data established a nucleus of meroterpenoid characterized by an andrastin scaffold, structurally similar to the citreohybriddione C (Figures S2 and S3) [28]. Analysis of the ^{1}H-^{1}H COSY data led to the identification of two isolated spin-systems of C-1/C-2 and C-5/C-6/C-7. The HMBC from H_2-1 to C-3, C-10, from H_1-5 to C-10, and from H_3-22 to C-3, C-4, C-5, C-23, ring A was formed. The HMBC correlations from H_3-24 to C-7, C-8, C-9, from H_1-11 to C-8, C-10 and from H_2-21 to C-5, C-10 completed ring B. Then, the HMBC cross-peaks from H_2-1, H_2-6 to C-10 and from H_3-23, H_2-7 to C-5 indicated ring A and ring B were fused at C-5 and C-10. Then, HMBC correlations from H_3-24 to C-14, C-15, from H_3-20 to C-11, C-12, C-13, C-17, from H_3-19 to C-12, C-13, C-14, C-17 and from H_3-18 to C-15, C-16, C-17, ring C and ring D were constituted, and they were blended at C-13 and C-14. The HMBC correlations from H_2-6, H_1-11 to C-8, from H_1-11 to C-10, suggested ring B and ring C were tightly connected. In addition, the HMBC correlation from H_3-26 to C-25 implied the presence of a methyl carboxylate. A weak HMBC correlation from H_3-26 to C-14 located the methyl carboxylate at C-14. Except one double bond, three carbonyls and four rings, ten degrees of unsaturation indicated that two new rings were required. According to the HMBC correlation from H_2-21 to C-3 (δ_C 99.8), a 6-membered ring was confirmed between C-1, C-2, C-3, C-10, and C-21. Finally, another new 5-membered ring was formed by intramolecular dehydration of hydroxyl groups at C-12 and C-16. Thus, the planar structure of **1** was established as shown in Figure 2.

Table 1. ^{1}H NMR (600 MHz) and ^{13}C NMR (150 MHz) of **1-3** in CD_3OD.

Position	1		2		3	
	δ_C	δ_H (*J* in Hz)	δ_C	δ_H (*J* in Hz)	δ_C	δ_H (*J* in Hz)
1	34.7 (CH_2)	1.42, m 2.33, m	35.3 (CH_2)	1.12, m 2.19, m	34.8 (CH_2)	1.15, m 2.20, m
2	30.2 (CH_2)	1.85, m 2.16, m	30.3 (CH_2)	1.74, m 2.12, m	30.1 (CH_2)	1.76, m 2.14, m
3	99.8 (C)		99.5 (C)		99.5 (C)	
4	41.4 (C)		41.3 (C)		41.3 (C)	
5	49.1 (CH)	1.47, m	50.9 (CH)	1.33, m	51.4 (CH)	1.21, m
6	19.4 (CH_2)	1.62, m 1.91, m	20.5 (CH_2)	1.55, m 1.75, m	20.4 (CH_2)	1.62, m 1.81, m
7	32.3 (CH_2)	1.94, m 2.61, m	32.5 (CH_2)	2.07, m 2.76, m	32.9 (CH_2)	1.94, m 2.23, m
8	40.2 (C)		42.4 (C)		41.9 (C)	
9	150.1 (C)		48.5 (CH)	1.89, t (2.7)	48.6 (CH)	1.98, t (2.7)
10	38.5 (C)		36.3 (C)		36.3 (C)	
11	127.2 (CH)	5.63, s	124.2 (CH)	5.42, m	126.0 (CH)	5.60, m
12	77.4 (C)		137.7 (C)		133.7 (C)	
13	54.4 (C)		57.5 (C)		60.6 (C)	
14	73.4 (C)		69.7 (C)		69.2 (C)	
15	203.7 (C)		190.7 (C)		171.9 (C)	
16	76.7 (C)		113.5 (C)		131.9 (C)	
17	203.8 (C)		201.4 (C)		202.1 (C)	
18	7.9 (CH_3)	1.19, s	6.6 (CH_3)	1.57, s	8.8 (CH_3)	1.55, s
19	11.0 (CH_3)	1.33, s	18.0 (CH_3)	1.18, s	17.4 (CH_3)	1.20, s
20	24.2 (CH_3)	1.23, s	20.2 (CH_3)	1.82, s	19.1 (CH_3)	1.75, s
21	74.4 (CH_2)	3.55, d (7.6) 4.39, d (8.7)	68.6 (CH_2)	3.81, d (9.0) 4.22, d (9.0)	68.5 (CH_2)	3.82, d (8.9) 4.21, d (9.0)
22	27.2 (CH_3)	1.07, s	27.9 (CH_3)	1.04, s	27.9 (CH_3)	1.07, s
23	18.9 (CH_3)	1.04, s	18.9 (CH_3)	1.01, s	18.8 (CH_3)	1.03, s
24	25.9 (CH_3)	1.49, s	16.7 (CH_3)	1.19, s	16.5 (CH_3)	1.24, s
25	169.3 (C)		172.6 (C)		170.9 (C)	
26	52.5 (CH_3)	3.60, s	51.9 (CH_3)	3.56,s	52.4 (CH_3)	3.59,s
Ac-CH_3					21.2 (CH_3)	2.36, s
Ac-OCO					167.3 (C)	

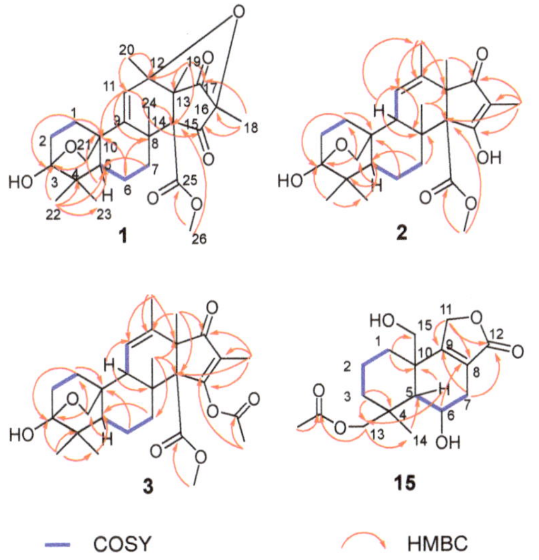

Figure 2. Key HMBC and COSY correlations of **1–3** and **15**.

The relative configuration of compound **1** was defined by the NOESY correlations. The correlations of H_2-21 with H_3-23, H_3-20 with H_3-19, H_3-24, and H_3-18 with H_3-19, H_3-26 were observed in the NOESY spectrum, which means H_3-18, H_3-19, H_3-20, H_2-21, H_3-23, H_3-24 and H_3-26 were on the same side. The NOESY correlations of H_1-5 with H_3-22 suggested that H_1-5 and H_3-22 were in the opposite face (Figure 3). The absolute configuration of **1** was determined by comparing the calculated ECD spectra generated by the time-dependent density functional theory (TDDFT) for two enantiomers 3*R*, 5*S*, 8*S*, 10*S*, 12*R*, 13*S*, 14*R*, 16*R*-**1a** and 3*S*, 5*R*, 8*R*, 10*R*, 12*S*, 13*R*, 14*S*, 16*S*-**1b** with the experimental one. Finally, the experimental ECD spectrum of **1** was nearly identical to the calculated ECD spectrum for **1a** (Figure 4), clearly suggesting the 3*R*, 5*S*, 8*S*, 10*S*, 12*R*, 13*S*, 14*R*, 16*R* absolute configuration for **1**.

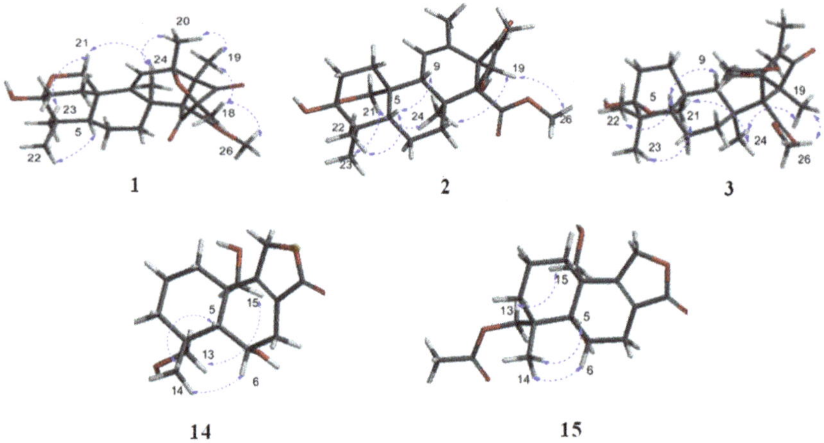

Figure 3. Key NOE correlations of **1–3** and **14–15**.

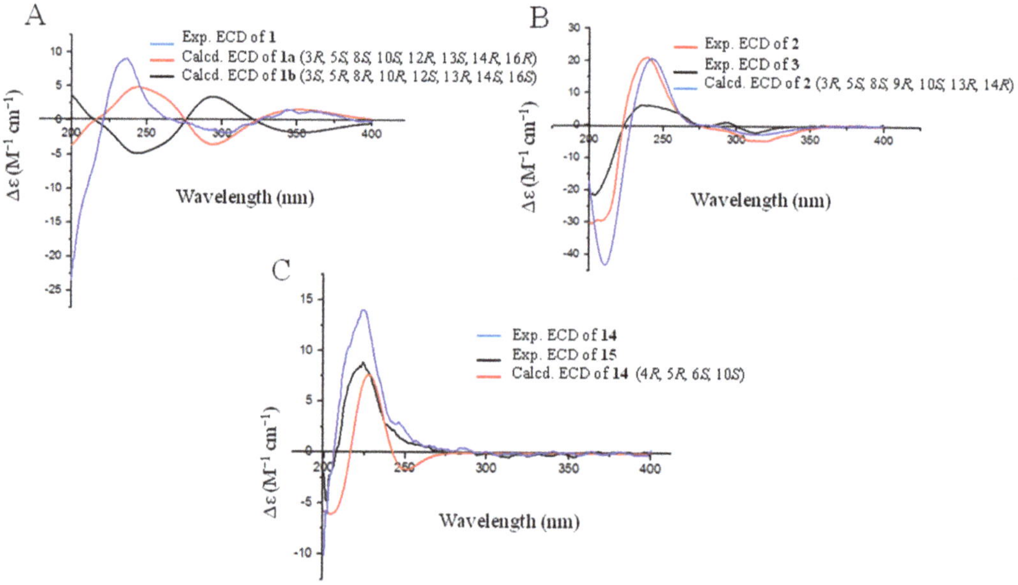

Figure 4. ECD spectra of compounds **1** (**A**), **2** and **3** (**B**), **14** and **15** (**C**) in CH_3OH.

Hemiacetalmeroterpenoid B (**2**) was isolated as a white powder and had a molecular formula of $C_{26}H_{36}O_6$, determined by HRESIMS data *m/z* 445.25772 $[M+H]^+$ (calcd. 445.25847) with nine degrees of unsaturation. The ^1H NMR spectrum of **2** displayed the signal for one olefinic proton (δ_H 5.42), one methoxyl (δ_H 3.56), two methines (δ_H 1.33 and 1.89), four methylenes (δ_H 1.12, 1.33, 1.55, 1.75, 2.07, 2.12, 2.19 and 2.76) and six methyls (δ_H 1.01, 1.04, 1.18, 1.19, 1.57 and 1.82). The ^{13}C NMR data revealed 26 carbon resonances, involving four olefinic carbons for two double bonds (δ_C 113.5, 124.2, 137.7, 190.7), two carbonyl carbons for one ketone (δ_C 201.4), one ester carbonyl (δ_C 172.6) (Table 1). According to 1D NMR and 2D NMR data, the planar structure of **2** was similar to the co-isolated andrastin B (**13**). The obvious difference is that the acetyl group at the C-3 position of compound **2** disappears. Meanwhile, the HMBC from H_2-21 to C-3 (δ_C 99.5) also indicated that a new 6-membered ring was formed between C-1, C-2, C-3, C-10 and C-21 (Figure 2).

The NOESY spectrum indicated that H_1-5, H_1-9 and H_3-22 were on the same side based on the correlations of H_1-5 with H_1-9 and H_3-22. On the contrary, it was suggested that H_3-19, H_3-21, H_3-23, H_3-24, and H_3-26 were on the other side based on the NOESY correlations of H_2-21 with H_3-23 and H_3-24, along with H_3-19 with H_3-24 and H_3-26 (Figure 3). Thus, the relative configuration of **2** was determined to be 3*R*, 5*S*, 8*S*, 9*R*, 10*S*, 13*R* and 14*R*. The absolute configuration of the stereogenic centers in **2** was assigned as 3*R*, 5*S*, 8*S*, 9*R*, 10*S*, 13*R* and 14*R* by comparing its experimental ECD spectrum with that of the calculated model molecule (Figure 4).

Hemiacetalmeroterpenoid C (**3**) was also purified as a white powder. The molecular formula was specified as $C_{28}H_{38}O_7$ (ten degrees of unsaturation) by HRESIMS (*m/z* 509.25015 $[M+Na]^+$), which is 42 mass units higher than that of **2** (Figure S17). Analysis of its NMR data (Table 1) revealed the presence of the same partial structure as that found in compound **2**. The only difference was **3** has an additional acetyl fragment. Finally, a weak HMBC correlation from Ac-CH_3 to C-15 suggested that the acetyl fragment was attached to C-15 (Figure 2).

Because compound **3** has the same chiral center as **2**, the NOESY correlation and experimental ECD spectrum of compound **3** were in agreement with those of **2** (Figures 3 and 4). Thus, the absolute configuration of **3** was identified as 3*R*, 5*S*, 8*S*, 9*R*, 10*S*, 13*R* and 14*R*.

Compound **14** was obtained as a yellow powder. Analysis of its ^1H NMR and ^{13}C NMR data showed that the planar structure of **14** was the same as compound V, which was the product of the alkaline hydrolysis of parasiticolide A [27]. However, the absolute configuration of compound V was ambiguous.

The relative configuration of **14** was also defined by the NOESY correlation. The correlations of H_3-14 with H_1-5 and H_1-6, and H_2-13 with H_2-15 were found in the NOESY spectrum, which means H_1-5, H_1-6, and H_3-4 were on the same side, and H_2-13 and H_2-15 were on the opposite face (Figure 3). Thus, the absolute configuration of the stereogenic centers in **14** was assigned as 4*R*, 5*R*, 6*S*, 10*S* by comparing its experimental ECD spectrum with that of the calculated model molecule (Figure 4). Finally, compound **14** was named as astellolide J.

Astellolide Q (**15**) was also acquired as a yellow powder. It molecular formula was determined as $C_{17}H_{24}O_6$ according to HRESIMS analysis at *m/z* 347.14578 $[M+Na]^+$ (calcd. 347.14651), indicating six degrees of unsaturation. The ^1H NMR of **15** showed two methyls (δ_H 1.15 and 2.08), four methylenes (δ_H 1.16, 1.45, 1.54, 1.78, 1.91, 2.05, 2.34 and 2.50), one methines (δ_H 1.74) one hydroxymethine (δ_H 4.55) and three hydroxy-methylenes (δ_H 3.34, 3.92, 4.09, 4.33, 4.84 and 5.04). In addition, according to the HSQC data, the ^{13}C NMR data showed the presence of 17 carbon signals, including two ester carbonyl carbons (δ_C 173.1, 177.0) and two olefinic carbons (δ_C 124.0, 169.0), one methyl, seven methylenes (three oxygenated), two methines (one oxygenated), two aliphatic quaternary carbons (Table 2). Analysis of its ^1H NMR and ^{13}C NMR data in association with the 2D NMR data established a nucleus of drimane sesquiterpenoid characterized by an astellolide scaffold, structurally similar to the co-isolated compound **14** (Figures S26 and S27). It can be clearly observed that compound **15** has an additional acetyl fragment. Furthermore, the HMBC from H_2-13 to Ac-OCO indicated that the acetyl fragment was linked to C-13 (Figure 3).

Table 2. ^{1}H NMR (400 MHz) and ^{13}C NMR (100 MHz) of **15** in CD_3OD.

Position	δ_C	δ_H (J in Hz)	Position	δ_C	δ_H (J in Hz)
1	35.3 (CH_2)	1.45, m 2.05, m	10	44.2 (C)	
2	19.5 (CH_2)	1.54, m 1.78, m	11	71.8 (CH_2)	5.00, d (17.7) 4.84, d (17.6)
3	37.0 (CH_2)	1.16, m 1.91, m	12	177.0 (C)	
4	39.3 (C)		13	68.1 (CH_2)	4.44, d (11.2) 4.62, d (5.4)
5	56.4 (CH)	1.74, s	14	28.1 (CH_3)	1.16, s
6	63.6 (CH)	4.61, d (11.0)	15	65.7 (CH_2)	3.76, d (12.0) 4.33, d (11.9)
7	33.0 (CH_2)	2.34, d, (18.9) 2.50, d (18.3)	Ac-CH_3	20.8 (CH_3)	2.08, s
8	124.0 (C)		Ac-OCO	173.1 (C)	
9	169.0 (C)				

Finally, the NOESY correlation and experimental ECD spectrum of compound **15** were identical to those of **14** (Figures 3 and 4). Thus, the absolute configuration of **15** was also assigned as 4*R*, 5*R*, 6*S*, 10*S*.

2.2. Antimicrobial Assay

Compounds **1–15** were investigated for their antimicrobial activities against two phytopathogenic fungi and four bacterial strains. As shown in Table 3, andrastin-type meroterpenoids have better antimicrobial activities against phytopathogenic fungus than against bacteria. Most of all the tested compounds (9 compounds out of total 15 compounds) displayed potent antimicrobial activities (MIC < 50 µg/mL). Among them, compounds 1, 5 and 10 exhibited remarkable antimicrobial activities against *Penicillium italicum* and *Colletrichum gloeosporioides* with MIC values of 6.25, 1.56, 6.25 and 6.25, 3.13, 6.25 µg/mL. Moreover, compound 1 showed inhibitory activities against *Bacillus subtilis* under concentration of 6.25 µg/mL. Compound 10 also displayed significant antimicrobial activity against *Salmonella typhimurium* with an MIC value of 3.13 µg/mL. Notably, compound 5 revealed potential antimicrobial activity against all the strains, the MIC values were lower than 25 µg/mL.

Table 3. Antimicrobial activity of compounds **1–15**.

Compound \ Microbial	Methicillin-Resistent *Staphyococcus aureus* (µg/mL) [a]	*Bacillus subtilis* (µg/mL) [a]	*Pseudomonas aeruginosa* (µg/mL) [a]	*Salmonella typhimurium* (µg/mL) [a]	*Penicillium italicum* (µg/mL) [a]	*Colletrichum gloeosporioides* (µg/mL) [a]
1	25	6.25	>50	>50	6.25	6.25
2	>50	>50	25	>50	50	>50
3	>50	>50	>50	>50	50	>50
4	>50	>50	>50	>50	>50	>50
5	50	25	25	>50	1.56	3.13
6	>50	25	50	>50	12.50	25
7	>50	>50	>50	>50	25	25
8	>50	>50	>50	>50	>50	>50
9	>50	>50	>50	>50	>50	>50

Table 3. *Cont.*

Compound \ Microbial	Methicillin-Resistent *Staphyococcus aureus* (μg/mL) [a]	*Bacillus subtilis* (μg/mL) [a]	*Pseudomonas aeruginosa* (μg/mL) [a]	*Salmonella typhimurium* (μg/mL) [a]	*Penicillium italicum* (μg/mL) [a]	*Colletrichum gloeosporioides* (μg/mL) [a]
10	25	12.50	25	3.13	6.25	6.25
11	>50	>50	>50	>50	>50	>50
12	>50	>50	>50	>50	>50	>50
13	50	>50	>50	>50	50	>50
14	>50	>50	>50	>50	>50	>50
15	>50	>50	>50	>50	25	25
Ampicillin	0.13	0.13	0.07	0.13	-	-
Ketoconazole	-	-	-	-	0.78	0.78

[a]: The deviation value of three parallel experiments; -: No test.

As for the study of the structure–activity relationship (SAR), it was found that the degree of oxidation at C-21 had different effects on the activities of the compounds. The compound with methyl (**10**) at C-21 has significantly antimicrobial activity, followed by the aldehyde group (**7**), and hydroxymethyl (**13**) was the weakest. In-depth analysis showed that apart from the degree of oxidation at C-21, keto-enol tautomerism at the cyclopentane ring also had obvious influences on the antimicrobial activities of compounds. Compared to compounds **7** and **13** (enol form), compounds **8** and **9** (keto form) showed no activities against all strains (Table 3).

3. Experimental Methods

3.1. General Experimental Procedures

The NMR were tested on a Bruker Avance 600 MHz spectrometer (Karlsruhe, Germany) at room temperature. Optical rotations data were recorded on an MCP300 (Anton Paar, Shanghai, China). UV were tested using a Shimadzu UV-2600 spectrophotometer (Shimadzu, Kyoto, Japan). IR spectra were recorded on IR Affinity-1 spectrometer (Shimadzu, Kyoto, Japan). HR-ESI-MS spectra were tested on a ThermoFisher LTQ-Orbitrap-LC-MS spectrometer (Palo Alto, CA, USA). LC-MS/MS data was performed on a Q-TOF manufactured by Waters and a Waters Acquity UPLC BEH C18 column (1.7 μm, 2.1 × 100 mm). Recoated silica gel plates (Qingdao Huang Hai Chemical Group Co., Qingdao, China, G60, F-254), Column chromatography (CC) and Sephadex LH-20 (Amersham Pharmacia, Stockholm, Sweden) were used to purify the compounds.

3.2. Fungal Material

Fungus N-5 was isolated from the rhizosphere soil of mangrove plant *Avicennia marina* (collected in October 2021 from Nansha Mangrove National Nature Reserve in Guangdong Province, China). It was identified as *Penicillum* sp. by the ITS region (deposited in GenBank, accession no ON926808), and fungus N-5 was deposited at Sun Yat-sen University, China.

3.3. Fermentation

The fungus *Penicillum* sp. N-5 was cultured in one hundred 1000 mL Erlenmeyer flasks at 25 °C for 30 days; these contained autoclaved rice solid-substrate medium composed of 50 g rice and 50 mL 3‰ saline water.

3.4. Extraction and Purification

After incubation, the mycelia and solid rice medium were extracted four times with EtOAc, and 75 g of residue was obtained. Next, the residue was separated by a gradient of petroleum ether/EtOAc from 9:1 to 0:10 (*v*/*v*) on silica gel CC and divided into six fractions (Fr.1–Fr.6). Fr. 2 (10 g) was separated to Sephadex LH-20 (methanol) to yield three sub-fractions (SFrs. 2.1–2.3). SFrs.2.3 (1.2 g) was applied to silica gel CC (DCM/MeOH *v*/*v*,

100:1) and further purified by reversed-phase (RP) high performance liquid chromatography (HPLC; 90–10% MeCN/H_2O for 25 min) to obtain compounds **1** (5 mg) and **3** (7 mg). Fr. 3 (16 g) was also separated to Sephadex LH-20 (methanol) to yield four sub-fractions (SFrs. 3.1–3.4). SFrs.3.1 (1.6 g) was separated to silica gel CC (DCM/MeOH *v*/*v*, 80:1) and further purified by reversed-phase (RP) high performance liquid chromatography (HPLC; 75–25% MeCN/H_2O for 22 min) to yield compounds **2** (11 mg) and **15** (6 mg).

Hemiacetalmeroterpenoid A (**1**): white powder, m.p. 122.8–124.1 °C; $[\alpha]_D^{25}$-52 (*c* 0.02, MeOH), UV (MeOH) λ_{max} (log ε): 206 (2.52) (Figure S33); ECD (MeOH) λ_{max} ($\Delta\varepsilon$): 240 (+5.47), 301 (−1.14), 362 (+0.61); ^1H (600 MHz, CD_3OD) and ^{13}C NMR (150 MHz, CD_3OD) data, see Table 1; HR-ESI-MS: *m*/*z* 459.23709 [M+H]$^+$ (calcd. for $C_{26}H_{35}O_7$, 459.23773).

Hemiacetalmeroterpenoid B (**2**): white powder, m.p. 106.9–108.4 °C; $[\alpha]_D^{25}$ −56 (*c* 0.02, MeOH), UV (MeOH) λ_{max} (log ε): 205 (2.83), 238 (1.36) (Figure S34); ECD (MeOH) λ_{max} ($\Delta\varepsilon$): 206 (−11.92), 248 (+3.38), 311 (−1.19); ^1H (600 MHz, CD_3OD) and ^{13}C NMR (150 MHz, CD_3OD) data, see Table 1; HR-ESI-MS: *m*/*z* 445.25772 [M+H]$^+$ (calcd. for $C_{26}H_{37}O_6$, 445.25847).

Hemiacetalmeroterpenoid C (**3**): white powder, m.p. 100.8–102.6 °C; $[\alpha]_D^{25}$ −66 (*c* 0.02, MeOH), UV (MeOH) λ_{max} (log ε): 205 (2.51), 260 (1.78) (Figure S35); ECD (MeOH) λ_{max} ($\Delta\varepsilon$): 206 (−18.31), 240 (+12.72), 310 (−2.18); ^1H (600 MHz, CD_3OD) and ^{13}C NMR (150 MHz, CD_3OD) data, see Table 1; HR-ESI-MS: *m*/*z* 509.25015 [M+Na]$^+$ (calcd. for $C_{28}H_{38}O_7Na$, 509.25097).

Astellolide J (**14**): yellow powder, m.p. 202.9–204.5 °C; $[\alpha]_D^{25}$ +16 (*c* 0.02, MeOH), UV (MeOH) λ_{max} (log ε): 216 (3.15) (Figure S36); ECD (MeOH) λ_{max} ($\Delta\varepsilon$): 203 (−1.82), 225 (+3.27); ^1H (400 MHz, CD_3OD) and ^{13}C NMR (100 MHz, CD_3OD) data; HR-ESI-MS: *m*/*z* 305.13565 [M+Na]$^+$ (calcd. for $C_{15}H_{22}O_5Na$, 305.13594).

Astellolide Q (**15**): yellow powder, m.p. 160.1–162.2 °C; $[\alpha]_D^{25}$ +12 (*c* 0.02, MeOH), UV (MeOH) λ_{max} (log ε): 218 (3.62) (Figure S37); ECD (MeOH) λ_{max} ($\Delta\varepsilon$): 232 (+5.19); ^1H (400 MHz, CD_3OD) and ^{13}C NMR (100 MHz, CD_3OD) data, see Table 2; HR-ESI-MS: *m*/*z* 347.14578 [M+Na]$^+$ (calcd. for $C_{17}H_{24}O_6Na$, 347.14578).

3.5. ECD Calculation

Firstly, ECD calculations of compounds **1–3** and **14–15** were performed by the Gaussian 09 program and Spartan'14. Next, the conformations with a Boltzmann population (>5%) were selected for optimization and calculation in methanol at B3LYP/6-31+G (d, p). Finally, the ECD spectra were generated by the program SpecDis 1.6 (University of Würzburg, Würzburg, Germany) and drawn by OriginPro 8.0 (OriginLab, Ltd., Northampton, MA, USA) from dipole-length rotational strengths by applying Gaussian band shapes with sigma = 0.30 eV [29,30].

3.6. Bioassays Antimicrobial Activity

Antimicrobial activity assay was performed as previously described in [31,32].

4. Conclusions

In summary, three new andrastin-type meroterpenoids (**1–3**), one new drimane sesquiterpenoid (**15**) and one sesquiterpenoid J (**14**) that was first isolated from a natural source, together with ten known compounds (**4–13**) were isolated from the cultures of the rhizosphere soil of mangrove plant *Avicennia marina* fungus *Penicillium* sp. N-5. Their structures were determined by the analysis of NMR, HR-MS and ECD spectra. All the isolated compounds were investigated for their antimicrobial activities against two phytopathogenic fungi and four bacterial strains. Among them, compounds **1**, **5** and **10** exhibited significant inhibition against *Penicillium italicum* and *Colletrichum gloeosporioides* with MIC values of 6.25, 1.56, 6.25 and 6.25, 3.13, 6.25 μg/mL. Notably, compound **5** showed potential antimicrobial activity against all the strains and the MIC values were lower than 25 μg/mL. Moreover, andrastin-type meroterpenoid antimicrobial activity against phytopathogenic fungi was reported for the first time.

Supplementary Materials: The following are available online at https://www.mdpi.com/article/10.3390/md20080514/s1, Figure S1: HRESIMS spectrum of compound **1**; Figure S2: ^{1}H NMR spectrum of compound **1**; Figure S3: ^{13}C NMR spectrum of compound **1**; Figure S4: DEPT135 spectrum of compound **1**; Figure S5. HSQC spectrum of compound **1**; Figure S6: H, H-COSY spectrum of compound **1**; Figure S7: HMBC spectrum of compound **1**; Figure S8: NOE spectrum of compound **1**; Figure S9: HRESIMS spectrum of compound **2**; Figure S10: ^{1}H NMR spectrum of compound **2**; Figure S11: ^{13}C NMR spectrum of compound **2**; Figure S12: DEPT135 spectrum of compound **2**; Figure S13: HSQC spectrum of compound **2**; Figure S14: H, H-COSY spectrum of compound **2**; Figure S15: HMBC spectrum of compound **2**; Figure S16: NOE spectrum of compound **2**; Figure S17: HRESIMS spectrum of compound **3**; Figure S18: ^{1}H NMR spectrum of compound **3**; Figure S19: ^{13}C NMR spectrum of compound **3**; Figure S20: HSQC spectrum of compound **3**; Figure S21: H, H-COSY spectrum of compound **3**; Figure S22: HMBC spectrum of compound **3**; Figure S23: NOE spectrum of compound **3**; Figure S24: NOE spectrum of compound **14**; Figure S25: HRESIMS spectrum of compound **15**; Figure S26: ^{1}H NMR spectrum of compound **15**; Figure S27: ^{13}C NMR spectrum of compound **15**; Figure S28. DEPT135 spectrum of compound **15**; Figure S29: HSQC spectrum of compound **15**; Figure S30: H, H-COSY spectrum of compound **15**; Figure S31: HMBC spectrum of compound **15**; Figure S32: NOE spectrum 3M of compound **15**; Figure S33: UV and ECD of compound **1**; Figure S34: UV and ECD of compound **2**; Figure S35: UV and ECD of compound **3**; Figure S36: UV and ECD of compound **14**; Figure S37: UV and ECD of compound **15**.

Author Contributions: T.C. performed the experiments and wrote the paper; W.Y. analyzed the data and discussed the results; T.L., Y.Y. and Y.L. participated in the experiments; B.W. and Z.S. reviewed the manuscript; Z.S. designed and supervised the experiments. All authors have read and agreed to the published version of the manuscript.

Funding: We thank the National Natural Science Foundation of China (U20A2001, 21877133), GDNRC [2022]35 and the Key-Area Research and Development Program of Guangdong Province (2020B1111030005) for their generous support.

Institutional Review Board Statement: Not applicable.

Informed Consent Statement: Not applicable.

Data Availability Statement: Data are contained within the article and Supplementary Material.

Conflicts of Interest: The authors declare no conflict of interest.

References

1. Matsuda, Y.; Quan, Z.; Mitsuhashi, T.; Li, C.; Abe, I. Cytochrome P450 for citreohybridonol synthesis: Oxidative derivatization of the andrastin scaffold. *Org. Lett.* **2016**, *18*, 296–299. [CrossRef] [PubMed]
2. Matsuda, Y.; Awakawa, T.; Abe, I. Reconstituted biosynthesis of fungal meroterpenoid andrastin A. *Tetrahedron* **2013**, *69*, 8199–8204. [CrossRef]
3. Cheng, X.; Liang, X.; Zheng, Z.H.; Zhang, X.X.; Lu, X.H.; Yao, F.H.; Qi, S.H. Penicimeroterpenoids A-C, meroterpenoids with rearrangement skeletons from the marine-derived fungus *Penicillium* sp. SCSIO 41512. *Org. Lett.* **2020**, *62*, 6330–6333. [CrossRef]
4. Qin, Y.Y.; Huang, X.S.; Liu, X.B.; Mo, T.X.; Xu, Z.L.; Li, B.C.; Qin, X.Y.; Li, J.; Schäberle, T.F.; Yang, R.Y. Three new andrastin derivatives from the endophytic fungus *Penicillium vulpinum*. *Nat. Prod. Res.* **2022**, *36*, 13. [CrossRef] [PubMed]
5. Ren, J.; Huo, R.; Liu, G.; Liu, L. New andrastin-type meroterpenoids from the marine-derived fungus *Penicillium* sp. *Mar. Drugs* **2021**, *19*, 189. [CrossRef]
6. Xie, C.L.; Xia, J.M.; Lin, T.; Lin, Y.J.; Lin, Y.K.; Xia, M.L.; Chen, H.F.; Luo, Z.H.; Shao, Z.Z.; Yang, X.W. Andrastone A from the deep sea-derived fungus *Penicillium allii-sativi* acts as an inducer of caspase and RXRa-dependant apoptosis. *Front. Chem.* **2019**, *7*, 692. [CrossRef] [PubMed]
7. Cheng, Z.; Xu, W.; Wang, Y.; Bai, S.; Liu, L.; Luo, Z.; Yuan, W.; Li, Q. Two new meroterpenoids and two new monoterpenoids from the deep sea-derived fungus *Penicillium* sp. YPGA11. *Fitoterapia* **2019**, *133*, 120–124. [CrossRef] [PubMed]
8. Powers, Z.; Scharf, A.; Cheng, A.; Yang, F.; Himmelbauer, M.; Mitsuhashi, T.; Barra, L.; Taniguchi, Y.; Kikuchi, T.; Fujita, M.; et al. Biomimetic synthesis of meroterpenoids by dearomatization-driven polycyclization. *Angew. Chem. Int. Ed.* **2019**, *58*, 16141–16146. [CrossRef] [PubMed]
9. Zong, Y.; Wang, W.J.; Xu, T. Total synthesis of bioactive marine mero-terpenoids: The cases of liphagal and frondosin B. *Mar. Drugs* **2018**, *16*, 115. [CrossRef] [PubMed]
10. Kuan, K.K.W.; Markwell-Heys, A.W.; Cruickshank, M.C.; Tran, D.P.; Adlington, R.M.; Baldwin, J.E.; George, J.H. Biomimetic synthetic studies on meroterpenoids from the marine sponge *Aka coralliphaga*: Divergent total syntheses of siphonodictyal B, liphagal and corallidictyals A–D. *Bioorgan. Med. Chem.* **2019**, *27*, 2449–2465. [CrossRef] [PubMed]

11. Wang, W.; Shi, Y.; Liu, Y.; Zhang, Y.; Wu, J.; Zhang, G.; Che, Q.; Zhu, T.; Li, M.; Li, D. Brasilterpenes A-E, bergamotane sesquiterpenoid derivatives with hypoglycemic activity from the deep sea-derived fungus *Paraconiothyrium brasiliense* HDN15-135. *Mar. Drugs* **2022**, *20*, 338. [CrossRef]
12. Chen, S.; Cai, R.; Liu, Z.; Cui, H.; She, Z. Secondary metabolites from mangrove-associated fungi: Source, chemistry and bioactivities. *Nat. Prod. Rep.* **2022**, *39*, 560. [CrossRef]
13. Zhao, Y.; Sun, C.; Huang, L.; Zhang, X.; Zhang, G.; Che, Q.; Li, D.; Zhu, T. Talarodrides A-F, nonadrides from the antarctic sponge-derived fungus *Talaromyces* sp. HDN1820200. *J. Nat. Prod.* **2021**, *84*, 3011–3019. [CrossRef]
14. Shun, C.; Liu, Q.; Shah, M.; Che, Q.; Zhang, G.; Zhu, T.; Zhou, J.; Rong, X.; Li, D. Talaverrucin A, heterodimeric oxaphenalenone from antarctica sponge-derived fungus *Talaromyces* sp. HDN151403, Inhibits Wnt/β-Catenin Signaling Pathway. *Org. Lett.* **2022**, *24*, 3993–3997. [CrossRef]
15. Ye, G.; Huang, C.; Li, J.; Chen, T.; Tang, J.; Liu, W.; Long, Y. Isolation, structural characterization and antidiabetic activity of new diketopiperazine Alkaloids from mangrove endophytic fungus *Aspergillus* sp. 16-5c. *Mar. Drugs* **2021**, *19*, 402. [CrossRef]
16. Wu, Q.; Chang, Y.; Che, Q.; Li, D.; Zhang, G.; Zhu, T. Citreobenzofuran D-F and phomenone A-B: Five novel sesquiterpenoids from the mangrove-derived fungus *Penicillium* sp. HDN13-494. *Mar. Drugs* **2022**, *20*, 137. [CrossRef]
17. Yang, W.; Tan, Q.; Yin, Y.; Chen, Y.; Zhang, Y.; Wu, J.; Gao, L.; Wang, B.; She, Z. Secondary metabolites with α-glucosidase inhibitory activity from mangrove endophytic fungus *talaromyces* sp. CY-3. *Mar. Drugs* **2021**, *19*, 492. [CrossRef]
18. Zang, Z.; Yang, W.; Cui, H.; Cai, R.; Li, C.; Zou, G.; Wang, B.; She, Z. Two antimicrobial heterodimeric tetrahydroxanthones with a 7, 7′-Linkage from mangrove endophytic fungus *Aspergillus flavus* QQYZ. *Molecules* **2022**, *27*, 2691. [CrossRef]
19. Zou, G.; Chen, Y.; Yang, W.; Zang, Z.; Jiang, H.; Chen, S.; Wang, B.; She, Z. Furobenzotropolones A, B and 3-Hydroxyepicoccone B with antioxidative activity from mangrove endophytic fungus *Epicoccum nigrum* MLY-3. *Mar. Drugs* **2021**, *19*, 395. [CrossRef]
20. Chen, Y.; Yang, W.; Zou, G.; Wang, G.; Kang, W.; Yuan, J.; She, Z. Cytotoxic bromine- and iodine-containing cytochalasins produced by the mangrove endophytic fungus *Phomopsis* sp. QYM-13 using the OSMAC approach. *J. Nat. Prod.* **2022**, *85*, 1229–1238. [CrossRef]
21. Jiang, H.; Cai, R.; Zang, Z.; Yang, W.; Wang, B.; Zhu, G.; Yuan, J.; She, Z. Azaphilone derivatives with anti-inflammatory activity from the mangrove endophytic fungus *Penicillium sclerotiorum* ZJHJJ-18. *Bioorg. Chem.* **2022**, *122*, 105721. [CrossRef] [PubMed]
22. Cai, R.; Jiang, H.; Xiao, Z.; Cao, W.; Yan, T.; Liu, Z.; Lin, S.; Long, Y.; She, Z. (−)- and (+)-Asperginulin A, a pair of indole diketopiperazine alkaloid dimers with a 6/5/4/5/6 pentacyclic skeleton from the mangrove endophytic fungus *Aspergillus* sp. SK-28. *Org. Lett.* **2019**, *21*, 9633–9636. [CrossRef] [PubMed]
23. Gao, S.S.; Shang, Z.; Li, X.M.; Li, C.S.; Gui, C.M.; Wang, B.G. Secondary metabolites produced by solid fermentation of the marine-derived fungus *Penicillium commune* QSD-17. *Biosci. Biotechnol. Biochem.* **2012**, *76*, 358–360. [CrossRef] [PubMed]
24. Kosemura, S.H.; Miyata, K.; Matsunaga, S. Yamamura, Biosynthesis of citreohybridones, the metabolites of a hybrid strain KO 0031 derived from *penicillium citreo-viride* B. IFO 6200 and 4692. *Tetrahedron Lett.* **1992**, *33*, 3883–3886. [CrossRef]
25. Shiomi, K.R.; Uchida, J.; Inokoshi, H.; Tanaka, Y.; Iwai, S. Ōmura, Andrastins A-C, new protein farnesyltransferase inhibitors, produced by *Penicillium* sp. FO-3929. *Tetrahedron Lett.* **1996**, *37*, 1265–1268. [CrossRef]
26. Yang, X.; Xie, C.; Xia, J.; He, Z. Andrastone Compound and its Preparation Method and Application in Preparation of Antiallergic Drug. CN 111217878, 2 June 2020.
27. Hamasaki, T.; Kuwano, H.; Isono, K.; Hatsuda, Y.; Fukuyama, K.; Tsukihara, T.; Katsube, Y. A new metabolite, parasiticolide A, from *Aspergillus parasiticus*. *Agric. Biol. Chem.* **1975**, *37*, 749–751. [CrossRef]
28. Kosemura, S. Meroterpenoids from *Penicillium citreo-viride* B. IFO 4692 and 6200 hybrid. *Tetrahedron* **2003**, *59*, 5055–5072. [CrossRef]
29. Cui, H.; Liu, Y.N.; Li, J.; Huang, X.S.; Yan, T.; Cao, W.H.; Liu, H.J.; Long, Y.H.; She, Z.G. Diaporindenes A-D: Four unusual 2,3-dihydro-1H-indene analogues with anti-inflammatory activities from the mangrove endophytic fungus *Diaporthe* sp. SYSU. *J. Org. Chem.* **2018**, *83*, 11804–11813. [CrossRef]
30. Frisch, M.J.; Trucks, G.W.; Schlegel, H.B.; Scuseria, G.E.; Robb, M.A.; Cheeseman, J.R.; Scalmani, G.; Barone, V.; Mennucci, B.; Petersson, G.A.; et al. *Gaussian 09*; Gaussian, Inc.: Wallingford, UK, 2016.
31. Yang, W.; Yuan, J.; Tan, Q.; Chen, Y.; Zhu, Y.; Jiang, H.; Zou, G.; Zang, Z.; Wang, B.; She, Z. Peniazaphilones A-I, produced by co-culturing of mangrove endophytic fungi, *Penicillium sclerotiorum* THSH-4 and *Penicillium sclerotio-rum* ZJHJJ-18. *Chin. J. Chem.* **2021**, *39*, 3404–3412. [CrossRef]
32. Chen, Y.; Yang, W.; Zou, G.; Chen, S.; Pang, J.; She, Z. Bioactive polyketides from the mangrove endophytic fungi *Phoma* sp. SYSU-SK-7. *Fitoterapia* **2019**, *139*, 10436. [CrossRef]

marine drugs

MDPI

Article

New Polyketides from Mangrove Endophytic Fungus *Penicillium* sp. BJR-P2 and Their Anti-Inflammatory Activity

Chen Chen, Geting Ye, Jing Tang, Jialin Li, Wenbin Liu, Li Wu and Yuhua Long *

GDMPA Key Laboratory for Process Control and Quality Evaluation of Chiral Pharmaceuticals, Guangzhou Key Laboratory of Analytical Chemistry for Biomedicine, School of Chemistry, South China Normal University, Guangzhou 510006, China
* Correspondence: longyh@scnu.edu.cn

Abstract: Four new polyketide compounds, including two new unique isocoumarins penicillol A (**1**) and penicillol B (**2**) featuring with spiroketal rings, two new citreoviridin derivatives citreoviridin H (**3**) and citreoviridin I (**4**), along with four known analogues were isolated from the mangrove endophytic fungus *Penicillium* sp. BJR-P2. Their structures were elucidated by extensive spectroscopic methods. The absolute configurations of compounds **1–4** based on electronic circular dichroism (ECD) calculations, DP4+ analysis, and single-crystal X-ray diffraction are presented. All the new compounds were evaluated for anti-inflammatory activity. An anti-inflammatory assay indicated that compound **2** inhibited lipopolysaccharide (LPS)-induced NO production in RAW 264.7 cells, with half-maximal inhibitory concentration (IC_{50}) values of 12 μM, being more potent than the positive control, indomethacin (IC_{50} = 35.8 ± 5.7 μM). Docking study showed that compound **2** was perfectly docking into the active site of murine inducible nitric oxide oxygenase (iNOS) via forming multiple typical hydrogen bonds.

Keywords: isocoumarins; polyketides; anti-inflammatory activity; *Penicillium* sp.

Citation: Chen, C.; Ye, G.; Tang, J.; Li, J.; Liu, W.; Wu, L.; Long, Y. New Polyketides from Mangrove Endophytic Fungus *Penicillium* sp. BJR-P2 and Their Anti-Inflammatory Activity. *Mar. Drugs* **2022**, *20*, 583. https://doi.org/10.3390/md20090583

Academic Editor: Dehai Li

Received: 27 August 2022
Accepted: 16 September 2022
Published: 18 September 2022

1. Introduction

Inflammation is an adaptive response triggered by harmful stimuli, including some conditions of infection and tissue damage [1,2]. Macrophages, neutrophils, and lymphocytes are important natural immune cells to participate in homeostasis and immune response, and play complicated action on the pathogenesis of inflammatory disease [3–5]. Stimulated immune cells regulate inflammation by producing proinflammatory factors and mediators, such as interleukin (ILS), tumor necrosis factor-α (TNF-α), NO, prostaglandin E2 (PGE2), and iNOS [6–8]. Therefore, inhibition of inflammatory cytokines, chemokines, and mediators can be potent therapeutic strategies for the prevention of inflammation-related diseases.

Penicillium species are among the most widespread fungal organisms on earth and contains more than 350 species. Many *Penicillium* species can produce plentiful secondary metabolites, such as alkaloids [9], polyketides [10], cyclic peptides [11], and terpenoids [12], that can ascribe specific structural characteristics and significant biological activities. Isocoumarins are important natural lactones with wide range of biological activities, such as neuroprotective, antibacterial, antivirus and antitumor activities, and distribute widely in various microorganisms and plants from natural sources [13–19]. Up to now, nearly 1000 naturally occurring isocoumarins were reported [14]. As our continuing interest in finding new compounds with potential anti-inflammatory activity, the chemical investigation of the endophytic fungus *Penicillium* sp. BJR-P2 yielded two new unique isocoumarins with spiroketal rings (**1–2**), two new citreoviridin derivatives (**3–4**), together with four known analogues (**5–8**) (Figure 1). In this paper, the stereochemistry of these compounds was determined for the first time by DP4+ analysis, ECD calculations, and single-crystal

X-ray diffraction. By screening of the inhibitory effects on NO production in the LPS-induced RAW 264.7 macrophages, the anti-inflammatory activities of these compounds were evaluated, and the results showed that compound **2** exhibited effective inhibitory activity with IC_{50} value of 12 μM. This article describes the isolation, structure elucidation, and NO production inhibition of the new compounds.

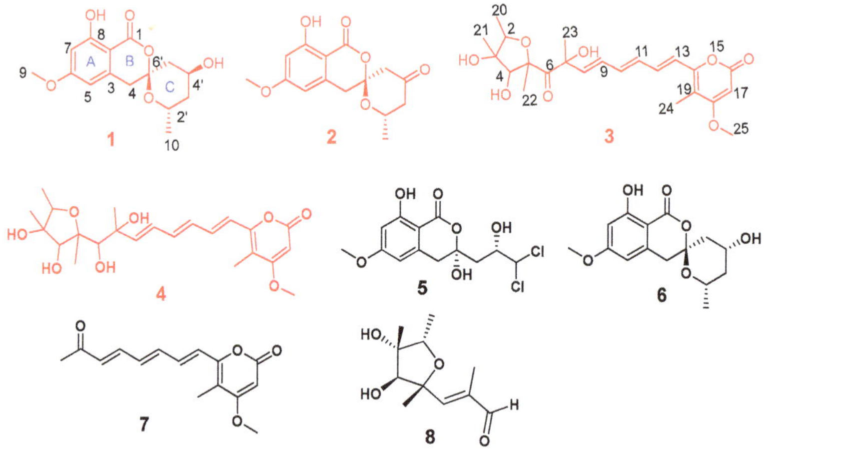

Figure 1. Chemical structures of compounds **1–8**.

2. Results and Discussion

2.1. Identification and Purification

The EtOAc extract of marine-derived fungus Penicillium sp. BJR-P2 was performed on repeated silica gel and Sephadex LH-20 column chromatography, followed by semipreparative HPLC to afford four new polyketides, penicillol A (**1**), penicillol B (**2**), and citreoviridin H (**3**), and citreoviridin I (**4**), along with four known polyketides **5**, **6**, **7**, and **8**.

2.2. Structural Elucidation

Compound **1** was obtained as a white powder. Its molecular formula was determined to be $C_{15}H_{17}O_6$ from the HRESIMS (*m/z* 293.1024 $[M - H]^-$ calcd for $C_{15}H_{17}O_6$, 293.1031), indicating 7 degrees of unsaturation. The IR spectrum of **1** at 3696, 1646, and 1585 cm^{-1} suggested the presence of hydroxyl, carbonyl, and aromatic ring groups. The ^{13}C NMR data in combination with HMQC spectra (Table 1) displayed one methyl carbon at δ_C 20.5, one methoxyl carbon at δ_C 55.2, three methylene carbons at δ_C 39.1, 38.9, and 38.5, four methine carbons at δ_C 106.8, 99.3, 63.4, 63.0, and six nonprotonated carbons at δ_C 169.2, 166.7, 164.4, 140.2, 104.6, 101.0. Analysis of the ^1H NMR spectrum of **1** revealed two aromatic protons at δ_H 6.35 (d, J = 2.1), 6.34 (s), indicating the presence of a tetrasubstituted phenyl. Two oxygenated methine protons signal at δ_H 4.39 (m) and 4.17 (m), three aliphatic methylenes at δ_H 1.53 (m), 1.78 (m), 1.87 (dd, J = 4.08, 15 Hz), 2.21 (dt, J = 15, 2.2 Hz), 2.95 (d, J = 16.5 Hz), and 3.17 (d, J = 16.5 Hz), one methoxyl at δ_H 3.82 (s), and one methyl at δ_H 1.07 (d, J = 6.3 Hz) were also recorded in this spectrum. The above spectroscopic features indicated that **1** belonged to the isocoumarin class. Further analysis of HMBC spectrum (Figure 2), the correlations from H-7 to C-2 (δ_C 101.0)/C-5 (δ_C 106.8)/C-6 (δ_C 166.7)/C-8 (δ_C 164.4), from H-5 to C-2/C-6, from H-4 to C-2/C-3 (δ_C 140.2)/C-6′ (δ_C 104.6), from H-9 to C-6 suggested that **1** was an isocoumarin derivative with a hydroxyl group at C-8 and a methoxy group at C-6. The ^1H-^1H COSY correlations between H-2′, H-3′, H-4′, H-5′, and H-10 combined with the HMBC correlations from H-5′ to C-3′ (δ_C 38.5)/C-4′ (δ_C 63.4), from H-3′ to C-2′ (δ_C 63.0)/C-10 (δ_C 20.5), from H-10 to C-2′/C-3′ suggested an aliphatic fragment of -CH_2-CH-CH_2-CH-CH_3. Furthermore, the key HMBC correlations from H-5′ to C-1/C-4/C-6′, from H-4′ to C-6′, and from H-10 to C-6′, together with the unsaturation of compound **1** and the chemical shift of C6′ (δ_C 104.6) indicated the presence of the spiroketal ring C. Therefore, the planar structure of **1** was shown in Figure 1. The absolute configuration of **1** was further verified by the X-ray diffraction analysis of a single crystal using Cu Kα as 6′*S*, 2′*S*, 4′*S*-**1** (Figure 3). Hence, the structure of compound **1** was identified as 6′*S*, 2′*S*, 4′*S*-**1** and named penicillol A.

Table 1. ^{1}H (600 MHz) and ^{13}C NMR (150 MHz) data for compounds **1** and **2**.

Position	1 [a] (δ_C, Type)	1 [a] (δ_H, Type)	2 [b] (δ_C, Type)	2 [b] (δ_H, Type)
1	169.2, C		167.5, C	
2	101.0, C		100.6, C	
3	140.2, C		138.3, C	
4	39.1, CH_2	2.95, d (16.5)	38.6, CH_2	3.14, d (16.4)
		3.17, d (16.5)		3.20, d (16.4)
5	106.8, CH	6.34, s	107.4, CH	6.29, s
6	166.7, C		166.4, C	
7	99.3, CH	6.35, d (2.1)	99.7, CH	6.38, d (2.3)
8	164.4, C		164.7, C	
9	55.2, CH_3	3.82, s	55.8, CH_3	3.84, s
10	20.5, CH_3	1.07, d (6.3)	21.6, CH_3	1.24, d (6.2)
2′	63.0, CH	4.39, m	67.9, CH	4.41, m
3′	38.5, CH_2	1.53, m	47.9, CH_2	2.29, dd (11.5, 15.1)
		1.78, m		2.53, m
4′	63.4, CH	4.17, m	202.5, C	
5′	38.9, CH_2	1.87, dd (4.1, 15.0)	49.5, CH_2	2.59, d (15.4)
		2.21, dt (2.2, 15.0)		2.85, dd (1.56, 15.3)
6′	104.6, C		104.7, C	
8-OH				11.0, s

[a] Measure in MeOD-d_4. [b] measure in $CDCl_3$.

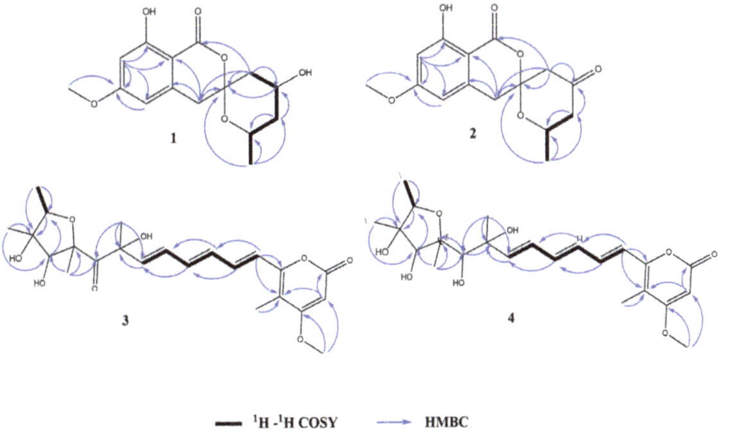

Figure 2. Key HMBC and ^{1}H-^{1}H COSY correlations of **1–4**.

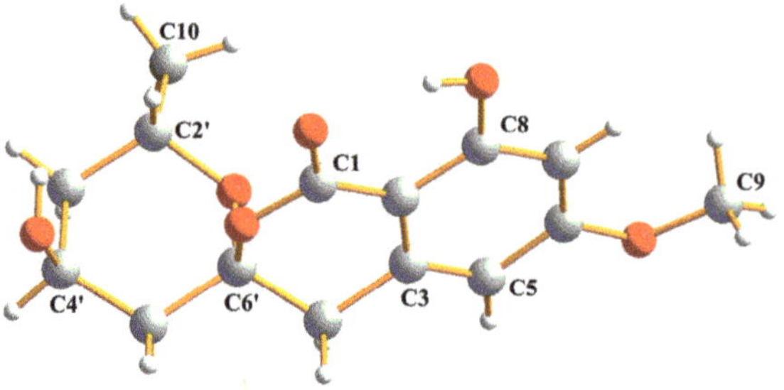

Figure 3. ORTEP representation of crystal structure of **1**.

Compound **2,** white powder, possesses the molecular formula $C_{15}H_{15}O_6$, as assigned by the HRESIMS ion at *m/z* 291.08727 $[M-H]^-$ (calcd for $C_{15}H_{15}O_6$, 291.08741), showing eight degrees of unsaturation. The IR spectrum suggested the presence of hydroxy (3680 cm^{-1}), aromatic ring (1580 cm^{-1}), and carbonyl (1736 cm^{-1}, 1671 cm^{-1}). Analysis of the ^{13}C NMR and HMQC (Table 1) data indicated the presence of 15 carbon atoms consisting of one methoxy group, one methyl, three methylenes, three methines (including two aromatic methines), 7 nonprotonated carbons (including one ester carbon, one carbonyl, four olefinic carbons). The ^1H NMR spectrum of **2** displayed two aromatic protons at δ_H 6.38 (d, J = 2.3 Hz), 6.29 (s), indicating the presence of a tetrasubstituted phenyl. One oxygenated methine proton signal at δ_H 4.41 (m), three aliphatic methylenes at δ_H 3.14 (d, J = 16.4), 3.20 d, J = 16.4 Hz), 2.29 (dd, J = 11.5, 15.1 Hz), 2.53 (m), 2.59 (d, J = 15.4 Hz), and 2.85 (dd, J = 1.6, 15.3 Hz), one methoxyl at δ_H 3.84 (s), one methyl at δ_H 1.24 (d, J = 6.2 Hz), and one active proton at δ_H 11.04 (s) were also recorded in this spectrum. Cumulative analyses of the ^1H and ^{13}C NMR spectra of compound **2** revealed that it possessed the similar planar structure as that of **1**. The main difference was the change of hydroxylated methine in **1** (δ_C 63.4) to a carbonyl in **2** (δ_C 202.5), indicating that compound **2** was an analog of **1**. Furthermore, the planar structure of **2** was further verified by the key HMBC correlation from H-5′ to C-4′ (δ_C 202.5), and H-3′ to C-4′ (Figure 2). The absolute configuration of **2** was determined by comparison of experimental and calculated ECD. Experimental data showed that the compounds **1**, **2**, and **6** had extremely similar ECD spectra, inferring that the cotton effect of compound **2** might be only affected by the chiral centers of C-6′ (δ_C 104.7) and C-2′ (δ_C 67.9) (Figure 4A). Then, the ECD spectra of the four possible configurations (6′*S*, 2′*S*-**2**; 6′*S*, 2′*R*-**2**; 6′*R*, 2′*S*-**2**; 6′*R*, 2′*R*-**2**) were calculated. The results disclosed that 6′*S*, 2′*S*-**2** and 6′*S*, 2′*R*-**2** displayed similar ECD spectra with the experimental one, showing a negative cotton effect (CE) at about 266 nm and a positive CE at about 285 nm (Figure 4B). Therefore, it was rationally speculated that the cotton effect of compound **2** was only affected by the chiral center of C-6′, and the configuration of C-6′ was determined to be 6′S as same as that of compound **1**. Based on the biosynthetic point of view, the absolute configuration at C- 2′ should be the same as that of compound **1**. Therefore, the structure of compound **2** was identified as 6′*S*, 2′*S*-**2** and named as penicillol B.

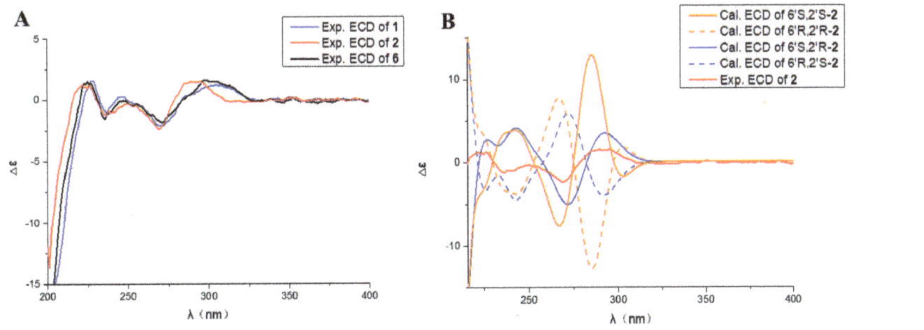

Figure 4. (**A**) Experimental ECD spectra of compounds **1**, **2**, and **6**. (**B**) Comparison between the experimental and calculated ECD spectra of **2**.

Compound **3** was obtained as pale-yellow oil with the molecular formula $C_{23}H_{30}O_8$ established from HRESIMS at *m/z* 415.17607 $[M - H_3O]^-$ (calcd for 415.17623), showing 9 degrees of unsaturation. The IR spectrum at ν_{max} 3685, 1725, 1693, and 1612 cm^{-1}, corresponded to a hydroxy, ester carbonyl, carbonyl and double bond group, respectively. Analysis of the ^{13}C NMR and HMQC spectra revealed 23 carbon resonances, corresponding to one ketone carbonyl (δ_C 204.7), one ester carbonyls (δ_C 163.9), three sp^2 nonprotonated carbons (δ_C 108.2, 154.5 and 170.8), three sp^3 nonprotonated carbon (δ_C 86.3, 84.9 and 82.3), seven sp^2 methine carbons (δ_C 88.9, −139.1), two sp^3 methine carbons (δ_C 80.2, 78.1), and five methyl carbons (δ_C 31.4, 16.9, 13.1, 12.5 and 9.0), one methoxyl (δ_C 56.3). Analysis of ^1H NMR spectrum of **3** exhibited a set of olefinic protons resonances at δ_H 5.49 (s), 6.34 (d, J = 15.0 Hz), 6.27 (dd, J = 15.5, 10.5 Hz), 6.36 (dd, J = 15.0, 10.5 Hz), 7.16 (dd, J = 15.0, 10.5 Hz), 6.43 (dd, J = 15.0, 10.5 Hz), 6.05 (d, J = 15.5 Hz,). Five methyls at δ_H 1.96 (s), 1.39 (s), 1.23 (d, J = 6.5 Hz), 1.35 (s), 1.46 (s), one methoxyl at δ_H 3.82 (s), and two oxygenated methine protons signal at δ_H 3.94 (s), 4.24 (dd, J = 6.5, 12.9 Hz) were also recorded. All the above data were indicative that compound **3** possessed a similar carbon skeleton as that of citreoviridin which was a polyketide derivative from a *Penicillium pulvillorum* [20]. The HMBC correlations from H-20 to C-2 (δ_C 80.2)/C-3 (δ_C 86.3), from H-2 to C-3, from H-21 to C-3/C-4 (δ_C 78.1), from H-4 to C-2, and from H-22 to C-4/C-5 (δ_C 84.9) suggested the presence of **3**, 4-dihydroxy-2, 3, 5-trimethyl-tetrahydrofuran moiety (Figure 2). The ^1H-^1H COSY correlation of H-2/ H_3-20 and H-8/ H-9/ H-10/ H-11/ H-12/ H-13 indicated two hydrocarbon fragments of -CH-CH_3 and -CH=CH-CH=CH-CH=CH-. Through the analysis of the ^1H NMR spectra of **3** to those of citreoviridin, combined with key HMBC correlation from H-17 to C-16/C-18/C-19, from H-24 to C-19, and from H-25 to C-17/C-18, suggested the presence of 4-methyl-5-methoxy-3, 4, 5-trisubstituted-α-pyrone moiety. Moreover, the HMBC correlation from H-13 to C-14 (δ_C 154.5)/C-19 (δ_C 108.2), indicated the unsaturated hydrocarbon fragment was lactated at C-14 of α-pyrone moiety. The main difference was the replacement of two double bond carbons in citreoviridin by a carbonyl carbon C-6 (δ_C 204.7) and an oxygenated quaternary carbon C-7 (δ_C 82.3) in **3**. This deduction was supported by the key HMBC correlations from H-23 to C-7/C-6 (δ_C 204.7)/C-8 (δ_C 139.1), from H-8 to C-7 (δ_C 82.3), and from H-22 to C-5/C-6. So, the planar structure of **3** was elucidated, as shown in Figure 1. The NOE correlations from H-2 to Me-21, from Me-21 to H-2/H-4, from Me-22 to H-4 revealed the relative configuration of the five-membered ring as 2*R**, 3*S**, 4*S**, 5*R** (Figure 5). The relative configuration of C-7 was subjected to DP4+ analysis. Therefore, a DP4+ analysis of two candidate structures (2*R**, 3*S**, 4*S**, 5*R**, 7*S**-**3**/2*R**, 3*S**, 4*S**, 5*R**, 7*R**-**3**) was performed by calculating for their theoretical 1D NMR chemical shifts. The result showed that the final score of the configuration 2*R**, 3*S**, 4*S**, 5*R**, 7*S**-**3** (100%) was the most probable (Figures S29 and S30). Thus, the structure of compound **3** was identified as 2*R**, 3*S**, 4*S**, 5*R**, 7*S**-**3**. Coupling constants between protons H-8 and H-9, H-10 and H-11, H-12 and H-13 ($^3J_{H\text{-}8,\,H\text{-}9}$ = 15.5 Hz; $^3J_{H\text{-}10,\,H\text{-}11}$ = 15.0 Hz; $^3J_{H\text{-}12,}$

H-13 = 15.0 Hz) inferred that the conformations of these three double bonds are all *trans* conformations, and compound **3** is named citreoviridin H.

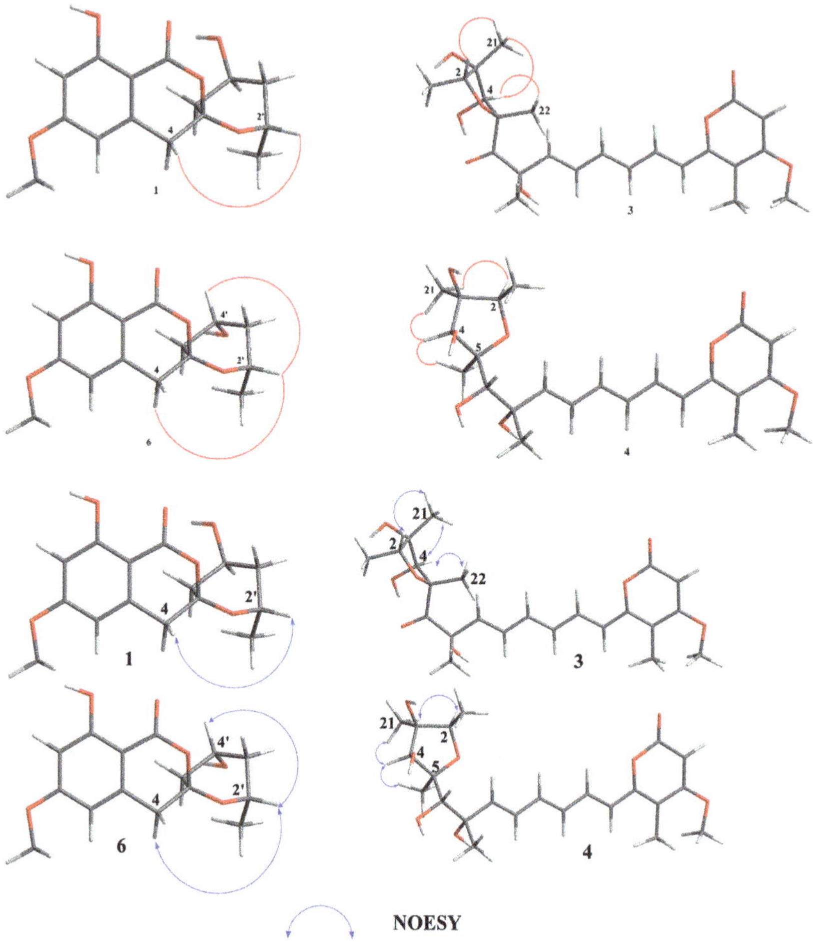

Figure 5. Key NOESY correlations of **1**, **3**, **4**, and **6**.

Compound **4** was isolated as pale-yellow oil, and its molecular formula was determined as $C_{23}H_{32}O_8$ on the basis of the pseudomolecular ion peak observed at *m/z* 417.19163 $[M - H_3O]^-$ (Calcd for 417.19188) in the HRESIMS spectrum, indicating 8 degrees of unsaturation. The IR spectrum showed absorption bands of hydroxyl, carbonyl, and double bond at νmax 3696, 1694, and 1623 cm^{-1}. The ^{13}C NMR and HMQC spectra (Table 2) displayed 23 carbon signals, consisting of six methyls (including one methoxyl), ten methines (including seven olefinic carbons), and seven quaternary carbons (including one ester carbon). The ^{1}H NMR spectrum of **4** revealed a set of olefinic protons resonances at δ_H 5.64 (s), 6.59 (d, *J* = 15.0 Hz), 6.43 (dd, *J* = 11.0, 15.4 Hz), 6.5 (dd, *J* = 11.0, 15.0 Hz), 7.16 (dd, *J* = 11.0, 15.0 Hz), 6.65 (dd, *J* = 11.0, 15.0 Hz), 6.01 (d, *J* = 15.4 Hz), five methyls at δ_H 2.02 (s), 1.32 (s), 1.20 (d, *J* = 6.4 Hz), 1.22 (s), 1.30 (s), one methoxyl at δ_H 3.92 (s), and three oxygenated methine protons signal at δ_H 3.79 (s), 4.08 (m), 3.69 (s). All the above data were indicative that compound **4** possessed a similar carbon skeleton of **3**. The main difference was the replacement of a carbonyl group in **3** (δ_C 204.7) by a hydroxymethine group in **4** (δ_C 91.2), indicating that compound **4** was a homologue of **3**. This inference was further confirmed by the key HMBC correlation from H-22 to C-5 (δ_C 86.3)/C-6 (δ_C 91.2), from

H-6 to C-5/C-7 (δ_C 73.3)/C-8 (δ_C 140.9), and from H-23 to C-7/C-6 (Figure 2). Therefore, the planar structure of **4** was established as shown in Figure 1. Furthermore, the relative configuration of compound **4** was partly determined by NOESY spectrum. The NOE correlations from H-2 to Me-21, from Me-21 to H-2/H-4, from Me-22 to H-4 indicated the relative configuration of the tetrahydrofuran ring to be 2*R**, 3*S**, 4*S**, 5*S** (Figure 5). In order to determine the relative configuration of C-6 and C-7, the DP4+ analysis of four candidate structures (2*R**, 3*S**, 4*S**, 5*S**, 6*S**, 7*S**-**4**/2*R**, 3*S**, 4*S**, 5*S**, 6*R**, 7*S**-**4**/2*R**, 3*S**, 4*S**, 5*S**, 6*R**, 7*R**-**4**/2*R**, 3*S**, 4*S**, 5*S**, 6*S**, 7*R**-**4**) was performed. The analysis results showed that the configuration of 2*R**, 3*S**, 4*S**, 5*S**, 6*S**, 7*S**-**4** was the correct structure with a 100% probability (Figures S31 and S32). Hence, the structure of compound **4** was identified as 2*R**, 3*S**, 4*S**, 5*S**, 6*S**, 7*S**-**4**. Coupling constants between protons H-8 and H-9, H-10 and H-11, H-12 and H-13 ($^3J_{H\text{-}8,\,H\text{-}9}$ = 15.4 Hz; $^3J_{H\text{-}10,\,H\text{-}11}$= 15.0 Hz; $^3J_{H\text{-}12,\,H\text{-}13}$ = 15.0 Hz) inferred that the conformations of these three double bonds are all *trans* conformations, and compound **4** is named citreoviridin I.

Table 2. ^{1}H (600 MHz) and ^{13}C NMR (150 MHz) data for compounds **3** and **4**.

Position	3 [b] (δ_C, Type)	3 [b] (δ_H, Type)	4 [a] (δ_C, Type)	4 [a] (δ_H, Type)
2	80.2, CH	4.24, q (6.4)	78.8, CH	4.08, q (6.4)
3	86.3, C		83.9, C	
4	78.1, CH	3.94, s	78.1, CH	3.79, s
5	84.9, C		86.3, C	
6	204.7, C		91.2, CH	3.69, s
7	82.3, C		73.3, C	
8	139.1, CH	6.05, d (15.5)	140.9, CH	6.01, d (15.4)
9	129.1, CH	6.27, dd (15.5, 10.5)	128.5, CH	6.43, dd (11.0, 15.4)
10	137.3, CH	6.43, dd (15.0, 10.5)	138.1, CH	6.65, dd (11.0, 15.0)
11	132.5, CH	6.36, dd (15.0, 10.5)	131.5, CH	6.50, dd (11.0, 15.0)
12	135.8, CH	7.16, dd (15.0, 10.9)	136.1, CH	7.16, dd (11.0, 15.0)
13	119.6, CH	6.34, d (15.0)	119.4, CH	6.59, d (15.0)
14	154.5, C		154.9, C	
16	163.9, C		165.4, C	
17	88.9, CH	5.49, s	88.1, CH	5.64, s
18	170.8, C		172.1, C	
19	108.2, C		108.8, C	
20	13.1, CH_3	1.23, d (6.5)	12.8, CH_3	1.20, d (6.4)
21	16.9, CH_3	1.35, s	12.8, CH_3	1.22, s
22	12.5, CH_3	1.39, s	26.3, CH_3	1.30, s
23	31.4, CH_3	1.46, s	12.2, CH_3	1.32, s
24	9.0, CH_3	1.96, s	7.9, CH_3	2.02, s
25	56.3, CH_3	3.82, s	56.3, CH_3	3.92, s

[a] Measure in MeOD-d_4; [b] measure in $CDCl_3$.

Four known compounds including dichlorodiaportal (**5**) [21], citreoviranol (**6**) [22], citreopyrone D (**7**) [23], citreoviral (**8**) [24], were isolated and identified from this fungus. Their structures were determined by comparing their NMR and MS data with those reported in the literature. Moreover, the absolute configuration of compound **6** was further determined to be 6′*S*, 2′*S*, 4′*R*-**6** by comparing its NOE correlations and ECD spectrum with compound **1**.

A plausible typical fungal polyketide synthetase (PKS) involved biosynthetic pathway for compounds **1**–**4** was proposed as shown in Figure S33. The condensation of one mole of acetyl coenzyme A and six moles of malonyl coenzyme A gives a mole linear polyketide chain. Subsequent keto-reduction, cyclization, methylation, and hydroxylation furnish compounds **1** and **2**. Similarly, one mole acetyl coenzyme A and eight moles malonyl coenzyme A condensed to form a mole linear polyketide chain, then keto-reduction, dehydration, cyclization, methylation, and hydroxylation form compounds **3** and **4** [25,26].

2.3. *Anti-Inflammatory Activity*

New compounds **1–4** were evaluated for anti-inflammatory activity in LPS-stimulated RAW 264.7 macrophages. Especially, Compound **2** significantly inhibited nitric oxide production with an IC_{50} value of 12 μM (Table 3).

Table 3. Inhibitory activities against LPS-induced NO production.

Compound	IC_{50} (μM)	Inhibition Ratio at 50 μM
1	-	<50%
2	12	97%
3	-	<50%
4	-	<50%
Indometacin [a]	35.8 ± 5.7	-

[a] Indometacin was used as positive control for the test.

2.4. *Molecular Docking Studies*

Inhibition of NO overproduction is usually the result of inhibition of iNOS enzyme expression or activity [27,28]. In order to investigate the inhibitory mechanism of compound **2** on NO production, the interaction and binding mode between compound **2** and iNOS (PDB: 3E6T) [29], molecular docking study was carried out using AUTODOCK 4.2.6 modeling software. Docking procedure was validated by docking of ligand indomethacin (positive drug) in the active site of iNOS, and root-mean square deviation (RMSD) of 0.12 Å to the X-ray structure. The results revealed that the lowest energy of compound **2** (−7.49 Kcal/mol) was lower than that of positive drug indomethacin (−7.45 Kcal/mol), and the lowest energy of compound **1** (−7.36 Kcal/mol) was higher than that of positive drug. Further observations showed indomethacin formed a hydrogen bond with key amino acid residues GLU-371, two hydrogen bonds with amino acid residues ARG-260, and a hydrogen bond with GLN-257 in the iNOS active pocket (Figure 6A). Compound **2** formed a hydrogen bond with the key amino acid residue GLU-371 through the methoxy group, two hydrogen bonds with the residue ASP-379 and ARG-382 by the hydroxyl group in the iNOS active pocket (Figure 6B), respectively. Notably, for compound **1**, while the carbonyl group in compound **2** at the 4′ position changed to the hydroxyl group, the optimized conformation of **1** was different from that of compound **2** and could not enter into the iNOS active pocket to form hydrogen bonds with the key amino acid residues (Figure 6C). As a result, compound **1** had no production inhibition activity (Table 3).

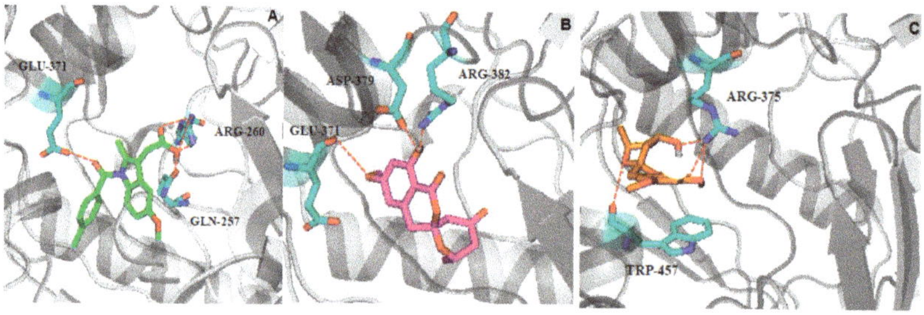

Figure 6. (**A**) Docking results of the binding pose of positive drug indomethacin in iNOS. (**B**) Predicted binding mode of compound **2** docked into iNOS (PDB: 3E6T). (**C**) Predicted binding mode of compound **1** docked into iNOS. (PDB: 3E6T) (indomethacin is in green stick; compound **2** is in purple stick; compound **1** is in yellow stick; red dashed lines represent H-bonds; the amino acids involved in hydrogen bond interactions are in blue stick).

3. Materials and Methods

3.1. General Experimental Procedures

HRESIMS data were measured on a Finnigan LTQ-Orbitrap Elite (Thermo Fisher Scientific, Waltham, MA, USA). NMR spectra were obtained by Bruker AVANCE NEO 600 MHz spectrometer (Bruker BioSpin, Switzerland). CD spectrum was reported on a Chirascan TM CD spectropolarimeter (Applied Photophysics, U.K). Optical rotation was obtained on an MCP 500 (Anton Paar, Austria). Single-crystal data was measured on an Oxford Gemini S Ultra diffractometer (Oxford Instrument, Oxfordshire, UK). Sephadex LH-20 (25–100 μm; GE Healthcare, Bio-Sciences AB, Stockholm, Sweden) and silica gel (200–300 mesh; Qingdao Marine Chemical Factory, Qingdao, China) were used for Column chromatography (CC). Thin layer chromatography (TLC) was detected on Silica gel GF254 plate (Qingdao Marine Chemical Ltd., Qingdao, China).

3.2. Fungal Material

The fungus BJR-P2 used in this research was isolated from the barks of *Avicennia marinav* (*Forsk.*) *Vierh*, a mangrove plant which collected from Yangjiang Hailing Island Mangrove Wetland Park. Using molecular biology methods, the fungus was identified through DNA amplification and ITS region sequencing. The sequence data of this strain has been deposited in Gen Bank with accession no. PRJNA793386. The BLAST search results show that the sequence is 100% similar to that of *Penicillium* sp.

3.3. Extraction and Isolation

The fungus *Penicillium* sp. BJR-P2 was fermented on solid autoclaved rice medium using one hundred 1 L Erlenmeyer flasks, each was containing 40 g rice and 40 mL 0.3% saline water, culturing in room temperature under static condition for 25 days. The solid rice and mycelia medium were extracted with methanol for three times. The organic solvents were evaporated under reduced pressure and extracted with ethyl acetate. We obtained 50 g of organic extract. Then, the extract was subjected to a silica gel column eluting with a gradient of petroleum ether/EtOAc from 1/0 to 0/1 to afford six fractions (Fractions 1–6). Fraction 2 (300 mg) was subjected to an open silica gel glass column eluted with DCM:MeOH (100:1) followed by Sephadex LH-20 open glass (30 mg) CC eluting with MeOH-H_2O (*v/v*, 7:3), and further purified by using HPLC on a semipreparative column (RP-18, 9.4 × 250 mm, 7 μm, 1.5 mL min^{-1}) eluted with MeOH:H_2O (6:4) to obtain compounds **1** (9.0 mg) and **4** (5.0 mg). Further separation of fraction 3 by preparative HPLC over silica gel (250 mm × 10 mm) with PE-EtOAc (4:2–1:1) and DCM:MO (200:1-80:1) as the eluent afforded **2** (10.0 mg) and **3** (5.0 mg). Fraction 4 (500 mg) was chromatographed on Sephadex LH-20 CC and silica gel CC to obtain compounds **5–8**.

Compound **1**: white powder; $[\alpha]_D^{25} - 88.9°$ (c 0.1, MeOH); IR (KBr): ν_{max} 3696, 1646, 1585 cm^{-1}; UV (MeOH) λ_{max} (logε): 216 (0.91) nm; 268 (0.53) nm; 303 (0.22); ECD (MeOH) λ_{max} (Δε), 228 (+1.56), 237 (−1.01), 247 (+2.02), 270 (−2.15), 305 (+1.20) nm; HRESIMS at *m/z* 293.1024 $[M - H]^-$ (calcd for 293.1031). ^1H and ^{13}C NMR see Table 1.

Compound **2**: white powder; $[\alpha]_D^{25} - 62.0°$ (c 0.1, MeOH); IR (KBr): ν_{max} 3680, 1736, 1671, 1580 cm^{-1}; UV (MeOH) λ_{max} (log ε): 216 (0.89) nm; 268 (0.52) nm; 303 (0.21); ECD (MeOH) λ_{max} (Δε), 222 (+1.25), 236 (−1.23), 250 (−0.31), 269 (−2.43), 295 (+1.48) nm; HRESIMS at *m/z* 291.08727 $[M - H]^-$ (calcd for 291.08741). ^1H and ^{13}C NMR see Table 1.

Compound **3**: yellow oil; $[\alpha]_D^{25} - 10°$ (c 0.1, MeOH); IR (KBr): ν_{max} 3685, 1725, 1693, 1612 cm^{-1}; UV (MeOH) λ_{max} (logε): 274 (0.65) nm; 367 (0.5) nm; HRESIMS at *m/z* 415.17607 $[M - H_3O]^-$ (calcd for 415.17623). ^1H and ^{13}C NMR see Table 2.

Compound **4**: yellow oil; $[\alpha]_D^{25} - 10°$ (c 0.1, MeOH); IR (KBr): ν_{max} 3696, 1694, 1623 cm^{-1}; UV (MeOH) λ_{max} (logε): 274 (0.96) nm; 368 (0.78) nm; HRESIMS at *m/z* 417.19163 $[M - H_3O]^-$ (calcd for 417.19188). ^1H and ^{13}C NMR see Table 2.

X-ray Crystal Data for **1**. Colorless crystal of **1** was obtained in methanol and EtOAc. Crystal data (CCDC 2131096) were collected with Cu Kα radiation. Monoclinic, space group P21, a = 7.18058(6) Å, b = 7.48777 (11) Å, c = 13.48676 (10) Å, α = 90°, β = 92.3992 (7) °, γ = 90,

V = 724.501(14) Å3, Z = 2, T = 149.99(10) K, μ (Cu Kα) = 0.963 mm^{-1}, ρcalc = 1.432 g/cm^3, F (000) = 332.0, R_1 = 0.0296, wR_2 =0.0745. Crystal dimensions 0.28 × 0.16 × 0.12 mm^3. Flack parameter = 0.03(4). The total number of independent reflections measured was 12,512, of which 2626 were observed, collected in the range of 6.56° ≤ 2θ ≤ 145.682°. The structure was determined and refined using full-matrix least-squares on F2 values for 1.109 I > = 2σ (I).

3.4. ECD and ^{13}C NMR Calculation

The conformational searches of the compounds were carried out by means of the Spartan'14 software and at Molecular Merck force field (MMFF) and DFT/TD-DFT calculations. Furthermore, Gaussian 05 program was used to generate and optimize the conformer at B3LYP/3-21G (d) level. Conformers with a Boltzmann distribution of over 1% were chosen for optimization at B3LYP/6-31+G (d, p), meanwhile, ECD calculation were conducted with the TD-DFT method at the B3LYP/6-31+G (d, p) level and the ^{13}C NMR calculation at mPW1PW91-SCRF/6-311+g (2d, p). The ECD spectra were generated using the SpecDis 3.0 (University of Würzburg, Würzburg, Germany) and Origin Pro 8.0 (Origin Lab, Ltd., Northampton, MA, USA) from dipole length rotational strengths by applying Gaussian band shapes with sigma = 0.30 eV. Then, the calculated and theoretical values of ^{13}C were analyzed by DP4+ [30].

3.5. Anti-Inflammatory Assays

The RAW 264.7 mouse macrophage cell line was purchased from the Cell Bank of Shanghai Institute of Biochemistry and Cell Biology (Chinese Academy of Sciences, Shanghai, China). Murine macrophage RAW 264.7 cells were cultured in DMEM (high glucose) medium supplemented with 10% (*v/v*) fetal bovine serum, 100 μg·mL^{-1} penicillin and streptomycin, and 10 mM HEPES buffer at 37oC in 5% CO_2 in air for 1 h. Cells were pretreated with different concentrations of samples (10, 5, 2.5, 1.25, and 0.625 μM) dissolved in serum-free medium containing 0.5% DMSO for 4 h, followed by stimulation with 1 μg·mL^{-1} LPS for 24 h. A total of 50 μL cell culture medium was mixed with 100 μL Griess reagents I and II and incubated horizontally at room temperature for 10 min. The absorbance was measured at 570 nm [31,32].

3.6. Molecular Docking Study

Virtual docking is implemented in the AutoDock tool of AutoDock4.2.6 software [33]. This is a common docking method that allows the ligand to have sufficient flexibility and maintain the rigidity of the target protein. The X-ray crystal structure of iNOS (PDB ID: 3E6T) [27] was obtained from the RCSB protein database (PDB) database. Before docking simulation, PyMOL was used to delete the original ligand and water molecules from the crystal structure, and the protein was saved in PDB format (receptor.pdb). The compound structure was drawn using ChemDraw 2D software, which was converted into three-dimensional (3D) structure by ChemDraw 3D software, and then stored as a file in PDB format. Furthermore, the molecular structure was optimized by Gaussian software. AutoDock tools converted both protein and ligand into PDBQT format for subsequent docking. Focusing on the protein, the parameters of the grid box were set to 126 × 126 × 126 points and the Lamarckian genetic algorithm was used to link the algorithm with 100 GA operations. Finally, PyMOL was used to visualize and analyze the results.

4. Conclusions

In conclusion, chemical investigation of the mangrove endophytic fungus BJR-P2 resulted in the isolation and identification of four new compounds (**1–4**), with four known analogs (**5–8**). Their structures were elucidated by extensive spectroscopic methods and quantum chemical calculations. The anti-inflammatory activity evaluation was carried out by screening their inhibition activity on NO production. The results showed compound

2 exhibited significant inhibitory activity with an IC_{50} value of 12 μM. This study may provide a new chemical lead candidate for the discovery of anti-inflammatory agents.

Supplementary Materials: The HRESIMS and NMR spectrum are available online at: https://www.mdpi.com/article/10.3390/md20090583/s1. Figure S1. HRESIMS of compound **1**. Figure S2. ^{1}H NMR spectrum of compound **1**. Figure S3. ^{13}C NMR spectrum of compound **1**. Figure S4. HMQC spectrum of compound **1**. Figure S5. HMBC spectrum of compound **1**. Figure S6. ^{1}H, ^{1}H-COSY spectrum of compound **1**. Figure S7. NOESY spectrum of compound **1**. Figure S8. HRESIMS of compound **2**. Figure S9. ^{1}H NMR spectrum of compound **2**. Figure S10.^{13}C NMR spectrum of compound **2**. Figure S11. HMQC spectrum of compound **2**. Figure S12. HMBC spectrum of compound **2**. Figure S13. ^{1}H, ^{1}H-COSY spectrum of compound **2**. Figure S14. NOESY spectrum of compound **2**. Figure S15. HR-ESI-MS spectrum of compound **3**. Figure S16. ^{1}H NMR spectrum of compound **3**. Figure S17. ^{13}C NMR spectrum of compound **3**. Figure S18. HMQC spectrum of compound **3**. Figure S19. HMBC spectrum of compound **3**. Figure S20. ^{1}H, ^{1}H-COSY spectrum of compound **3**. Figure S21. NOESY spectrum of compound **3**. Figure S22. HR-ESI-MS spectrum of compound **4**. Figure S23. ^{1}H NMR spectrum of compound **4**. Figure S24. ^{13}C NMR spectrum of compound **4**. Figure S25. HMQC spectrum of compound **4**. Figure S26. HMBC spectrum of compound **4**. Figure S27. ^{1}H, ^{1}H-COSY spectrum of compound **4**. Figure S28. NOESY spectrum of compound **4**. Figure S29. Experimental (Exp.) and calculated (Cal.) ^{1}H and ^{13}C chemical shift values of **3** and its possible isomers. Figure S30. DP4+ analysis of **3**. Figure S31. Experimental (Exp.) and calculated (Cal.) ^{1}H and ^{13}C chemical shift values of **4** and its possible isomers. Figure S32. DP4+ analysis of **4**. Figure S33. Proposed biosynthetic pathways for compounds **1–4**.

Author Contributions: C.C. and G.Y. performed the experiments for the isolation, structure elucidation, and biological evaluation and prepared the manuscript; J.T., J.L., W.L., L.W. contributed to part work of fermentation, extraction, structure characterization of all the compounds; Y.L. supervised the research work and revised the manuscript. Conceptualization, Y.L.; methodology, C.C. and G.Y.; software, C.C. and G.Y.; validation, Y.L.; formal analysis, C.C., G.Y., J.T.; investigation, C.C., G.Y., J.L., W.L., L.W.; resources, Y.L.; data curation, C.C., G.Y., J.T.; writing—original draft preparation, C.C. and G.Y.; writing—review and editing, Y.L. and C.C.; supervision, Y.L.; project administration, Y.L.; funding acquisition, Y.L. All authors have read and agreed to the published version of the manuscript.

Funding: This research was funded by the National Natural Science Foundation of China, grant number 41876153; Guangdong Marine Economy Development Special Project (No. GDNRC [2022]35).

Institutional Review Board Statement: Not applicable.

Informed Consent Statement: Not applicable.

Acknowledgments: The authors gratefully acknowledge grant from the National Natural Science Foundation of China (No. 41876153) and Guangdong Marine Economy Development Special Project (No. GDNRC [2022]35).

Conflicts of Interest: The authors declare no conflict of interest.

References

1. Medzhitov, R. Origin and physiological roles of inflammation. *Nature* **2008**, *454*, 428–435. [CrossRef] [PubMed]
2. Serhan, C.N.; Savill, J. Resolution of inflammation: The beginning programs the end. *Nat. Immunol.* **2005**, *6*, 1191–1197. [CrossRef]
3. Pierce, G.F. Macrophages: Important physiologic and pathologic sources of polypeptide growth factors. *Am. J. Resp. Cell Mol.* **1990**, *2*, 233–234. [CrossRef]
4. Moghaddam, A.S.; Mohammadian, S.; Vazini, H.; Taghadosi, M.; Esmaeili, M.; Mardani, F.; Seifi, B.; Mohammadi, A.; Afshari, J.T.; Sahebkar, A. Macrophage plasticity, polarization, and function in health and disease. *J. Cell. Physiol.* **2018**, *233*, 6425–6440. [CrossRef] [PubMed]
5. Carl, N. Points of control in inflammation. *Nature* **2002**, *420*, 846–852.
6. Glancy, R.M.; Amin, A.R.; Abramson, S.B. The role of nitric oxide in inflammation and immunity. *Arthritis Rheum.* **1998**, *41*, 1111–1151.
7. Yang, Y.Z.; Wei, Z.; Teichmann, A.T.; Wieland, F.H.; Wang, A.; Lei, X.G.; Zhu, Y.; Yin, J.X.; Fan, T.T.; Zhou, L.; et al. Development of a novel nitric Oxide (NO) production inhibitor with potential therapeutic effect on chronic inflammation. *Eur. J. Med. Chem.* **2020**, *193*, 112216. [CrossRef] [PubMed]

8. Chen, J.N.; Mejia, E.G.D.; Wu, J.S. Inhibitory effect of a glycoprotein isolated from golden oyster mushroom (pleurotus citrinopileatus) on the lipopolysaccharide-induced inflammatory reaction in RAW 264.7 macrophage. *J. Agric. Food Chem.* **2011**, *59*, 7092–7097. [CrossRef]
9. Dalsgaard, P.W.; Blunt, J.W.; Munro, M.H.; Frisvad, J.C.; Christophersen, C. Communesins G and H, new alkaloids from the psychrotolerant fungus *Penicillium rivulum*. *J. Nat. Prod.* **2005**, *68*, 258–261. [CrossRef] [PubMed]
10. Orfali, R.; Perveen, S.; Al-Taweel, A.; Ahmed, A.F.; Majrashi, N.; Alluhay, K.; Khan, A.; Luciano, P.; Taglialatela-Scafati, O. Penipyranicins A-C: Antibacterial methylpyran polyketides from a hydrothermal spring sediment *Penicillium* sp. *J. Nat. Prod.* **2020**, *83*, 3591–3597. [CrossRef]
11. Dalsgaard, P.W.; Larsen, T.O.; Frydenvang, K.; Christophersen, C. Psychrophilin A and cycloaspeptide D, novel cyclic peptides from the psychrotolerant fungus *Penicillium ribeum*. *J. Nat. Prod.* **2004**, *67*, 878–881. [CrossRef] [PubMed]
12. Zhuravleva, O.I.; Sobolevskaya, M.P.; Leshchenko, E.V.; Kirichuk, N.N.; Denisenko, V.A.; Dmitrenok, P.S.; Dyshlovoy, S.A.; Zakharenko, A.M.; Kim, N.Y.; Afiyatullov, S. Meroterpenoids from the alga-derived fungi *Penicillium thomii Maire* and *Penicillium lividum Westling*. *J. Nat. Prod.* **2014**, *77*, 1390–1395. [CrossRef] [PubMed]
13. Gu, W.; Wang, W.; Li, X.; Zhang, Y.; Wang, L.; Yuan, C. A novel isocoumarin with anti-influenza virus activity from strobilanthes cusia. *Fitoterapia* **2015**, *107*, 60–62. [CrossRef] [PubMed]
14. Xu, Y.H.; Lu, C.H.; Zheng, Z.H. A new 3,4-dihydroisocoumarin isolated from *Botryosphaeria* sp. F00741. *Chem. Nat. Comp.* **2012**, *48*, 205–207. [CrossRef]
15. Zhao, Y.; Liu, D.; Proksch, P.; Yu, S.; Lin, W.H. Isocoumarin derivatives from the sponge-associated fungus *Peyronellaea glomerata* with antioxidant activities. *Chem. Biodivers.* **2016**, *3*, 1186–1193. [CrossRef] [PubMed]
16. Girich, E.V.; Yurchenko, A.N.; Smetanina, O.F.; Trinh, P.T.H.; Ngoc, M.T.D.; Pivkin, M.V.; Popov, R.S.; Pislyagin, E.A.; Menchinskaya, E.S.; Chingizova, E.A.; et al. Neuroprotective metabolites from vietnamese marine derived fungi of *Aspergillus* and *Penicillium* genera. *Mar. Drugs* **2020**, *18*, 608. [CrossRef]
17. Gu, B.B.; Wu, Y.; Tang, J.; Jiao, W.H.; Li, L.; Sun, F.; Wang, S.P.; Yang, F.; Lin, H.W. Azaphilone and isocoumarin derivatives from the sponge-derived fungus *Eupenicillium* sp. 6A-9. *Tetrahedron Lett.* **2018**, *59*, 3345–3348. [CrossRef]
18. Shabir, G.; Saeed, A.; El-Seedi, H.R. Natural isocoumarins: Structural styles and biological activities, the revelations carry on. *Phytochemistry* **2021**, *181*, 112568. [CrossRef] [PubMed]
19. Noor, A.O.; Almasri, D.M.; Bagalagel, A.A.; Abdallah, H.M.; Mohamed, S.G.A.; Mohamed, G.A.; Ibrahim, S.R.M. Naturally Occurring Isocoumarins Derivatives from Endophytic Fungi: Sources, Isolation, Structural Characterization, Biosynthesis, and Biological Activities. *Molecules* **2020**, *25*, 395. [CrossRef] [PubMed]
20. Shizuri, Y.; Shigemori, H.; Sato, R.; Yamamura, S.; Kawai, K.; Furukawa, H. Four new metabolites produced by *Penicillium citreo-viride B.* on addition of NaBr. *Chem. Lett.* **2006**, *17*, 1419–1422. [CrossRef]
21. Nagel, D.W.; Steyn, P.S.; Scott, D.B. Production of citreoviridin by *Penicillium pulvillorum*. *Phytochemistry* **1972**, *11*, 627–630. [CrossRef]
22. Yang, M.H.; Li, T.X.; Wang, Y.; Liu, R.H.; Luo, J.; Kong, L.Y. Antimicrobial metabolites from the plant endophytic fungus *Penicillium* sp. *Fitoterapia* **2017**, *116*, 72–76. [CrossRef]
23. Kosemura, S.; Kojima, S.I.; Yamamura, S. Citreopyrones, new metabolites of two hybrid strains, KO 0092 and KO 0141, derived from the *Penicillium* species. *Chem. Lett.* **1997**, *26*, 33–34. [CrossRef]
24. Shizuri, Y.; Nishiyama, S.; Imai, D.; Yamamura, S. Isolation and stereostructures of citreoviral, citreodiol, and epicitreodiol. *Tetrahedron Lett.* **1984**, *25*, 4771–4774. [CrossRef]
25. Oikawa, H. Biosynthesis of structurally unique fungal metabolite GKK1032A2: Indication of novel carbocyclic formation mechanism in polyketide biosynthesis. *J. Org. Chem.* **2003**, *68*, 3552–3557. [CrossRef]
26. Li, H.; Jiang, J.; Liu, Z.; Lin, S.; Xia, G.; Xia, X.; Ding, B.; He, L.; Lu, Y.; She, Z. Peniphenones A−D from the mangrove fungus *Penicillium* dipodomyicola HN4-3A as inhibitors of mycobacterium tuberculosis phosphatase MptpB. *J. Nat. Prod.* **2014**, *77*, 800–806. [CrossRef] [PubMed]
27. Vallance, P.; Leiper, J. Blocking NO synthesis: How, where and why? *Nat. Rev. Drug Discov.* **2002**, *1*, 939–950. [CrossRef]
28. Eiserich, J.P.; Hristova, M.; Cross, C.E.; Jones, A.D.; Freeman, B.A.; Halliwell, B.; Vliet, A. Formation of nitric oxide-derived inflammatory oxidants by myeloperoxidase in neutrophils. *Nature* **1998**, *391*, 393–397. [CrossRef]
29. Garcin, E.D.; Arvai, A.S.; Rosenfeld, R.J.; Kroeger, M.D.; Crane, B.R.; Andersson, G.; Andrews, G.; Hamley, P.J.; Mallinder, P.R.; Nicholls, D.J.; et al. Anchored plasticity opens doors for selective inhibitor design in nitric oxide synthase. *Nat. Chem. Biol.* **2008**, *4*, 700–707. [CrossRef] [PubMed]
30. Cui, H.; Liu, Y.N.; Li, J.; Huang, X.S.; Yan, T.; Cao, W.H.; Liu, H.J.; Long, Y.H.; She, Z.G. Diaporindenes A–D: Four unusual 2, 3-dihydro-1H -indene analogues with anti-inflammatory activities from the mangrove endophytic fungus *Diaporthe* sp. SYSU-HQ3. *J. Org. Chem.* **2018**, *83*, 11804–11813. [CrossRef] [PubMed]
31. Morikawa, T.; Matsuda, H.; Toguchida, I.; Ueda, K.; Yoshikawa, M. Absolute stereostructures of three new sesquiterpenes from the fruit of alpiniaoxyphylla with inhibitory effects on nitric oxide production and degranulation in RBL-2H3 Cells. *J. Nat. Prod.* **2002**, *65*, 1468–1474. [CrossRef] [PubMed]

32. Chen, Y.; Liu, Z.M.; Liu, H.J.; Pan, Y.H.; Li, J.; Liu, L. Dichloroisocoumarins with potential anti-inflammatory activity from the mangrove endophytic fungus *Ascomycota* sp. CYSK-4. *Mar. Drugs* **2018**, *16*, 54. [CrossRef] [PubMed]
33. Garrett, M.M.; Huey, R.; Lingstrom, W.; Sanner, M.F.; Belew, R.K.; Goodsell, D.S.; Olson, A.J. AutoDock4 and AutoDockTools4: Automated docking with selective receptor flexibility. *J. Comput. Chem.* **2009**, *30*, 2785–2791.

marine drugs

MDPI

Article

Unusual Tetrahydropyridoindole-Containing Tetrapeptides with Human Nicotinic Acetylcholine Receptors Targeting Activity Discovered from Antarctica-Derived Psychrophilic *Pseudogymnoascus* sp. HDN17-933

Xuewen Hou [1], Changlong Li [1], Runfang Zhang [1], Yinping Li [1], Huadong Li [1], Yundong Zhang [1], Han-Shen Tae [2], Rilei Yu [1], Qian Che [1], Tianjiao Zhu [1], Dehai Li [1,3,4,*] and Guojian Zhang [1,3,5,*]

1 Key Laboratory of Marine Drugs, Chinese Ministry of Education, School of Medicine and Pharmacy, Ocean University of China, Qingdao 266003, China
2 Illawarra Health and Medical Research Institute (IHMRI), University of Wollongong, Wollongong, NSW 2522, Australia
3 Laboratory for Marine Drugs and Bioproducts, Pilot National Laboratory for Marine Science and Technology, Qingdao 266237, China
4 Open Studio for Druggability Research of Marine Natural Products, Pilot National Laboratory for Marine Science and Technology, Qingdao 266237, China
5 Marine Biomedical Research Institute of Qingdao, Qingdao 266101, China
* Correspondence: dehaili@ouc.edu.cn (D.L.); zhangguojian@ouc.edu.cn (G.Z.); Tel.: +86-532-8203-1619 (D.L.); +86-532-82032971 (G.Z.)

Citation: Hou, X.; Li, C.; Zhang, R.; Li, Y.; Li, H.; Zhang, Y.; Tae, H.-S.; Yu, R.; Che, Q.; Zhu, T.; et al. Unusual Tetrahydropyridoindole-Containing Tetrapeptides with Human Nicotinic Acetylcholine Receptors Targeting Activity Discovered from Antarctica-Derived Psychrophilic *Pseudogymnoascus* sp. HDN17-933. *Mar. Drugs* **2022**, *20*, 593. https://doi.org/10.3390/md20100593

Academic Editor: Rosa Maria Vitale

Received: 26 August 2022
Accepted: 19 September 2022
Published: 22 September 2022

Abstract: Chemical investigation of the psychrophilic fungus *Pseudogymnoascus* sp. HDN17-933 derived from Antarctica led to the discovery of six new tetrapeptides psegymamides A–F (**1–6**), whose planar structures were elucidated by extensive NMR and MS spectrometric analyses. Structurally, psegymamides D–F (**4–6**) possess unique backbones bearing a tetrahydropyridoindoles unit, which make them the first examples discovered in naturally occurring peptides. The absolute configurations of structures were unambiguously determined using solid-phase total synthesis assisted by Marfey's method, and all compounds were evaluated for their inhibition of human (h) nicotinic acetylcholine receptor subtypes. Compound **2** showed significant inhibitory activity. A preliminary structure–activity relationship investigation revealed that the tryptophan residue and the C-terminal with methoxy group were important to the inhibitory activity. Further, the high binding affinity of compound **2** to h$\alpha 4\beta 2$ was explained by molecular docking studies.

Keywords: psychrophilic fungus; tetrapeptides; tetrahydropyridoindoles unit; solid-phase synthesis; nicotinic acetylcholine receptors; molecular docking

1. Introduction

Psychrophilic fungi are a group of cold-adapted fungi residing in polar regions, alpine permafrost, glaciers, deep oceans, and other habitats [1–6], which are known for long-term low temperature, strong ultraviolet radiation, low nutrient and water availability, and frequent freeze and thaw cycles [7–9]. In order to adapt to such harsh conditions, these fungi have evolved special strategies in their metabolism and physiology [7,10–12], which endows the ability to produce diversified secondary metabolites, and makes psychrophilic fungi competitive microorganism group to serve structurally novel and bioactive natural products for drug development [13–16].

So far, a large number of novel natural products have been isolated from psychrophilic fungi, such as trisorbicillone A, a novel sorbicillin trimer that showed cytotoxic against HL60 cell lines (IC_{50} 3.14 μM) from deep-sea fungal strain *Phialocephala* sp. [17]; brevione A, the first breviane spiroterpenoid family from *Penicillium brevicompactum* [18]; penilactones

A and B, novel polyketides with NF-kB inhibitory activity (inhibitory rate of 40% at 10 Mm) from *Penicillium crustosum* PRB-2 [19]; 16-membered trichobotryside A inhibiting the larvae settlement of Balamus amphitrite (EC_{50} 2.5 μg/mL) and 18-membered trychobotrysides B-C from *Trichobotrys effuse* [20]; and cytotoxic diterpenes conidiogenones B–G isolated from *Penicillium* sp, and conidiogenone C displayed significant cytotoxicity against HL60 and BEL-7047 cell lines with IC_{50} values of 0.038 mM and 0.9 mM respectively [21].

During our ongoing research on searching bioactive structures from Antarctic-derived fungi, a psychrophilic strain *Pseudogymnoascus* sp. HDN17-933, isolated from sand samples collected from Fildes Peninsula, was chosen for study based on its unique HPLC-UV profile. Detailed chemical investigation on its fermentation products afforded six tetrapeptides psegynamides A–F with human nicotinic acetylcholine receptors (nAChRs) targeting activity. Among those structures, psegynamides D–F (**4–6**) represented the first group of naturally occurring tetrapeptides carrying tetrahydropyridoindole units in the backbones. Herein, we will describe the isolation, structural elucidation, solid-phase total synthesis, and biological activity evaluation of these new compounds.

2. Results and Discussion

The psychrophilic fungal strain *Pseudogymnoascus* sp. HDN17-933 was isolated from Antarctica. They can grow in an environment of −5 °C, and the optimum growth temperature is 15–18 °C. The studies of their secondary metabolites are infrequent, and only 44 secondary metabolites have been reported until the time of writing. In this paper, the strain was cultured under static conditions at 15 °C for 30 days in a rice medium. The EtOAc extract of strain *Pseudogymnoascus* sp. HDN17-933 (Figure S2) was fractionated and purified by vacuum chromatography on silica gel, octadecyl-silica (ODS), Sephadex LH-20, and HPLC to afford six tetrapeptides psegynamides A–F (**1–6**) (Figure 1).

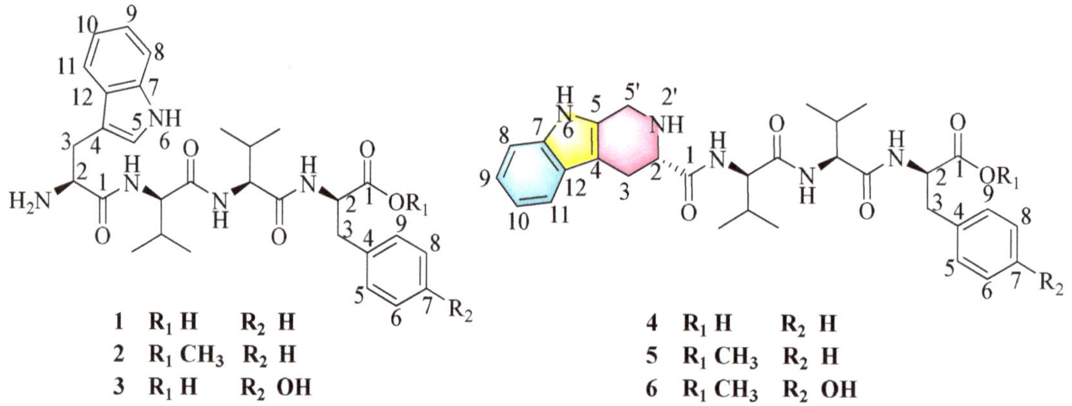

Figure 1. The structures of compounds **1–6** isolated from *Pseudogymnoascus* sp. HDN17-933.

Compound **1** named psegynamide A was obtained as a white amorphous powder. Its molecular formula was determined to be $C_{30}H_{39}N_5O_5$ based on the molecular ion peak at *m/z* 550.3025 $[M + H]^+$ (calcd 550.3024) in the HRESIMS analysis, indicating 14 degrees of hydrogen deficiency. In the ^{1}H-NMR spectrum of **1** (Table 1), the characteristic signals of NH were displayed at low-field δ_H 8.68 (1H, d, J = 9.5 Hz), 8.37 (1H, d, J = 7.8 Hz), and 8.13 (1H, d, J = 9.1 Hz). The signals of α-proton, the characteristic for amino acid residues, were exhibited at δ_H 4.56 (1H, m), 4.44 (1H, m), 4.30 (1H, m), and 4.18 (1H, m). Four methyl groups were displayed at δ_H 0.72 (3H, d, J = 6.8 Hz), 0.68 (3H, d, J = 6.8 Hz), 0.61 (3H, d, J = 6.8 Hz), and 0.48 (3H, d, J = 6.8 Hz). Analysis of the ^{13}C-NMR and HSQC spectra indicated the presence of four carbonyl peaks assignable to amide carbonyl groups at δ_C 173.6, 171.1, 170.7, and 161.9. Four groups of α-proton signals were showed at δ_C 57.3, 57.1, 53.9, and 52.9. The appearance of these spectra was typical of peptides. Combined analyses of ^{1}H-^{1}H COSY, HMBC, HSQC, NOESY, and TOCSY spectra further revealed the presence of 1 × Trp, 2 × Val, and 1 × Phe residues (Table 1), which were in accordance with the result of Marfey's analysis (Figure S58). In the HMBC spectrum (Figure 2), NH of Val (1) has a correlation signal with C-1 on Trp; NH of Val (2) has a correlation signal with C-1 on Val (1); NH of Phe has a correlation signal with C-1 on Val (2). Meanwhile, in the NOESY spectrum, the cross-peaks of NH in Val (1) and H-2 in Trp, NH in Val (2) and H-2 in Val (1), NH in Phe and H-2 in Val (2) further confirmed the linkage of amino acid residues by HMBC spectrum. Thus, the planar structure of compound **1** was established as Trp^1-Val^2-Val^3-Phe^4.

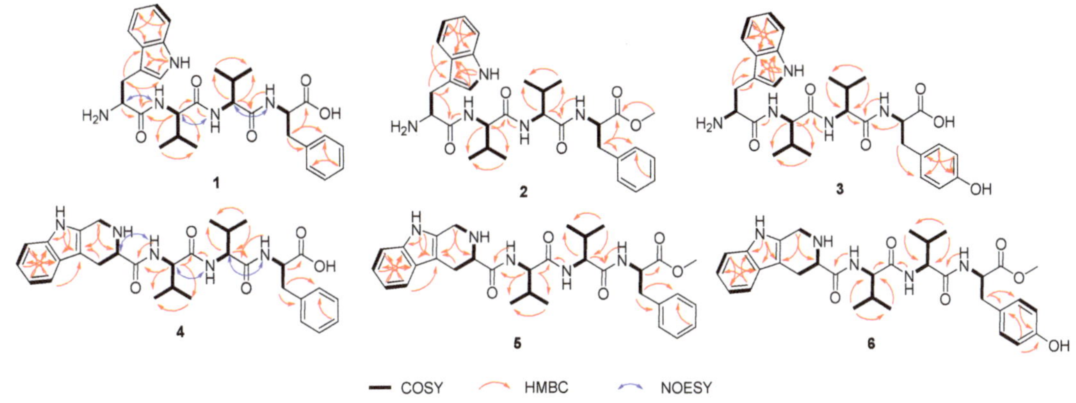

Figure 2. Key COSY, HMBC, and NOESY correlations of **1–6**.

The advanced Marfey's acidolytic method was considered to determine the absolute configurations of compound **1** [22]. After FDAA derivatization and assisted by HPLC analysis, the presence of *L*-Val, *D*-Val, and *D*-Phe in compound **1** (Figure S58) was confirmed. However, due to the destruction of Trp residue during the strong acid environment, we failed to detect the Trp-FDAA derivative from the hydrolytic mixture. Therefore, we turned to using alkaline hydrolytic to determine the absolute configuration of Trp residue. With the alkaline hydrolytic condition of 5 M LiOH at 110 °C for 16 h was finally adopted, compound **1** was successfully hydrolyzed, and the absolute configuration of Trp was determined as *L*. As analyzed from the HPLC profile, both *L*-Val and *D*-Val derivatives were detected in the acidolytic mixture of compound **1**, which makes it another challenge to determine the accurate sequence of the tetrapeptide. Hence, two possible stereoisomers, **1a** (*L*-Trp^1-*L*-Val^2-*D*-Val^3-*D*-Phe^4) and **1b** (*L*-Trp^1-*D*-Val^2-*L*-Val^3-*D*-Phe^4), were prepared using solid-phase total synthesis (Scheme 1A). By comparing the experimental NMR spectrum with the synthesized NMR data, the absolute configuration of compound **1** (Scheme 1B, Figure S64) was determined to be *L*-Trp^1-*D*-Val^2-*L*-Val^3-*D*-Phe^4, which was inconsistent with the synthesized stereoisomer **1b**.

Table 1. ^{1}H and ^{13}C NMR data in DMSO-d_6 for compounds **1–3** ([a] measured at 600 MHz; [b] measured at 150 MHz).

	No	Compound 1		Compound 2		Compound 3	
		δ_H [a], Mult (J in Hz)	δ_C [b]	δ_H [a], Mult (J in Hz)	δ_C [b]	δ_H [a], Mult (J in Hz)	δ_C [b]
Trp	1	-	169.1	-	169.2	-	169.2
	2	4.18 (1H, m)	52.9	4.20 (1H, m)	52.9	4.17 (1H, m)	53.0
	3	3.21 (1H, dd, J = 7.1, 14.7 Hz) 2.98 (1H, dd, J = 8.4, 14.7 Hz)	28.5	3.22 (1H, dd, J = 7.5, 15.5 Hz) 3.03 (1H, overlap, m)	28.4	3.19 (1H, dd, J = 6.8, 15.2 Hz) 2.96 (1H, dd, J = 8.3, 15.2 Hz)	28.6
	4	-	107. 4	-	107.4	-	107.4
	5	7.21 (1H, s)	125.5	7.23 (1H, overlap, s)	125.5	7.19 (1H, s)	125.6
	6 (NH)	10.98 (1H, s)	-	11.03 (1H, s)	-	10.98 (1H, s)	-
	7	-	136.8	-	136.8	-	136.9
	8	7.34 (1H, d, J = 7.8 Hz)	111.9	7.33 (1H, d, J = 7.6 Hz)	111.8	7.32 (1H, d, J = 7.0 Hz)	111.9
	9	7.07 (1H, t, J = 7.8 Hz)	121.6	7.07 (1H, t, J = 7.6 Hz)	121.6	7.05 (1H, t, J = 7.0 Hz)	121.7
	10	6.99 (1H, t, J = 7.8 Hz)	118.7	7.00 (1H, t, J = 7.6 Hz)	118.7	6.97 (1H, overlap, m)	118.8
	11	7.77 (1H, d, J = 7.8 Hz)	119.2	7.76 (1H, d, J = 7.6 Hz)	119.2	7.76 (1H, d, J = 7.0 Hz)	119.3
	12	-	127.4	-	127.5	-	127.5
	2-NH_2	-	-	-	-	7.98 (1H, s)	-
Val-1	1	-	170.7	-	170.7	-	170.8
	2	4.56 (1H, m)	57.3	4.52 (1H, overlap, m)	57.5	4.57 (1H, m)	57.4
	3	1.86 (1H, m)	31.9	1.87 (1H, m)	31.9	1.85 (1H, m)	32.1
	4	0.72 (3H, d, J = 6.8 Hz)	17.8	0.71 (3H, d, J = 6.9 Hz)	17.8	0.71 (3H, d, J = 6.4 Hz)	17.9
	5	0.68 (3H, d, J = 6.8 Hz)	19.5	0.66 (3H, d, J = 6.2 Hz)	19.4	0.66 (3H, d, J = 6.3 Hz)	19.6
	NH	8.68 (1H, d, J = 9.5 Hz)	-	8.69 (1H, d, J = 9.6 Hz)	-	8.69 (1H, d, J =8.6 Hz)	-
Val-2	1	-	171.1	-	171.4	-	171.2
	2	4.30 (1H, m)	57.1	4.28 (1H, m)	57.4	4.31 (1H, overlap, m)	57.2
	3	1.72 (1H, m)	31.3	1.76 (1H, m)	31.2	1.73 (1H, m)	31.5
	4	0.48 (3H, d, J = 6.8 Hz)	17.7	0.52 (3H, d, J = 6.8 Hz)	17.8	0.50 (3H, d, J = 6.8 Hz)	17.8
	5	0.61 (3H, d, J = 6.8 Hz)	19.7	0.62 (3H, d, J = 6.7 Hz)	19.6	0.62 (3H, d, J = 6.7 Hz)	19.8
	NH	8.13 (1H, d, J = 9.1 Hz)	-	8.13 (1H, d, J = 9.4 Hz)	-	8.13 (1H, d, J = 9.2 Hz)	-

Table 1. *Cont.*

	No	Compound 1		Compound 2		Compound 3	
		δ_H [a], Mult (J in Hz)	δ_C [b]	δ_H [a], Mult (J in Hz)	δ_C [b]	δ_H [a], Mult (J in Hz)	δ_C [b]
Phe/Phe-OCH_3/Tyr	1	-	173.6	-	172.6	-	173.8
	2	4.44 (1H, m)	53.9	4.48 (1H, overlap, m)	53.9	4.34 (1H, overlap, m)	54.2
	3	3.07 (1H, dd, J = 4.4, 13.8 Hz) 2.77 (1H, dd, J = 10.6, 13.8 Hz)	37.4	3.03 (1H, overlap, m) 2.83 (1H, dd, J = 10.0,13.6 Hz)	37.2	2.91 (1H, m) 2.63 (1H, m)	36.8
	4	-	138.0	-	137.5	-	128.0
	5	7.22 (1H, overlap, s)	129.5	7.23 (1H, overlap, m)	129.5	6.97 (1H, overlap, m)	130.5
	6	7.22 (1H, overlap, s)	128.5	7.23 (1H, overlap, m)	128.6	6.59 (1H, overlap, m)	115.4
	7	7.16 (1H, m)	126.8	7.17 (1H, overlap, m)	127.0	-	156.5
	8	7.22 (1H, overlap, s)	128.5	7.23 (1H, overlap, m)	128.6	6.59 (1H, overlap, m)	115.4
	9	7.22(1H, overlap, s)	129.5	7.23 (1H, overlap, m)	129.5	6.97 (1H, overlap, m)	130.5
	NH	8.37 (1H, d, J = 7.8 Hz)	-	8.58 (1H, d, J = 8.0 Hz)	-	8.34 (1H, d, J = 8.3 Hz)	-
	1-OCH_3	-	-	3.56 (3H, s)	52.4	-	-
	7-OH	-	-	-		9.18 (1H, s)	-

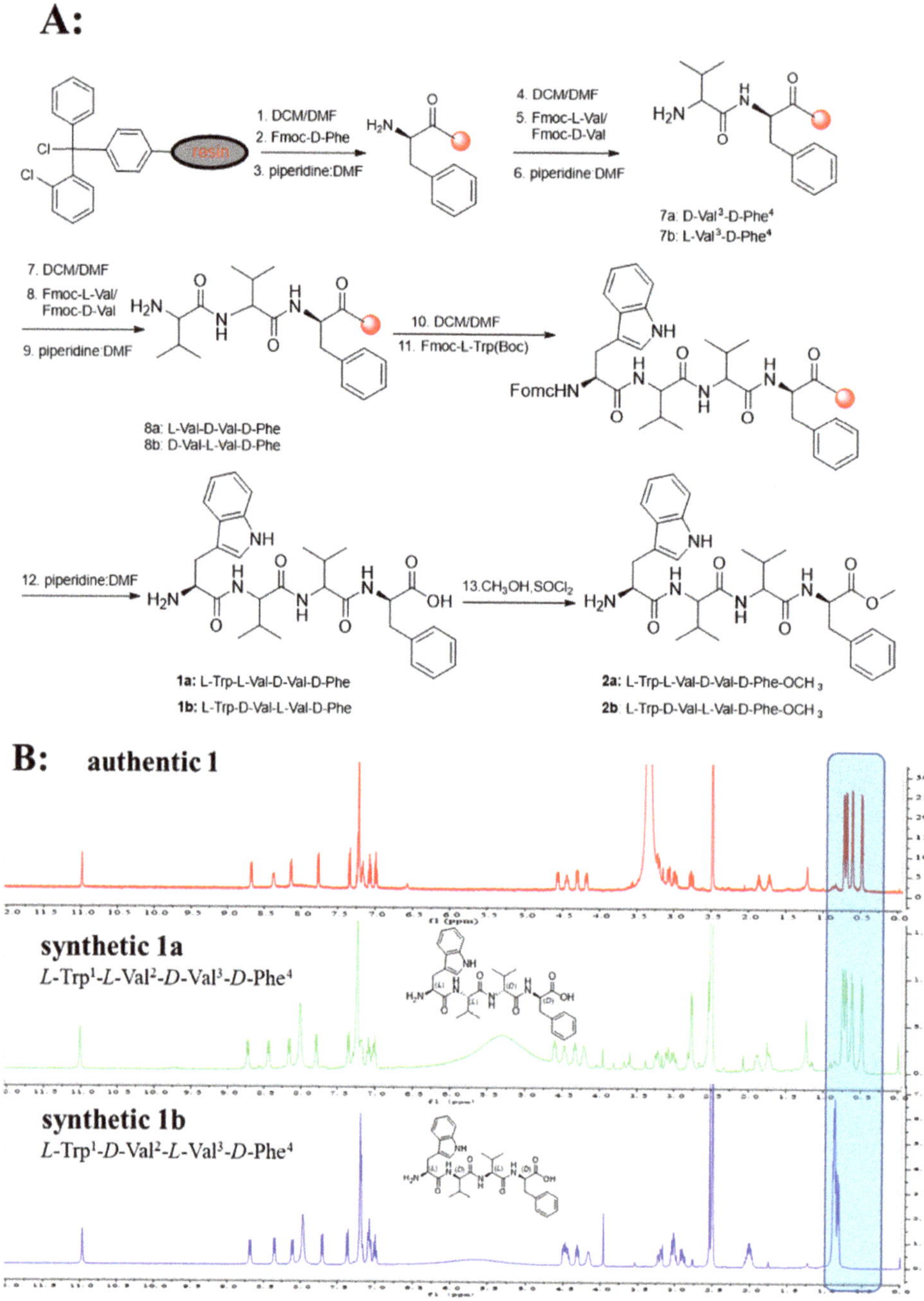

Scheme 1. (**A**): Solid-Phase Synthesis of **1a**, **1b** and **2a**, **2b**. (**B**): Comparison of ^{1}H-NMR spectra of authentic **1**, synthetic **1a**, and **1b**.

Compound **2** named psegynamide B was obtained as a white amorphous powder. Its molecular formula was determined to be $C_{31}H_{41}N_5O_5$ based on the molecular ion peak at *m/z* 564.3172 $[M + H]^+$ (calcd 564.3180) in the HRESIMS analysis, which showed 14 Da molecular weight (MW) surplus to compound **1**. The ^{1}H and ^{13}C NMR of **2** were similar to those of **1** (Table 1), except for the presence of a methoxy group. The ^{1}H, ^{13}C NMR, and HSQC spectra showed a signal at δ_C 52.4/δ_H 3.56, which was assignable to a methoxy group. In addition, the methoxy site was established unambiguously by an HMBC experiment, in which a long-range correlation between CH_3 (δ_H 3.56) and C-1 (δ_C 172.6) in the Phe unit was observed. Combined analyses of ^{1}H-^{1}H COSY, HMBC, HSQC, and TOCSY spectra assigned the linkage of amino acid residues as Trp^1-Val^2-Val^3-Phe^4-OCH_3. Marfey's acidolytic analysis confirmed the presence of *L*-Val, *D*-Val, and *D*-Phe in compound **2**, which was a similar case to compound **1**. Then, the solid-phase total synthesis and Marfey's alkaline hydrolytic method were also used to determine the absolute configurations of compound **2**. Finally, the absolute configuration of **2** was assigned to be *L*-Trp^1-*D*-Val^2-*L*-Val^3-*D*-Phe^4-OCH_3.

Compound **3** named psegynamide C was isolated as a white amorphous powder. The HRESIMS analysis of compound **3** gave a hydrogen adduct $[M + H]^+$ at m/z 566.2977 (calcd 566.2973), corresponding to the molecular formula of $C_{30}H_{39}N_5O_6$, which showed one more oxygen than compound **1**. In the ^{1}H NMR spectrum (Table 1), two aromatic H-atom signals appearing at δ_H 6.97 (2H, overlap, m) and 6.59 (2H, overlap, m) were attributable to a 1,4-substituted phenyl. In addition, a characteristic signal of OH was observed at low-field δ_H 9.18 (1H, s). Careful analysis of NMR of compound **3** revealed a similar structure as compound **1**, except for the presence of Tyr residue instead of Phe residue. Thus, the planar structure of compound **3** was established as Trp^1-Val^2-Val^3-Tyr^4. Based on Marfey's acid/alkaline hydrolytic analysis, as well as solid-phase total synthesis, the absolute configuration of **3** was assigned to be *L*-Trp^1- *D*-Val^2-*L*-Val^3-*D*-Tyr^4.

Compound **4** was obtained as a white amorphous powder. Its molecular formula was determined to be $C_{31}H_{39}N_5O_5$ on the basis of the molecular ion peak at *m/z* 562.3032 $[M + H]^+$ (calcd 562.3024) in the HRESIMS analysis, indicating 15 degrees of hydrogen deficiency. Compared the 1D-NMR data with compound **1**, compound **4** has an additional methylene (δ_C 40.8) and the CH-5 of Trp in compound **1** was changed as a quaternary carbon. In addition, one more degree of unsaturation indicated a cyclic structure for compound **4**. With the assistance of COSY correlations of H_2-5′/NH-2′, H-2/NH-2′ and H-2/H_2-3, and the HMBC correlations from H_2-5′ to C-4/C-2, the tetrahydropyridoindole residue was deduced. Compound **4** represented the first example naturally occurring peptide with a tetrahydropyridoindole moiety. On the basis of Marfey's acid hydrolytic analysis and solid-phase total synthesis, the absolute configuration of **4** was assigned to be as alternating LDLD chirality and named psegynamide D.

Psegymamides E-F (**5–6**) were obtained as white amorphous powders with the molecular formulas of $C_{32}H_{41}N_5O_5$ and $C_{32}H_{41}N_5O_6$ by HRESIMS, respectively. The 1D NMR spectra of **5** and **6** (Table 2) indicated a skeleton similar to psegynamide D (**4**). The difference between **4** and **5** was the replacement of the hydroxide group by a methoxy group (δ_C 52.3/δ_H 3.57) in the Phe unit, which was confirmed by HMBC correlation from OCH_3 (δ_H 3.57) and C-1 (δ_C 172.7). The difference between **4** and **6** was the presence of Tyr-OCH_3 residue instead of Phe residue. Thus, the planar structures of **5–6** were established. Similarly, the absolute configuration of **5–6** was assigned as alternating LDLD chirality by Marfey's method and solid-phase total synthesis and finally named psegynamides E-F, respectively.

Table 2. ^{1}H and ^{13}C NMR data in DMSO-d_6 for compounds **4–6** (^{a} measured at 600 MHz; ^{b} measured at 150 MHz).

	No	Compound 4		Compound 5		Compound 6	
		δ_H ^{a}, Mult (J in Hz)	δ_C ^{b}	δ_H ^{a}, Mult (J in Hz)	δ_C ^{b}	δ_H ^{a}, Mult (J in Hz)	δ_C ^{b}
	1	-	168.4	-	168.8	-	168.9
	2	4.26 (1H, m)	55.5	4.21 (1H, overlap, m)	55.8	4.20 (1H, overlap, m)	55.9
	3	3.38 (1H, m) 2.83 (1H, m)	23.6	3.38 (1H, overlap, m) 2.82 (1H, overlap, m)	24.1	3.37 (1H, overlap, m) 2.81(1H, t, J = 13.7 Hz)	24.1
	4	-	105.0	-	105.3	-	105.3
	5	-	125.8	-	127.0	-	127.0
	6 (NH)	11.1 (1H, s)	-	11.18 (1H, s)	-	11.10 (1H, s)	-
	7	-	136.4	-	136.8	-	136.8
	8	7.37 (1H, d, J = 8.1 Hz)	111.5	7.33 (1H, d, J = 8.1 Hz)	112.0	7.33 (1H, d, J = 8.2 Hz)	122.3
	9	7.11 (1H, t, J = 7.3 Hz)	121.8	7.07 (1H, t, J = 7.2 Hz)	122.2	7.07 (1H, t, J = 7.4 Hz)	112.0
	10	7.47 (1H, d, J = 7.8 Hz)	118.0	7.43 (1H, d, J = 7.8 Hz)	119.6	7.43 (1H, d, J = 7.9 Hz)	119.6
	11	7.02 (1H, t, J = 7.4 Hz)	119.2	6.97 (1H, t, J = 7.3 Hz)	118.3	7.00 (1H, overlap)	118.4
	12	-	125.8	-	126.2	-	126.2
	2′ (NH)	9.68 (1H, s)	-	9.67 (1H, s)		-	
	5′	4.44 (1H, overlap, m) 4.28 (1H, overlap, m)	40.2	4.41 (1H, overlap, m) 4.26 (1H, overlap, m)	40.8	4.40 (1H, overlap, m) 4.25 (1H, overlap, m)	40.8
Val-1	1	-	170.4	-	170.9	-	170.9
	2	4.68 (1H, m)	57.1	4.61 (1H, dd, J = 6.0, 9.4 Hz)	57.6	4.63 (1H, dd, J = 6.1, 6.4 Hz)	57.5
	3	2.01 (1H, m)	31.6	2.00 (1H, m)	32.0	1.99 (1H, m)	32.1
	4	0.89 (3H, d, J = 6.8 Hz)	17.8	0.85 (1H, d, J = 6.8 Hz)	18.2	0.85 (3H, d, J = 6.8 Hz)	18.2
	5	0.91 (3H, d, J = 6.8 Hz)	19.3	0.87 (1H, d, J = 6.8 Hz)	19.9	0.88 (3H, d, J = 6.7 Hz)	19.9
	NH	8.83 (1H, d, J = 9.5 Hz)	-	8.80 (1H, d, J = 9.4 Hz)	-	8.78 (1H, d, J = 9.4 Hz)	-
Val-2	1	-	170.8	-	171.4	-	171.4
	2	4.36 (1H, m)	56.8	4.28 (1H, dd, J = 5.8, 9.2 Hz)	57.5	4.31 (1H, dd, J = 5.7, 9.2 Hz)	57.3
	3	1.73 (1H, m)	31.1	1.75 (1H, m)	31.4	1.77 (1H, m)	31.5
	4	0.50 (3H, d, J = 6.8 Hz)	17.4	0.52 (1H, d, J = 6.8 Hz)	17.9	0.56 (3H, d, J = 6.8 Hz)	17.9
	5	0.64 (3H, d, J = 6.8 Hz)	19.4	0.62 (1H, d, J = 6.8 Hz)	19.7	0.66 (3H, d, J = 6.8 Hz)	19.8
	NH	8.24 (1H, d, J = 9.1 Hz)	-	8.20 (1H, d, J = 9.2 Hz)	-	8.20 (1H, d, J = 9.1 Hz)	-

Table 2. *Cont.*

	No	Compound 4		Compound 5		Compound 6	
		δ_H [a], Mult (*J* in Hz)	δ_C [b]	δ_H [a], Mult (*J* in Hz)	δ_C [b]	δ_H [a], Mult (*J* in Hz)	δ_C [b]
Phe/Phe-OCH_3/Tyr	1	-	173.3	-	172.7	-	172.9
	2	4.49 (1H, m)	53.4	4.48 (1H, m)	54.0	4.40 (1H, overlap, m)	54.3
	3	3.10 (1H, dd, *J* = 4.4, 13.8 Hz) 2.79 (1H, dd, *J* = 10.6,13.8 Hz)	37.1	3.03 (1H, dd, *J* =5.0, 13.9 Hz) 2.82 (1H, overlap, m)	37.3	2.89 (1H, dd, *J* = 5.2, 13.9 Hz) 2.69 (1H, dd, *J* = 10.0, 13.7 Hz)	36.7
	4	-	137.6	-	137.6	-	127.5
	5	7.25 (1H, overlap, m)	129.2	7.21 (1H, overlap, m)	129.6	6.96 (1H, d, *J* = 8.5 Hz)	130.5
	6	7.25 (1H, overlap, m)	128.3	7.21 (1H, overlap, m)	128.7	6.60 (1H, d, *J* = 8.3 Hz)	115.5
	7	7.19 (1H, m)	126.5	7.16 (1H, overlap, m)	127.1	-	156.6
	8	7.25 (1H, overlap, m)	128.3	7.21 (1H, overlap, m)	128.7	6.60 (1H, d, *J* = 8.3 Hz)	115.5
	9	7.25 (1H, overlap, m)	129.2	7.21 (1H, overlap, m)	129.6	6.96 (1H, d, *J* = 8.5 Hz)	130.5
	NH	8.47 (1H, d, *J* = 7.8 Hz)	-	8.58 (1H, d, *J* =8.0 Hz)	-	8.49 (1H, d, *J* = 7.9 Hz)	-
	1-OCH_3	-	-	3.57 (3H, s)	52.5	3.56 (3H, s)	52.4
	7-OH	-	-	-	-	9.21(1H, s)	-

Nicotinic acetylcholine receptors (nAChRs) belong to the ligand-gated ion channel superfamily [23,24]. There are a large number of nAChR subunits, such as α1-10, β1-4, γ, δ, and ε, which may mediate analgesic effects and adverse reactions [25,26]. Targeting specific nAChRs subtypes may reduce adverse reactions while maintaining efficient analgesia and becoming a real new analgesic target. All new compounds (**1**–**6**) were evaluated for their activity at ACh-evoked currents mediated by human (h) α1β1εδ, α1β1γδ, α3β2, α3β4, α4β2, α7, and α9α10 nAChRs, among which compound **2** showed significant inhibitory activity at all subtypes (>70% inhibition) except α7 (Figure 3, Table S1). Interestingly, compound **2** selectively inhibited (>98% inhibition, n = 8–11) the α1β1εδ and α1β1γδ subtypes. For comparison, the α-conotoxin GI peptide from the marine cone snail *Conus geographus* venom antagonizes ACh-evoked currents mediated by hα1β1εδ with a half-maximal inhibitory concentration (IC_{50}) of 20 nM [27]. According to the structural features of **1**–**6**, the preliminary structure–activity relationship (SAR) of inhibitory activities was tentatively discussed. Generally, compounds **1**–**3** exhibited stronger activities than compounds **4**–**6**, indicating that the presence of tetrahydropyridoindoles unit decreases inhibitory activities. Psegynamide B (**2**) showed stronger activities than psegynamide A (**1**), suggesting that C-terminal replacement with the methoxy group was beneficial to the activity.

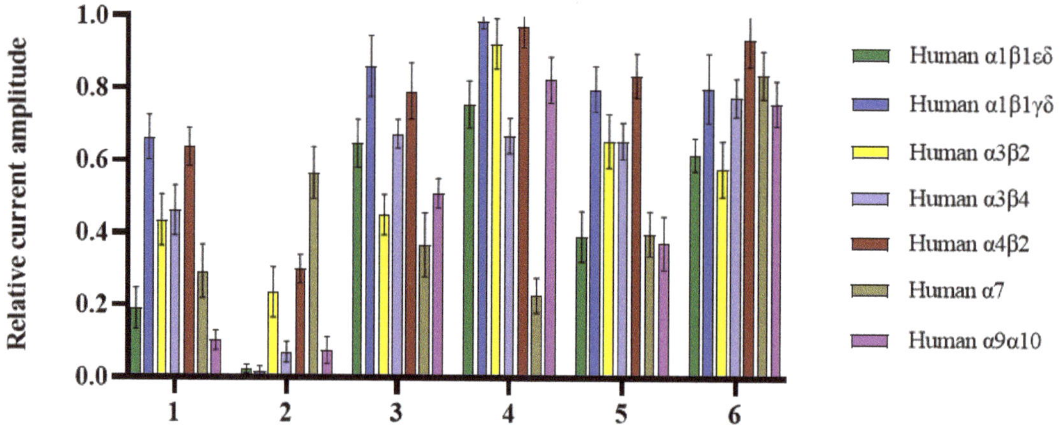

Figure 3. Bar graph of compounds (**1**–**6**) (100 μM) inhibition of ACh-evoked peak current amplitude mediated by human (h) α1β1εδ, α1β1γδ, α3β2, α3β4, α4β2, α7 and α9α10 nAChRs. Whole-cell currents at hα1β1εδ and hα1β1γδ were activated by 5 μM ACh, hα3β2 and hα9α10 were activated by 6 μM ACh, and hα3β4, hα4β2, and hα7 were activated by 300, 3, and 100 μM ACh, respectively (mean ± SD, n = 6–11).

To explain the different inhibitory activities and obtain further insight into the mechanism of nAChRs inhibition, molecular docking studies were carried out to explore the possible binding modes of compounds **1**–**6** and key interactions with nAChRs. The crystal structure of hα4β2 (PDB code: 5KXI) was used for further docking. The results in the docking study matched well with the inhibition activities (Table S2), compounds with lower calculated docking scores are considered to have higher binding affinities with the target. Among them, compound **2** showed the highest binding affinity to hα4β2 with the most negative free binding energy (−4.2 kcal/mol). Analysis for optimized binding conformation of compound **2** displayed that the hydrogen at N-6 in Trp interacts with Ser A180 through a hydrogen bond with a distance of 2.2 Å, and the Trp formed H-pi stacked bonds with Gln B50 with a distance of 4.0 Å (Figure 4). While the Trp residue was replaced by the tetrahydropyridoindoles unit, the H-H and H-pi bond was lost. In addition, the NH in Val (2) of compound **2** interacts with Asp A49 through a hydrogen bond with a distance of 2.5 Å. It is interesting to find that the Arg B45 may be important for the activity because of that NH_2 in Trp and the C=O in Phe-OCH_3 of compound **2** simultaneously interact with Arg B45 through a hydrogen bond with distance of 2.2 Å and 3.3 Å, respectively. Furthermore, considering non-steroidal anti-inflammatory drugs and narcotics (opioids) are currently the most commonly used analgesic drugs, these drugs exhibit limitations in efficacy, unwanted side effects, and the problem of drug abuse [28,29]. The above findings provided that compound **2** might be helpful in developing different analgesic drugs by inhibiting nicotinic acetylcholine receptors.

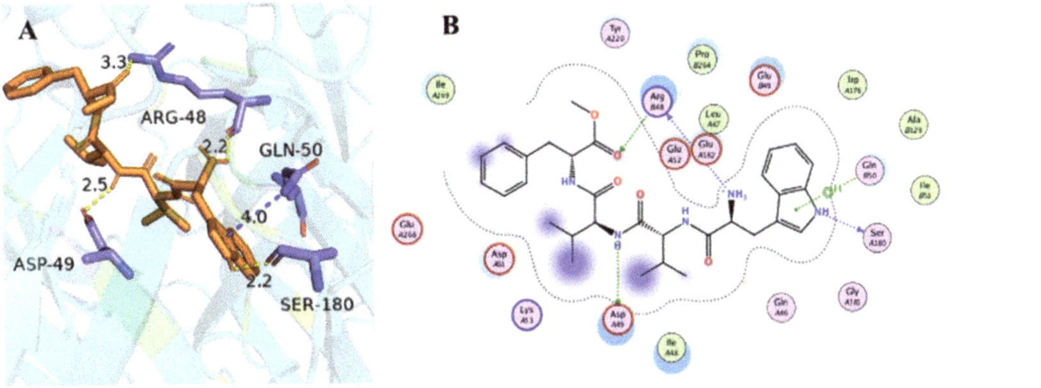

Figure 4. Docking models of compound **2** with the crystal structure of h$\alpha4\beta2$ (PDB code: 5kxi). (**A**) H-bonding interactions of compound **2** with h$\alpha4\beta2$. (**B**) 2D schematic diagram of compound **2** with h$\alpha4\beta2$. The H-bonds and H-pi stacked bonds are shown in yellow dashed lines and blue dashed lines, respectively.

3. Materials and Methods

3.1. General Experimental Procedures

NMR spectra were obtained on JEOLJN M-ECP 600 MHz spectrometers, of which TMS was an internal standard. The optical rotation of new compounds was calculated in MeOH on a JASCOP-1020 digital polarimeter. By using an LTQ Orbitrap XL (Thermo Fisher Scientific, Waltham, MA, USA) mass spectrometer, HRESIMS data were obtained. The spray voltage, capillary voltage, and tube lens were 4.0 kv, 16 v, and 35 v, respectively. The capillary temperature was 275 °C with a sheath gas flow rate of 10 arb. unit, and FT full mass spectra were acquired in the positive ionization mode at a resolution of 30,000 with 100–1500 Da mass range. The crude extract of *Pseudogymnoascus* sp. HDN17-933 were analyzed by reversed-phase HPLC (5 × 250 mm YMC C18 column, 5 µm) with a linear gradient of MeOH (A) and 0.1% aqueous TFA (B) from 5% to 100% A over 60 min at a flow rate of 1 mL/min. Column chromatography was carried out using the following chromatographic substrates: silica gel (300–400 mesh; Qingdao Marine Chemical Industrials, Qingdao, China), Sephadex LH-20 (GE Healthcare, Bio-Sinences Corp, Piscataway, NJ, USA). HPLC of the Waters company equipped with a 2998 PDA detector was performed on an ODS column (YMC-Pack ODS-A, 10 × 250 mm, 5 µm, 3 mL/min). UV spectra were carried out on Waters 2487 developed by Waters Corporation, Milford, USA. All Fmoc-amino acids were purchased from GL Biochem Ltd. (Shanghai, China). 2-chlorotrityl chloride resin at 1 mmol scale was purchased from Tianjin Nankai Hecheng S&T Co., Ltd. (Tianjin, China).

3.2. Fungal Material and Fermentation

The fungal strain HDN17-933 was isolated from Fildes Peninsula, Antarctica, and identified as *Pseudogymnoascus* sp. based on internal transcribed spacer DNA sequencing. The sequence is available with the accession number MZ268166 at Genbank. It has been submitted to the Key Laboratory of Marine Drugs, working under the Ministry of Education of China, School of Medicine and Pharmacy, Ocean University of China. The fungus was cultured under the static condition at 15 °C in winter for 30 days in 1 L Erlenmeyer flasks, each containing rice (80 g) and naturally collected seawater (Huiquan Bay, Yellow Sea) (120 mL).

3.3. Extraction and Purification

All fermentation broth (40 L) was extracted with MeOH, filtered, concentrated, and partitioned between EtOAc and H_2O. The EtOAc extract was evaporated under reduced pressure to give a crude gum (18.5 g). Moreover, the extract was subjected vacuum chromatography on silica gel (200–300 mesh) and eluted with stepped gradient elution via DCM:MeOH (100:1–1:1) to yield eight combined fractions (Fr.1 to Fr.8). Fr.6 was further separated using ODS (MeOH: H_2O; 30:80–100:0) to obtain Fr.6-1 to Fr.6-6. Then Fr.6-3 and Fr.6-4 were further subjected to a Sephadex LH-20 column and eluted with MeOH to provide subfractions (from Fr.6-3-1 to Fr.6-3-4 and from Fr.6-4-1 to Fr.6-4-5). Fr.6-4-2 was purified by HPLC eluted with MeCN-H_2O (38:62) to give compounds **1** (4.2 mg, t_R = 11.5 min), **2** (3.4 mg, t_R = 7.2 min), and **5** (3.5 mg, t_R = 8.0 min), respectively. Fr.6-3-2 was purified by HPLC eluted with MeOH-H_2O (65:35) to give compounds **3** (5.1 mg, t_R = 10.2 min), **6** (4.3 mg, t_R = 11.0 min), and **4** (3.2 mg, t_R = 13.5 min), respectively.

3.4. Physical and Chemical Data

Psegynamide A (**1**): white amorphous powder; $[\alpha]^{20}_D$ + 30.2 (*c* 0.2, MeOH); UV (DAD) λ_{max} 218 nm, 283 nm; ^{1}H and ^{13}C NMR (DMSO-d_6), see Table 1; HREIMS *m/z* 550.3025 $[M + H]^+$ (cacld for $C_{30}H_{40}N_5O_5$, 550.3024).

Psegynamide B (**2**): white amorphous powder; $[\alpha]^{20}_D$ + 23.8 (*c* 0.3, MeOH); UV (DAD) λ_{max} 217 nm, 283 nm; ^{1}H and ^{13}C NMR (DMSO-d_6), see Table 1; HREIMS *m/z* 564.3172 $[M + H]^+$ (cacld for $C_{31}H_{42}N_5O_5$, 564.3180).

Psegynamide C (**3**): white amorphous powder; $[\alpha]^{20}_D$ + 25.3 (*c* 0.3, MeOH); UV (DAD) λ_{max} 217 nm, 282 nm; ^{1}H and ^{13}C NMR (DMSO-d_6), see Table 1; HREIMS *m/z* 566.2977 $[M + H]^+$ (cacld for $C_{30}H_{40}N_5O_6$, 566.2973).

Psegynamide D (**4**): white amorphous powder; $[\alpha]^{20}_D$ − 14.7 (*c* 0.3, MeOH); UV (DAD) λ_{max} 217 nm, 282 nm; ^{1}H and ^{13}C NMR (DMSO-d_6), see Table 2; HREIMS *m/z* 562.3032 $[M + H]^+$ (cacld for $C_{31}H_{40}N_5O_5$, 562.3024).

Psegynamide E (**5**): white amorphous powder; $[\alpha]^{20}_D$ − 17.8 (*c* 0.2, MeOH); UV (DAD) λ_{max} 216 nm, 282 nm; ^{1}H and ^{13}C NMR (DMSO-d_6), see Table 2; HREIMS *m/z* 576.3186 $[M + H]^+$(cacld for $C_{32}H_{42}N_5O_5$, 576.3180).

Psegynamide F (**6**): white amorphous powder; $[\alpha]^{20}_D$ − 20.3 (*c* 0.2, MeOH); UV (DAD) λ_{max} 216 nm, 283 nm; ^{1}H and ^{13}C NMR (DMSO-d_6), see Table 2; HREIMS *m/z* 592.3133 $[M + H]^+$ (cacld for $C_{32}H_{42}N_5O_6$, 592.3130).

3.5. Advanced Marfey's Analysis of Acid Hydrolytic for Val, Phe, Tyr

Compounds **1–6** (1.0 mg each) were reacted with 6 M HCl (1.5 mL) at 110 °C for 12 h; the hydrolysates were concentrated to dryness. The hydrolysates and standard amino acids (150 μM) were then successively treated with water (300 μL), FDAA (10 mg/mL solution in acetone, 100 μL), acetone (300 μL), and $NaHCO_3$ (1 M, 150 μL) at 45 °C water bath heating for 2.0 h. Then, the reaction was stopped with HCl (2 M, 75 μL) prior to HPLC analysis. Amino acid standards were similarly derivatized with FDAA. The resulting FDAA derivatives of compounds **1–6**, *L*- and *D*-Val, *L*- and *D*- Phe, *L*- and *D*- Tyr were analyzed by HPLC eluted with a linear gradient of MeOH (A) and 0.10% aqueous TFA (B) from 40% to 100% in an over 45 min with UV detection at 340 nm.

3.6. Modified Marfey's Analysis of Alkaline Hydrolytic for Trp

Considering that Trp FDAA derivatives were not detected in the resulting FDAA derivatives of compounds **1–3**, it might result from that Trp amino acid was hydrolytic and destroyed in strong hydrochloric acid. Therefore, Marfey's alkaline hydrolytic method was considered to determine the absolute configuration of Trp residue. Through continuous attempts, the alkaline hydrolytic condition of 5 M LiOH at 110 °C for 16 h was finally adopted, and the absolute configuration of Trp was determined as *L*.

3.7. Solid-Phase Total Synthesis of 1a, 1b

Compounds **1a**, **1b** were synthesized by 2-chlorotrityl chloride resin (loading 1 mmol/g) as described in Scheme 1. Then, 4 equiv of Fmoc-*D*-Phe (2 mmol) was added to a suspension of 1 equiv of 2-chlorotrityl chloride resin (0.50 g, 0.5 mmol), 4 equiv of HCTU (2 mmol), 8 equiv of DIEA (4 mmol), and DCM/DMF (*v:v* = 1:1). After stirring for 2 h, the resin was filtered and washed with DCM/DMF (1:1) for three times. The reaction mixture was treated with piperidine:DMF (*v:v* = 1:4) for stirring 30 min to deprotect the Fmoc group. After filtering and washing, the resin was used as resin-*D*-Phe^4-NH_2 for the next coupling reaction.

A solution of HCTU (2 mmol) and DIEA (4 mmol) in DCM/DMF (1:1) was added to a mixture of resin-*D*-Phe^4-NH_2, Fmoc-*L*-Val/ Fmoc-*D*-Val (2 mmol). After the reaction mixture was stirred for 60 min, the resin was filtered and washed. To remove the Fmoc group, the same procedure was repeated as above. The resin was used as resin-*D*-Phe^4-*L*-Val^3-NH_2/*D*-Phe^4-*D*-Val^3-NH_2 for the subsequent next coupling reaction, respectively. A solution of HCTU (2 mmol) and DIEA (4 mmol) in DCM/DMF (1:1) was added to a mixture of resin-*D*-Phe^4-*L*/*D*-Val^3-NH_2, Fmoc-*D*-Val/ Fmoc-*L*-Val (2 mmol). After the reaction mixture was stirred for 60 min, the resin was filtered and washed. To remove the Fmoc group, the same procedure was repeated as above. The resin was used as resin-*D*-Phe^4-*L*-Val^3-*D*-Val^2-NH_2/*D*-Phe^4-*D*-Val^3-*L*-Val^2-NH_2 for the next coupling reaction, respectively.

A solution of HCTU (2 mmol) and DIEA (4 mmol) in DCM/DMF (1:1) was added to a mixture of resin-*D*-Phe^4-*L*/*D*-Val^3-*D*/*L*-Val^2-NH_2, Fmoc-*L*-Trp (Boc)-OH (2 mmol). After the reaction mixture was stirred for 60 min, the resin was filtered and washed. To remove the Fmoc group, the same procedure was repeated as above. The resin was used as resin-*D*-Phe^4-*L*-Val^3-*D*-Val^2-*L*-Trp^1(Boc)-NH_2/*D*-Phe^4-*D*-Val^3-*L*-Val^2-*L*-Trp^1(Boc)-NH_2, respectively. Subsequent to TFA cleavage (TFA: TIPS: H_2O, 90:5:5, 3 h), the reaction solution was treated with cold ether to obtain precipitation. The precipitate was furtherly purified by HPLC to give the compounds **1a** (*L*-Trp^1-*L*-Val^2-*D*-Val^3-*D*-Phe^4) and **1b** (*L*-Trp^1-*D*-Val^2-*L*-Val^3-*D*-Phe^4), respectively.

3.8. Xenopus Laevis Oocyte Preparation and Microinjection

All procedures were approved by the University of Wollongong Animal Ethics Committees (project number: AE2003). Female *Xenopus laevis* were sourced from Nasco (Fort Atkinson, WI, USA), and a maximum of four frogs were kept in a 15 L aquarium at 20–26 °C with 12 h light/dark cycle. Oocytes were obtained from five-year-old frogs anesthetized with 1.7 mg/mL ethyl 3-aminobenzoate methanesulfonate (pH 7.4 with $NaHCO_3$). Stage V-VI oocytes (Dumont's classification; 1200–1300 μm diameter) were defolliculated with 1.5 mg/mL collagenase Type II (Worthington Biochemical Corp., Lakewood, NJ, USA) at room temperature for 1–2 h in OR-2 solution containing (in mM): 82.5 NaCl, 2 KCl, 1 $MgCl_2$ and 5 HEPES at pH 7.4.

The human muscle nAChR clones ($\alpha 1$, $\beta 1$, γ, δ and ε) were purchased from Integrated DNA Technologies (Coralville, IA, USA), whereas the human $\alpha 3$, $\alpha 9$, $\alpha 10$, $\beta 2$, and $\beta 4$ clones were purchased from OriGene (Rockville, MD, USA), and all were subsequently inserted into the pT7TS vector. The human $\alpha 4$ and $\alpha 7$ clones were obtained from Prof. Jon Lindstrom (University of Pennsylvania, Philadelphia, PA, USA). Plasmid constructs of the human nAChR clones were linearized for in vitro mRNA synthesis using mMessage mMachine transcription kit (AMBION, Forster City, CA, USA).

Oocytes were injected with 5 ng cRNA for h$\alpha 1\beta 1\gamma\delta$, h$\alpha 1\beta 1\varepsilon\delta$, h$\alpha 3\beta 2$, h$\alpha 3\beta 4$ and h$\alpha 4\beta 2$, 10 ng cRNA for h$\alpha 7$ nAChR, and 35 ng cRNA for h$\alpha 9\alpha 10$ nAChR (concentration confirmed spectrophotometrically and by gel electrophoresis). The muscle subunit cRNA ratio was 2:1:1:1 ($\alpha 1$:$\beta 1$: γ/ε:δ), whereas the heteromeric α and β subunit cRNA ratio was 1:1, injected using glass pipettes pulled from glass capillaries (3-000-203 GX, Drummond Scientific Co.,

Broomall, PA, USA). Oocytes were incubated at 18 °C in sterile ND96 solution composed of (in mM): 96 NaCl, 2 KCl, 1 $CaCl_2$, 1 $MgCl_2$, and 5 HEPES at pH 7.4, supplemented with 5% fetal bovine serum, 50 mg/L gentamicin (GIBCO, Grand Island, NY, USA) and 10,000 U/mL penicillin-streptomycin (GIBCO).

3.9. Oocyte Two-Electrode Voltage Clamp Recording and Data Analysis

Electrophysiological recordings were carried out 2–5 days post cRNA microinjection. Two-electrode voltage clamp recordings of *X. laevis* oocytes expressing human nAChRs were performed at room temperature (21–24 °C) using a GeneClamp 500B amplifier and pClamp9 software interface (Molecular Devices, Sunnyvale, CA, USA) at a holding potential −80 mV. Voltage-recording and current-injecting electrodes were pulled from GC150T-7.5 borosilicate glass (Harvard Apparatus, Holliston, MA) and filled with 3 M KCl, giving resistances of 0.3–1 MΩ. Due to the Ca^{2+} permeability of α9α10 nAChRs, 100 μM BAPTA-AM incubation was carried out before recording to prevent the activation of *X. laevis* oocyte endogenous Ca^{2+}-activated chloride channels. Oocytes expressing hα9α10 nAChRs were perfused with ND115 solution containing (in mM): 115 NaCl, 2.5 KCl, 1.8 $CaCl_2$, and 10 HEPES at pH 7.4, whereas oocytes expressing all other nAChR subtypes were perfused with ND96 solution using a continuous Legato 270 push/pull syringe pump perfusion system (KD Scientific, Holliston, MA, USA) at a rate of 2 mL/min in an OPC-1 perfusion chamber of <20 μL volume (Automate Scientific, Berkeley, CA, USA).

Initially, oocytes were briefly washed with ND115/ND96 solution, followed by 3 applications of ACh at a half-maximal excitatory ACh concentration (EC_{50}) for the nAChR subtypes (3 μM for hα4β2, 5 μM for hα1β1γδ and hα1β1εδ, 6 μM for hα3β2 and hα9α10, 100 μM for hα7 and 300 μM for hα3β4) [27]. Washout with bath solution was performed for 3 min between ACh applications. Oocytes were incubated with compounds for 5 min with the perfusion system turned off, followed by co-application of ACh and compound with flowing bath solution. All compound solutions were prepared in ND115/ND96 + 0.1% bovine serum albumin (BSA). Incubation with 0.1% BSA was performed to ensure that the BSA and the pressure of the perfusion system had no effect on nAChRs. Peak current amplitudes before (ACh alone) and after (ACh + compound) compound incubation were measured using Clampfit version 10.7.0.3 software (Molecular Devices, Sunnyvale, CA, USA), where the ratio of ACh + compound-evoked current amplitude to ACh alone-evoked current amplitude was used to assess the activity of the compounds at the nAChRs. All electrophysiological data were pooled (n = 6–11) and represent means ± standard deviation (SD). Data analysis was performed using GraphPad Prism 5 (GraphPad Software, La Jolla, CA, USA).

3.10. Molecular Docking

Molecular docking studies were performed using MOE 2016. The crystal structure of human α4β2 nAChR (PDB identifier: 5kxi) was obtained from the Protein Data Bank (http://www.rcsb.org, accessed on 28 September 2016). Prior to docking, heteroatoms and water molecules in the protein were removed. In the meantime, compounds were minimized. The result of each ligand was furtherly analyzed by using PyMOL.

4. Conclusions and Discussion

In conclusion, the genus *Pseudogymnoascus* is a group of psychrophilic fungi that possess good potential in serving structurally unique structures but not widely investigated. In this study, chemical investigation of secondary metabolites from the psychrophilic fungus *Pseudogymnoascus* sp. HDN17-933 led to the discovery of six new tetrapeptides psegynamides A–F (**1–6**), among which psegynamides D–F (**4–6**) was the first naturally occurring peptide bearing a tetrahydropyridoindoles moiety. To the best of our knowledge, only a few synthetic peptides bearing the tetrahydropyridoindoles motif have been reported [30,31], for example, tetrahydropyridoindoles as cholecystokinin and gastrin antagonists in 1992 [30]. Solid synthesis techniques assisted by Marfey's method were employed to work out the absolute configuration, especially in solving the challenges of determining the order of *L*-Val and *D*-Val isomers in the peptide sequences. Moreover, compound **2** was found to have bioactivity by inhibiting human nAChRs. Considering that non-steroidal anti-inflammatory drugs and narcotics (opioids) are currently the most commonly used analgesic drugs, these drugs exhibit limitations in efficacy, unwanted side effects, and the problem of drug abuse. To overcome these problems, the discovery of different molecular participants in the pain pathways could bring new opportunities for therapeutic intervention. Compound **2** might be helpful in developing potential natural short peptide inhibitor of nAChRs from Antarctica-derived fungus. The above findings illustrate and highlight the validity of exploiting psychrophilic fungus for the discovery of structurally novel and bioactive natural products.

Supplementary Materials: The following are available as supplementary materials https://www.mdpi.com/article/10.3390/md20100593/s1: Figure S1: 18S rRNA sequences data of *Pseudogymnoascus* sp. HDN17-933; Figure S2: HPLC analysis of the crude extract of HDN17-933; Figures S3–S57: 1D and 2D NMR spectra, HRESIMS spectra, IR spectra of compounds **1–6**; Figures S58–S88: Advanced Marfey's acid/alkaline hydrolytic analysis of compounds **1–6**, and NMR spectra of solid-phase total synthesis. Table S1. Table of compounds (**1–6**) inhibition of ACh-evoked peak current amplitude mediated by human (h) $\alpha1\beta1\varepsilon\delta$, $\alpha1\beta1\gamma\delta$, $\alpha3\beta2$, $\alpha3\beta4$, $\alpha4\beta2$, $\alpha7$ and $\alpha9\alpha10$ nAChRs. Table S2. Free binding energy estimation and molecular interactions of compounds **1–6.**

Author Contributions: The contributions of the respective authors are as follows: X.H.: conceptualization, resources, investigation, writing—original draft, and writing—review and editing. C.L., R.Z., Y.L., H.L. and Y.Z.: investigation and data curation. H.-S.T. and R.Y.: assistance of nicotinic acetylcholine receptors activity and molecular docking. D.L., G.Z., T.Z. and Q.C.: funding acquisition, and writing—review and editing. All authors have read and agreed to the published version of the manuscript.

Funding: This research was funded by Marine S&T Fund of Shandong Province for Pilot National Laboratory for Marine Science and Technology (2022QNLM030003-1, 2022QNLM030003-2), National Natural Science Foundation of China (81991522, 41976105), the NSFC-Shandong Joint Fund (U1906212), the Hainan Provincial Joint Project of Sanya Yazhou Bay Science and Technology City (2021CXLH0012), Taishan Scholar Youth Expert Program in Shandong Province (tsqn 202103153, tsqn 201812021), Major Basic Research Programs of Natural Science Foundation of Shandong Province (ZR2019ZD28) and an Australian Research Council (ARC) Discovery Project Grant (DP150103990) awarded to D.J. Adams.

Institutional Review Board Statement: Not applicable.

Informed Consent Statement: Not applicable.

Data Availability Statement: Not applicable.

Acknowledgments: We thank Adams for the facilities and support of the functional studies carried out in *Xenopus* oocytes at IHMRI, University of Wollongong.

Conflicts of Interest: The authors declare no conflict of interest.

References

1. Hassan, N.; Rafiq, M.; Hayat, M.; Shah, A.A.; Hasan, F. Psychrophilic and psychrotrophic fungi: A comprehensive review. *Rev. Environ. Sci. Bio/Technol.* **2016**, *15*, 147–172. [CrossRef]
2. Golubev, W.I. New species of basidiomycetous yeasts, Rhodotorula creatinovora and R. yakutica, isolated from permafrost soils of Eastern-Siberian Arctic. *Mikol. Fitopatol.* **1998**, *32*, 8–13.
3. Tosi, L.; Carbognin, L.; Teatini, P.; Rosselli, R.; Stori, G.G. The ISES project subsidence monitoring of the catchment basin south of the Venice Lagoon. In *Land Subsidence*; La Garangola: Padova, Italy, 2000; pp. 113–126.
4. Ma, L.; Catranis, C.M.; Starmer, W.T.; Rogers, S.O. Revival and characterization of fungi from ancient polar ice. *Mycologist* **1999**, *13*, 70–73. [CrossRef]
5. Broady, P.A.; Weinstein, R.N. Algae, lichens and fungi in La Gorce Mountains, Antarctica. *Antarct. Sci.* **1998**, *10*, 376–385. [CrossRef]
6. Tojo, M.; Newshamb, K.K. Snow moulds in polar environments. *Fungal Ecol.* **2012**, *5*, 395–402. [CrossRef]
7. Robinson, C.H. Cold adaptation in Arctic and Antarctic fungi. *New Phytol.* **2001**, *151*, 341–353. [CrossRef]

8. McKenzie, R.L.; Björn, L.O.; Bais, A.; Ilyasd, M. Changes in biologically active ultraviolet radiation reaching the Earth's surface. *Photochem. Photobiol. Sci.* **2003**, *2*, 5–15. [CrossRef]
9. Selbmann, L.; Onofri, S.; Fenice, M.; Federici, F.; Petruccioli, M. Production and structural characterization of the exopolysaccharide of the Antarctic fungus *Phoma herbarum* CCFEE 5080. *Res. Microbiol.* **2002**, *153*, 585–592. [CrossRef]
10. Brown, A. Compatible solutes and extreme water stress in eukaryotic micro-organisms. *Adv. Microb. Physiol.* **1978**, *17*, 181–242.
11. Lewis, D.; Smith, D. Sugar alcohols (polyols) in fungi and green plants. I. Distribution, physiology and metabolism. *New Phytol.* **1967**, *66*, 143–184. [CrossRef]
12. Weinstein, R.N.; Montiel, P.O.; Johnstone, K. Influence of growth temperature on lipid and soluble carbohydrate synthesis by fungi isolated from fellfield soil in the maritime Antarctic. *Mycologia* **2000**, *92*, 222–229. [CrossRef]
13. Margesin, R.; Schinner, F.; Marx, J.C.; Gerday, C. Bacteria in snow and glacier ice. In *Psychrophiles: From Biodiversity to Biotechnology*; Springer: Berlin/Heidelberg, Germany, 2008; pp. 381–387.
14. Georlette, D.; Damien, B.; Blaise, V.; Depiereux, E.; Uversky, V.N. Structural and functional adaptations to extreme temperatures in psychrophilic, mesophilic, and thermophilic DNA ligases. *J. Biol. Chem.* **2003**, *278*, 37015–37023. [CrossRef]
15. Krohn, D. Endophytic fungi: A source of novel biologically active secondary metabolites. *Mycol. Res.* **2002**, *106*, 996–1004.
16. Singh, P.; Raghukumar, C.; Verma, P.; Shouche, Y. Fungal community analysis in the deep-sea sediments of the Central Indian Basin by culture-independent approach. *Microb. Ecol.* **2011**, *61*, 507–517. [CrossRef]
17. Li, D.; Wang, F.; Xiao, X.; Fang, Y.; Zhu, T.; Gu, Q.; Zhu, W. Trisorbicillinone A, a novel sorbicillin trimer, from a deep sea fungus, *Phialocephala* sp. FL30. *Tetrahedron Lett.* **2007**, *48*, 5235–5238. [CrossRef]
18. MacíAs, F.; Varela, R.M.; Simonet, A.M.; Cutler, H.G.; Cutler, S.J.; Ross, S.A.; Dunbar, D.C.; Dugan, F.M.; Hill, R.A. (+)-Brevione A. The first member of a novel family of bioactive spiroditerpenoids isolated from *Penicillium brevicompactum* Dierckx. *Tetrahedron Lett.* **2000**, *41*, 2683–2686. [CrossRef]
19. Wu, G.; Ma, H.; Zhu, T.; Jing, L.; Gu, Q.; Li, D. Penilactones A and B, two novel polyketides from Antarctic deep-sea derived fungus *Penicillium crustosum* PRB-2. *Tetrahedron* **2012**, *68*, 9745–9749. [CrossRef]
20. Sun, Y.L.; Zhang, X.Y.; Nong, X.H.; Xu, X.Y.; Qi, S.H. New antifouling macrodiolides from the deep-sea-derived fungus *Trichobotrys effuse* DFFSCS021. *Tetrahedron Lett.* **2016**, *57*, 366–370. [CrossRef]
21. Du, L.; Li, D.; Zhu, T.; Cai, S.; Wang, F.; Xiao, X.; Gu, Q. New alkaloids and diterpenes from a deep ocean sediment derived fungus *Penicillium* sp. *Tetrahedron* **2009**, *65*, 1033–1039. [CrossRef]
22. Fujii, K.; Ikai, Y.; Oka, H.; Suzuki, M.; Harada, K.I. A nonempirical method using LC/MS for determination of the absolute configuration of constituent amino acids in a peptide: Combination of Marfey's method with mass spectrometry and its practical application. *Anal. Chem.* **1997**, *69*, 549–557. [CrossRef]
23. Albuquerque, E.X.; Pereira, E.F.R.; Alkondon, M.; Rogers, S.W. Mammalian nicotinic acetylcholine receptors: From structure to function. *Physiol. Rev.* **2009**, *89*, 73–120. [CrossRef]
24. Thompson, A.J.; Lester, H.A.; Lummis, S.C.R. The structural basis of function in Cys-loop receptors. *Q. Rev. Biophys.* **2010**, *43*, 449–499. [CrossRef]
25. Anand, R.; Conroy, W.G.; Schoepfer, R.; Whiting, P.; Lindstrom, J. Neuronal nicotinic acetylcholine receptors expressed in Xenopus oocytes have a pentameric quaternary structure. *J. Biol. Chem.* **1991**, *266*, 11192–11198. [CrossRef]
26. Sherry, L.; Daniel, B. Neuronal nicotinic receptors: From structure to function. *Nicotine Tob. Res.* **2001**, *3*, 203–223.
27. Tae, H.S.; Gao, B.; Jin, A.H.; Alewood, P.F.; Adams, D.J. Globular and ribbon isomers of Conus geographus α-conotoxins antagonize human nicotinic acetylcholine receptors. *Biochem. Pharmacol.* **2021**, *190*, 114638–114646. [CrossRef] [PubMed]
28. Woodcock, J. A difficult balance-pain management, drug safety, and the FDA. *N. Engl. J. Med.* **2009**, *361*, 2105–2107. [CrossRef] [PubMed]
29. Ballantyne, J.C.; Shin, N.S. Efficacy of opioids for chronic pain: A review of the evidence. *Clin. J. Pain* **2008**, *24*, 469–478. [CrossRef]
30. Molino, B.F.; Darkes, P.R.; Ewing, W.R. Tetrahydro-Pyrido-Indoles as Cholecystokinin and Gastrin Antagonists. U.S. Patent 5,162,336, 10 November 1992.
31. Steffen, W.; Torsten, D.; Degenhard, M.; Beate, S.; Thomas, S.; Dieter, F.; Ulrich, K.; Daniela, H.; Jörg, D.; Christiaans, J.A.M. Christiaans, 6-Benzyl-2, 3, 4, 7-Tetrahydro-Indolo [2, 3-c] Quinoline Compounds Useful as Pde5 Inhibitors. U.S. Patent 12/449,173, 13 May 2010.

marine drugs

MDPI

Review

Application of Gene Knockout and Heterologous Expression Strategy in Fungal Secondary Metabolites Biosynthesis

Yaodong Ning †, Yao Xu †, Binghua Jiao and Xiaoling Lu *

Department of Biochemistry and Molecular Biology, College of Basic Medical Sciences, Naval Medical University, Shanghai 200433, China
* Correspondence: luxiaoling80@126.com
† These authors contributed equally to this work.

Abstract: The in-depth study of fungal secondary metabolites (SMs) over the past few years has led to the discovery of a vast number of novel fungal SMs, some of which possess good biological activity. However, because of the limitations of the traditional natural product mining methods, the discovery of new SMs has become increasingly difficult. In recent years, with the rapid development of gene sequencing technology and bioinformatics, new breakthroughs have been made in the study of fungal SMs, and more fungal biosynthetic gene clusters of SMs have been discovered, which shows that the fungi still have a considerable potential to produce SMs. How to study these gene clusters to obtain a large number of unknown SMs has been a research hotspot. With the continuous breakthrough of molecular biology technology, gene manipulation has reached a mature stage. Methods such as gene knockout and heterologous expression techniques have been widely used in the study of fungal SM biosynthesis and have achieved good effects. In this review, the representative studies on the biosynthesis of fungal SMs by gene knockout and heterologous expression under the fungal genome mining in the last three years were summarized. The techniques and methods used in these studies were also briefly discussed. In addition, the prospect of synthetic biology in the future under this research background was proposed.

Keywords: fungi; secondary metabolites; gene knockout; heterologous expression

Citation: Ning, Y.; Xu, Y.; Jiao, B.; Lu, X. Application of Gene Knockout and Heterologous Expression Strategy in Fungal Secondary Metabolites Biosynthesis. *Mar. Drugs* **2022**, *20*, 705. https://doi.org/10.3390/md20110705

Academic Editor: Dehai Li

Received: 27 September 2022
Accepted: 9 November 2022
Published: 10 November 2022

1. Introduction

Microorganisms can produce a wide variety of SMs (e.g., polyketides, terpenoids, saponins, and non-ribosomal peptides), most of which show good biological activity, such as antibacterial, anti-tumor, and immunoregulation properties [1–3]. The research on SMs from prokaryotes (such as *Actinomyces*, *Streptomyces*, and other bacteria) started earlier and more thoroughly, while the research on SMs from eukaryotes is relatively scarce. However, with the in-depth study of fungi in recent years, it has been shown that fungi possess more potential to produce SMs than bacteria, and the products have better biological activity, which has attracted more extensive attention and research [4,5]. At the same time, with the ongoing advancement of gene sequencing technology and bioinformatics, a large amount of genomic information on bacteria and fungi has been analyzed and annotated [6,7]. After "mining" this gene information, numerous "silent" biosynthetic gene clusters (BGCs) of SMs from bacteria and fungi still have not been characterized, while fungi show more powerful production potential for SMs than bacteria because of their larger and more complex genomes [8]. The continuous progress of molecular biology technology, such as gene knockout and heterologous expression, as well as the application of combinatorial biosynthesis strategies, makes the manipulation of genes increasingly convenient, which greatly expands the research on microbial SMs [9,10]. All of these promote the biosynthetic pathway study for microbial SMs and show great advantages in the field of biosynthesis.

Gene knockout and heterologous expression are common and mature strategies for the study of the biosynthesis of microbial SMs under the genome mining. The common gene knockout methods mainly include PEG/$CaCl_2$-mediated homologous recombination, *Agrobacterium*-mediated transformation, and the CRISPR/Cas technology [11]. In the heterologous expression, the model strains mainly include *Escherichia coli* and *Bacillus subtilis* in the prokaryotes and *Saccharomyces cerevisiae* and filamentous fungi such as *Aspergillus nidulans* and *Aspergillus oryzae* in the eukaryotes [12]. For the prokaryotes, gene knockout and heterologous expression techniques have been well experienced and established because the BGCs of the SMs are usually distributed in clusters and exist in the form of operons, which are convenient for the investigation and operation of target genes (clusters) [13]. For eukaryotes, because of the intact nucleus and genetic system and the larger and more complex genome of fungi, the BGCs of the SMs are usually scattered, which makes it very challenging to dig out and analyze the fungal target genes [14]. Additionally, due to the high level of fungal evolution, it is difficult to establish genetic operating systems, which consequently started late and are relatively scarce in the study of the biosynthesis of fungal SMs [15]. Nowadays, more and more scholars are dedicated to the genome mining of fungal SMs and have obtained a series of achievements. Therefore, it is important and necessary to summarize the research on the molecular biology technology in the secondary metabolites of fungi, especially the gene knockout and the heterologous expression techniques.

In this paper, we primarily concentrated on the representative studies of the biosynthesis of fungal SMs by gene knockout and heterologous expression under the fungal genome mining in the last three years, of which the techniques and methods were briefly introduced. The purpose of this paper is to classify and summarize the strains which have been studied, elucidating the fact that some gene knockout and heterologous expression methods are indeed applicable to the gene manipulation of certain species of fungi, which can provide some ideas and references for future research. In addition, we look forward to the prospect and direction of biosynthesis in the future and to providing new ideas for the biosynthesis.

2. Traditional Strategies of Diversity of SMs from Fungi

Microorganisms are important sources of natural products, most of which are isolated from bacteria and fungi. The SMs from fungi have attracted extensive attention and research because of their novel structure (e.g., terpenoids, polyketides, anthraquinones, steroids, and non-ribosomal polypeptides), diverse biological activity (e.g., antibacterial, anti-inflammatory, anti-tumor, and immunoregulation), and rich yield [1–3].

Studies on the diversity of SMs from fungi primarily followed the traditional natural product discovery strategy before information on fungal genomes became available (Figure 1). To obtain more SMs with diverse structures and biological activity from the same fungus, the most representative method is the OSMAC strategy. Generally, the same strain may produce various amounts or even distinct kinds of natural products from the different culture conditions (such as the composition of the culture medium, the fermentation conditions, the added precursors, etc.). The scientists have discovered numerous novel and bioactive natural products by the OSMAC method, showing that different culture conditions may activate the silent genes to produce new secondary metabolites [16–18]. The OSMAC strategy will continue to be used as a valuable method to exploit the biosynthetic potential of strains. Co-culture, the other traditional natural product mining method, can activate silent genes or clusters through interspecific interactions [19]. In the two-way chemical communications between the co-cultured strains, the signal molecules are transmitted back and forth to interfere with the compound library of co-cultured strains to enrich the quantity of the compounds. Epigenetic regulation is to activate the silent gene clusters through DNA methylation and histone modification without changing the DNA sequence to regulate the secondary metabolic pathway of the strain to obtain new products [18,20]. However, no matter what the traditional method is, the essence is a random selection of the de-silencing of the secondary metabolic pathway, and any conditions affecting the response of a strain to the external conditions may be used to change the transcriptome and then to change the proteome; finally, it can be read out in the variable SMs. This "blind" selection and these changes make it neither possible to accurately understand the law of the biosynthesis of the strains from the perspective of genes, nor to directionally discover the SMs that interested us under the guidance of the genomic information, which still has great limitations in the process of mining natural products.

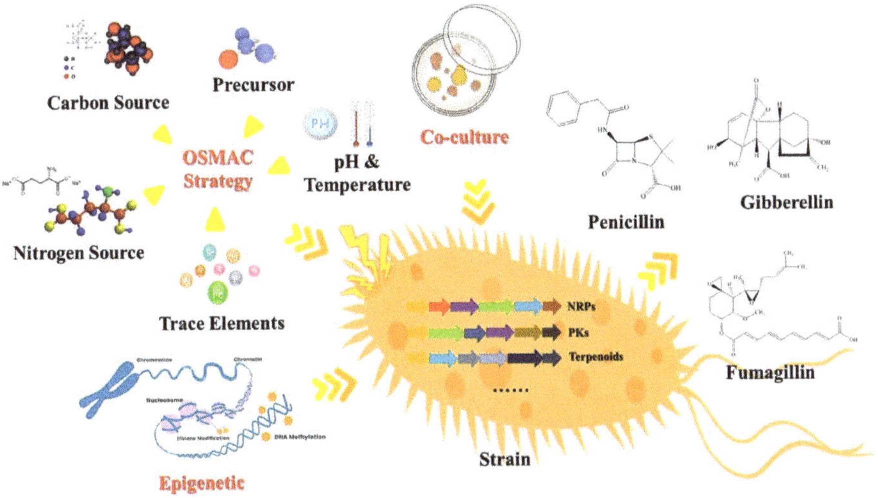

Figure 1. Traditional strategies for fungal secondary metabolites discovery. These mainly include OSMAC strategy, strains co-culture, and epigenetic regulation, aiming to stimulate the expression of biosynthetic gene clusters of SMs through changes of external conditions or exchange of signal molecules between strains in order to obtain more SMs.

3. Gene Mining and Bioinformatics Broaden the Discovery of SMs of Fungi

With the rapid development of genome sequencing technology, many fungal genome data have been identified and reported, which makes it possible to predict which kinds of compounds may be produced. More and more bioinformatics tools have been developed with the continuous improvement of bioinformatics analysis. BLAST and FASTA are currently the most commonly used database search programs based on local similarity and are tools for the sequence similarity search, which can be used for homologous gene retrieval in public databases [21]. TOUCAN [22] and ARTS [23] can be used for gene mining; antiSMASH [24–26] and cluster finder [27,28] were used for analyzing and predicting BGCs; these have greatly promoted the research on fungal SM BGCs. A new stage in the study of the SMs of fungi has been entered. The fungi have large and complex genomes, restricting the analysis of gene clusters. After mining the BGCs of the SMs of fungi, it has been shown that most of the gene clusters (>90%) were unknown [29–31]. The genomes of the marine fungi *Calcarisporium* sp. and *Pestalotiopsis* sp. possessed 60 and 67 BGCs, respectively, by bioinformatics analysis, of which the new clusters accounted for 98% and 97%, respectively, and only a small number of BGCs were expressed after RNA-seq verification [30]. Gao et al. reported the high-quality genome sketch sequence of the endophytic fungus *Neonectria* sp. DH2, of which 14,163 genes are predicted to encode proteins, and 557 of the genes are unique. According to the neighborhood-linked phylogenetic tree of the ITS region, there were 47 BGCs in the DH2 genome, of which only 5 BGCs were previously reported, showing the huge production potential of the SMs of fungi [31].

Unlike the traditional research strategy, the study of the fungal SMs based on the gene mining is to associate the SMs with the BGCs by bioinformatics analysis, predict the potential biosynthesis pathway, and then verify the prediction by molecular biology techniques and to analyze the biosynthesis pathway of the SMs. In this research strategy, it is more definite to discover what we want, master the law of biosynthesis, and to develop it more accurately. Gene knockout and heterologous expression have been widely used in the study of fungal secondary metabolic biosynthesis and have played an important role.

4. Application of Gene Knockout Strategy in Biosynthesis of SMs of Fungi

Gene knockout is one of the important methods for studying the biosynthesis pathway of fungal SMs; it mainly focuses on the in vivo verification of the gene function of the studied strains (Figure 2). The studied genes (clusters) are usually predicted to be expressed in the original strain, rather than silent genes. The research idea is usually to determine the existence of interesting compounds in the fermentation broth of the original strain first, then to sequence the gene of the strains, predict the biosynthesis pathway (gene clusters) of the interesting compounds after homologous gene mining and bioinformatics analysis of the genomes, and then verify the correlation between the gene cluster and the biosynthesis of the natural products by the gene knockout. The function of the studied genes in the natural product biosynthesis can be roughly judged by the construction of single or multiple gene mutants and the accumulation of intermediate metabolites, which preliminarily analyzed the biosynthetic pathway of the products.

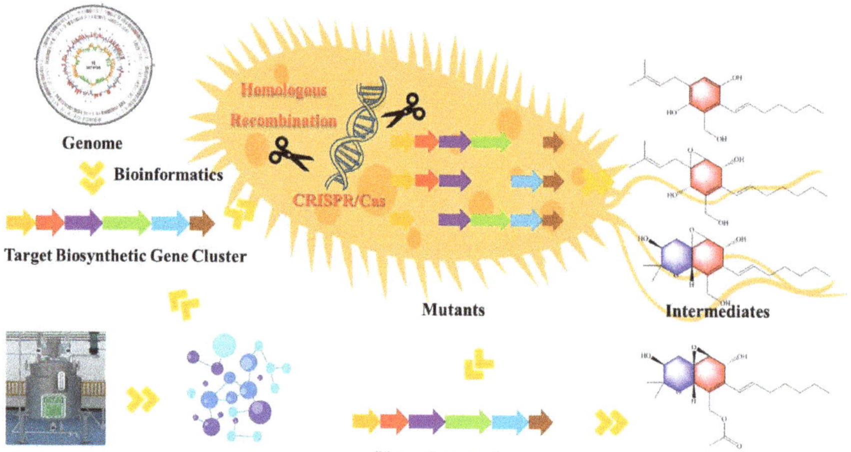

Figure 2. Application of gene knockout strategy in the study of fungal secondary metabolites. It mainly includes the determination of target BGCs, construction of mutants, and identification of intermediate products, which verifies the biosynthetic pathway.

4.1. *Application of Different Gene Knockout Methods*

Different strains have formed different physiological and biochemical characteristics in the long process of evolution. Even if the evolutionary tree shows that the strains are in the same genus, their morphological, physiological, and biochemical characteristics are also quite different. Therefore, the establishment of the genetic transformation system needs to be investigated and selected according to the actual situation of the different strains. Currently, PEG-mediated homologous recombination, *Agrobacterium*-mediated transformation, and PEG-mediated CRISPR/Cas technology are the main methods used in the research on the gene knockout related to the biosynthesis of fungal secondary metabolites.

4.1.1. PEG-Mediated Homologous Recombination

PEG-mediated homologous recombination is a classic method for fungal gene research, based on the preparation of high-quality protoplasts, which induces foreign DNA into cells by PEG/CaCl2 and other methods. PEG is a cell fusion agent which can interfere with the recognition between cells by causing the disorder of the surface charge of the cell membrane, thus facilitating the intercellular fusion and the entry of foreign DNA molecules into the protoplast. The homologous recombination reaction strictly depends on the homology between the DNA molecules. The homologous recombination reaction is usually based on the formation and resolution of cross molecules or the Holliday junction structure, that is, the precursor stage, the formation of the synaptonemial complex, and the resolution of the Holliday structure.

It has been widely used in the biosynthesis of SMs derived from *Aspergillus*. In knocking out the key genes *GedF* and *GedK* of the anthraquinones biosynthesis from *A. fumigatus*, a revised questin ring-opening mechanism was elucidated; this caused a classic Baeyer–Villiger oxidation hypothesis, which has been challenged [32]. By comparing the metabolites of wildtype *A. fumigatus* and *crmA*, a deleted strain grown under Cu^{2+}, it was found that at the level of trace Cu^{2+} CrmA participated in two different biosynthetic pathways to improve the adaptability under environmental pressure [33]. After knocking out the oxepinamides biosynthesis gene derived from *A. ustus*, the necessary intermediates were obtained, and the biosynthesis pathway was analyzed for the first time [34]. Analyzing the azaphilones biosynthesis gene derived from *A. terreus*, synthesized by two independent gene clusters, provides a new idea for the biological mechanism of complex compounds synthesized by filamentous fungi [35]. At the same time, it has also been applied in the study of the biosynthesis pathways of natural products, such as the oxygenated phenethyl derivative from *A. ustus* [36] and the hopane-type triterpenoid glycoside from *A. fumigatus* [37], indicating that it is generally applicable to *Aspergillus* fungi. In addition, for xylomyrocins from *Paramyrothecium* sp., their biosynthesis pathway was identified through gene knockout and stable isotope feeding, which clarified the fusion coordination between carbohydrate metabolism and NRPS skeleton synthesis and enriched the biosynthesis sources of the special assembly units of non-ribosomal peptides [38]. Liu et al. confirmed the biosynthetic gene cluster of sordarin in the *Sordaria araneosa*, proving that four P450 oxidases play an important role in the rearrangement process [39]. The PEG-mediated homologous recombination method has also been used in the study of biosynthetic genes in epidithiodiketopiperazines derived from *Trichoderma hypoxylon* [40] and indolizidine alkaloids derived from *Curvularia* sp. [41], showing that this method has been widely used in gene knockout.

4.1.2. PEG-Mediated CRISPR/Cas Technique

The CRISPR/Cas technique uses RNA to guide the Cas protein to modify the targeted sequences; this has been widely used in various fields as a hot spot. The CRISPR-Cas9 gene editing technology is to identify the target genome sequence through the artificially designed sgRNA (guide RNA) and to guide the Cas9 protease to effectively cut the double strands of DNA to form double strand breaks. The damage repair will cause gene knockout or knock-in and finally achieve the goal of modifying the genome DNA.

In the fungi gene study, CRISPR/Cas technology often requires PEG-mediated protoplast transformation. Scientists knocked out the meroterpenoids biosynthesis gene from marine fungus *Talaromyces purpureogenus* and evaluated two NHI proteins from a heterodimer for catalysis, analyzing the biosynthesis of heteroterpene [42]. Several biosynthetic genes of aculenes derived from *A. aculeatus* were inactivated, which provided reference for the synthesis and derivation of daucane sesquiterpenes [43]. In addition, CRISPR/Cas technology has played an important role in the functional research on phomoxanthone A, derived from marine fungus *Diaporthe* sp. [44], and flavoprotein monooxygenase, derived from *A. terreus* [45]. This technique is becoming mature, with more and more applications in the future study of fungi genes.

4.1.3. *Agrobacterium*-Mediated Transformation

There are relatively few studies on fungal gene knockout mediated by *Agrobacterium* because of the difficulty of the genetic operation and the low transformation efficiency of the fungi. Mycotoxin patulin isolated from *Penicillium expansum* can cause fruit and product pollution. Li et al. used *Agrobacterium*-mediated transformation to study the patulin biosynthesis gene cluster and confirmed the function of all the genes involved in its biosynthesis, providing the support for the prevention and treatment of pathogenic microorganisms [46]. Zhang et al. knocked out a pyrone meroterpenoid oxalicine B gene cluster from *P. oxalicum* and further elucidated its biosynthesis pathway and oxidase catalytic mechanism through in vitro biochemical verification. Oxalicine B possessed good anti-influenza virus activity [47]. Research on a "super" gene cluster of *Metarhizium robertsii* shows that this cluster contains three secondary metabolic gene clusters. It is confirmed that different gene deletions do not affect the insecticidal virulence of *M. robertsii* but do significantly affect the ability of *M. robertsii* to resist different bacteria because of the different gene deletions leading to the production of different structural compounds [48].

4.2. Other Applications of Gene Knockout Strategy

Gene knockout could also be applied to the diverse study of SMs in fungi, except for the verification of the function of genes, such as for the activating of silent gene clusters and increasing product diversity. Wei et al. knocked out the key genes in the biosynthesis of rubratoxins, the main product of *P. dangeardi*, which makes it easier to inhibit the production of the main compounds and competitively obtain common precursors of polyketide synthesis and isolate novel skeleton compounds [49]. Qi et al. realized the abundant accumulation of emodin, the precursor of physcion, by knocking out the key emodin-1-OH-O-methyltransferase gene in *A. terreus* [50]. This method could also apply in the agricultural pathogenic bacteria, such as the biosynthesis of toxin ustilaginoidins derived from *Ustilaginoidea virens* [51], the infection of fusaoctaxin B derived from *Fusarium graminearum* on the plant virulence factor [52], and the verification of the multiple gene function in the biosynthesis pathway of penifulvin, an anti-insect compound derived from *P. griseofulvum*, which provides an important reference for the prevention and control of agricultural pathogenic microorganisms and the development of new green biological pesticides [53].

4.3. Limitations of Gene Knockout Strategy

There are also many defects and limitations in the study of fungal SM biosynthesis by gene knockout technology. First of all, because of the high level of evolution and incomplete genetic system of fungi, it is difficult for most strains to establish a genetic transformation system by conventional methods. Regardless of the PEG-mediated homologous recombination or CRISPR/Cas technology, they both require the high-quality protoplasts, while the low transformation efficiency is common in the protoplasts [54]. In addition, there were usually a lot of verification works needed in the screening of mutant strains. Secondly, it is difficult to analyze and identify all the metabolites in the fermentation by the current separation and identification techniques because of the unknown metabolic pathways of most of the studied strains and the complex metabolite compositions. Based on this, the gene deletion is easy to ignore in the regulation of biological metabolites. In the process of studying the biosynthesis pathway of mycotoxin flavipucine derived from *A. nidulans*, gene knockout could only determine the protein involved in the synthesis of the toxin. When the single gene in the BGC was knocked out, there were no intermediates observed; this needs to use the heterologous expression strategy to clarify its specific biosynthesis pathway [55]. Gene knockout strategies play the role of "verifier" in most studies focusing on whether genes (clusters) are involved in the synthesis of certain SMs, but they cannot explain how genes participate in biosynthesis exactly. This rough verification can be used as a guiding tool to study the primary stage of fungal SMs.

5. Application of Heterologous Expression Strategy in Biosynthesis of SMs of Fungi

Because of the disadvantages and limitations of the gene knockout strategy, the heterologous expression strategy has become the other important method in the study of the biosynthesis of the SMs of fungi. Gene data mining reveals that many fungi possess cryptic BGCs that appear to be silent when cultivated in the conventional fermentation conditions [8]. It is difficult to study the function of these genes by gene knockout when carrying out a detailed study of the biosynthetic pathway of complex SMs, while it is necessary to use heterologous hosts with a clear genetic background and mature genetic transformation system to express these silent or complex biosynthetic genes (clusters) (Figure 3). The general research idea of heterologous expression mainly includes the whole genome sequencing of the original strain, blasting and searching for the homologous genomic data to obtain the gene clusters encoding the biosynthesis of the SMs of interest. After the bioinformatics analysis of these genes (clusters), the meaningful genes (clusters) are cloned. Finally, the genes were heterologously expressed by genetic transformation, and the corresponding biosynthetic pathway was clarified by the identification and analysis of the products. There could be the heterologous expression not only of a single gene, but also of the whole gene cluster. When a single gene was heterologously expressed, it mainly combined with the in vitro enzyme experiments to characterize the expressed proteins in detail; when the gene cluster was heterologously expressed, it mainly combined with precursor feeding to verify the hypothetical biosynthesis pathway.

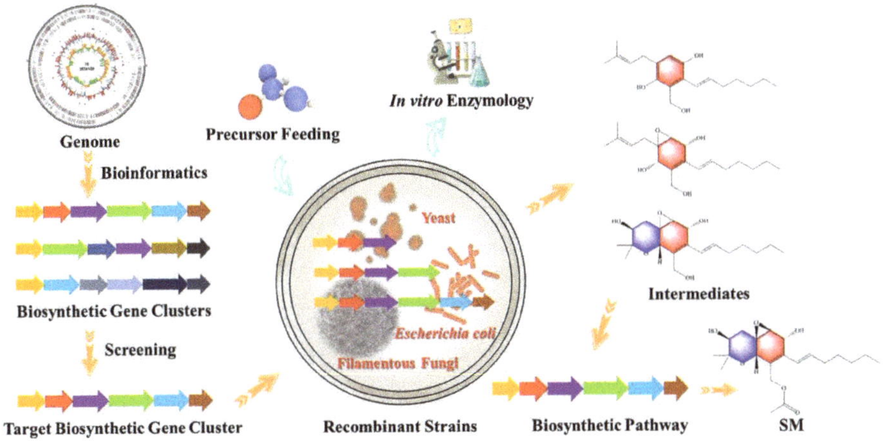

Figure 3. Application of heterologous expression strategy in the study of fungal secondary metabolites. It mainly includes the screening of target BGCs, heterologous expression of recombinant strain, feeding of precursor, in vitro enzymatic verification, and identification of intermediate products, in order to predict the biosynthetic pathway.

5.1. Application of Different Heterologous Hosts

The selection of the heterologous host is the key to the successful application of heterologous expression. Firstly, it is important to determine the clear genetic background of the heterologous expression strain. Secondly, the heterologous expression strain should have the advantages of simple culture and fast growth. Finally, the simple and easy manipulation and high transformation efficiency of the genetic transformation were chosen. At present, the mature and commonly used expression systems in the biosynthesis of SMs of fungi include *Escherichia coli* and *Bacillus subtilis* in prokaryotes and *Saccharomyces cerevisiae* and filamentous fungi in eukaryotes, such as *Aspergillus nidulans* and *Aspergillus oryzae*. The selection of model strains should consider the sources of different genes (clusters) and the physical and chemical properties of the expressed products.

5.1.1. Application of Filamentous Fungi as Heterologous Hosts

With the in-depth study of filamentous fungi, there are more outstanding advantages than the model strains with, for example, prokaryotes and yeast as the heterologous hosts. The filamentous fungi can correctly splice introns from heterogenous fungi and express multiple biosynthetic genes in fungi concurrently, which plays a role in protein translation and post-modification to obtain the target product. At present, the filamentous fungi expression systems commonly used in fungal heterologous expression were mainly *A. nidulans* and *A. oryzae*.

As a filamentous fungus, *A. nidulans* has a relatively clear genetic background after years of research, and a relatively mature genetic transformation system has been established. It has been widely used as a model strain in the study of the heterologous expression of the SMs of fungi. With the combined heterologous expression of the biosynthesis gene cluster of sordarinane, derived from *Sordaria araneosa* with enzyme experiments in vitro, a group of P450 multi-enzyme systems was confirmed. The new catalytic function of the P450 family oxidases was demonstrated, which laid a foundation for the development of P450 multi-enzyme synergistic catalysis for the synthesis of new chemical entities [56]. Wei et al. identified the functions of some P450 enzymes and sesquiterpene synthetases by the heterologous expression of the BGC of asperaculin A, isolated from *A. aculeatus* [57]. Due to the failure of the gene knockout strategy, Zhong et al. heterogeneously expressed the BGC of rumbrin, isolated from *Auxarthron umbrinum*, and found an autoantibody gene, verifying that the product had anti-HIV activity [58]. In addition, the BGCs of flavunoidine-1 derived from *A. flavus* were recombinantly arranged and heterologously expressed in *A. nidulans* in different combinations, and the BGCs containing both TC and NRPS core enzymes were identified [59]. In addition, the BGCs of harzianic acid [60] and trichoxide [61] from *Trichoderma*, citridone [62] and ilicicolin H [63] from *Penicillium*, and decarestricitine from *Beauveria bassiana* were heterologously expressed in *A. nidulans*, combined with precursor feeding, which clarified the biosynthesis pathway to some extent, respectively [64].

A. oryzae, a heterologous host, has similar characteristics and advantages to *A. nidulans*, which is also widely used as a model strain of biosynthetic genes in the SMs of fungi. In the reconstruction of the biosynthetic pathway of phlegmacins derived from *Talaromyces* sp., an unprecedented laccase-involved unsymmetrically regioselective oxidative coupling reaction was shown, which provides a new reference for the synergistic catalytic mechanism of laccases and other proteins [65]. The BGC of funiculolides derived from *A. funiculosus* were heterogeneously expressed to elucidate the fact that α-ketoglutarate-dependent dioxygenase FncG catalyzed spirocyclopentanone [66]. By the heterologous expression of CJ-20557 biosynthetic gene clusters from *A. duricaulis*, in combination with in vitro enzyme experiments, it was clarified that the SAT domain of polyketide synthase DrcA was responsible for the formation of the depside bond, which enriched the understanding of the mechanism of fungal NR-PKS biosynthesis [67]. Chen et al. confirmed that the enzyme IlIS, a sesquiterpene synthase from *Irpex lactous*, was responsible for the synthesis of the tremulane skeleton through heterologous expression and in vitro enzymatic reaction, and four new tremulane sesquiterpene products were isolated [68]. The specific PKS synthase GrgF in the BGC of gregatin A derived from *Penicillium* sp. [69] and the trichobrasilenol terpene cyclase derived from *Trichoderma atroviride* [70] were verified by the heterologous expression of *A. oryzae*. In addition, the NRPS-PKS gene cluster from *A. candidus* was heterologously expressed to obtain a pyrrolobenzazepine alkaloid [71]. Additionally, conidiogenone [72], brevianamide A [73], and brevione E [74] derived from *Penicillium* were heterologously expressed in *A. oryzae*, combined with the precursor feeding, which is helpful in analyzing the biosynthesis pathway.

5.1.2. Application of *Saccharomyces cerevisiae* as Heterologous Host

Because of the early study of the genetic background and the mature genetic manipulation system, *Saccharomyces cerevisiae* was widely used as the heterologous expression model strain in the study of the SMs of fungi in the past. Eukaryotic *S. cerevisiae* could translate proteins correctly and post-modify. Zhang et al. heterologously expressed *P. funiculosum* source chrodrimanin-type meroterpenoids BGCs and verified them in in vitro catalytic experiments to clarify the function of CdnC protein, obtaining a series of new chrodrimanins compounds [75]. In the biosynthetic path of cyclohexanoid terpenoids derived from *Aspergillus* sp., two key enzymatic functions were characterized by heterologous expression and in vitro enzymology experiments [76]. In addition, *S. cerevisiae* has also been applied for the expression of entire BGCs. *S. cerevisiae* was used as an expression vector in the heterologous expression of the BGC of the formation central C ring in the tetracyclic ergoline derived from *A. fumigatus* [77]. In the study of the biosynthesis pathway of shimalactones derived from *Emericella variecolor* [78], and flavunoidine 1 [59] and diorcinol [79] derived from *Aspergillus, S. cerevisiae* was also the heterologous expression vector. However, compared with filamentous fungi as expression vectors, *S. cerevisiae* still has some shortcomings, such as the lack of an advanced mRNA splicing system and the difficult in expressing complex BGCs; it is only suitable for expression of single or simple BGCs, which restricts its further application in the study of fungal SMs.

5.1.3. Application of *Escherichia coli* as Heterologous Host

As the earliest heterologous host, *Escherichia coli* has been widely used in genetic engineering, metabolic engineering, and other fields. Because of its advantages, such as simple cultivable, rapid growth, high transformation efficiency, and so on, *E. coli* has a good effect on the heterologous expression of BGCs from prokaryotes. However, it has similar shortcomings with *S. cerevisiae*, lacking the introns splicing of eukaryotes and the post-translational modification process, (glycosylation, phosphorylation et al.), which restrict the heterologous expression of the fungal biosynthetic gene cluster. It is generally used to heterologously express a single gene of BGC and is combined with in vitro enzyme experiments to characterize the protein function. The key genes in the BGCs of nanangelenin A [80], diorcinol [79], and asperaculin A [57], derived from *Aspergillus*, fumiquinazoline [81], and brevianamide A [73], derived from *Penicillium*, and trichobrasilenol, derived from *T. atroviride* [70], were heterogeneously expressed in *E. coli.*, combined with in vitro catalysis and other experiments, which clarified the functions of a series of key enzymes in the biosynthesis pathway.

5.2. Application of Heterologous Expression Strategy in Mining Silent BGCs

The rapid development of gene mining and bioinformatics has brought new breakthroughs to the discovery of fungal SMs. A large number of terpenoid, PKS, and NRPS gene clusters have been predicted from the genome sequences of many fungi, most of which are unknown. The actual kinds of obtained secondary metabolites of the strains were far less than the predicted kinds based on the BGCs' functions because of the gene silence under conventional experimental conditions. How to activate these silent genes to obtain lots of SMs is the current research focus. At present, there are generally two ways to activate the silent gene clusters. One is to overexpress the silent gene clusters by adding strong promoters to or regulating the transcription factors of the original strains. Seven new compounds were obtained by overexpressing the specific transcription factor tenS of *Beauveria bassiana*-derived silent gene clusters [82]. However, it has not been widely used in fungi because of the difficulty in establishing genetic system in the original strain. The other is to clone these silent gene clusters and transform them into model strains for heterologous expression, which is a common research method at present. Nine new sesquiterpenes were discovered after the heterologous expression of silent gene clusters derived from *A. ustus* in *A. oryzae*, which strongly supports the application of heterologous expression in the gene mining of silent BGCs [83].

5.3. Limitations of Heterologous Expression Strategy

Although heterologous expression technology has significant advantages in the study of fungal secondary metabolite biosynthesis, it still has some limitations, especially in terms of the compatibility of heterologous strains with foreign genes. The proteins could be expressed and modified well by heterologous host filamentous fungi, but filamentous fungi still have some defects. When the silent hancockinone A BGC, derived from *A. hancockii*, was heterologously expressed in *A. nidulans*, the target product was not obtained due to intron cleavage and inactive expressed protein [84]. Moreover, the genetic stability of the recombinant strain is very important. How to maintain stable genetic characteristics is a problem to be solved. In short, the heterologous expression system still needs continuous technical optimization in order to better serve the biosynthesis of SMs of fungi.

6. Conclusions

With the advent of the post-genome era, the research on SMs has gradually deepened from the characterization of phenomena to the process of mining essence, which gene sequencing and bioinformatics have strongly promoted. The research on the SMs of fungi has entered a new stage. In the study of fungal SMs biosynthesis, gene mining provides a theoretical foundation; molecular biology sets up the essential circumstances, and strategies such as gene knockout and heterologous expression were the methodologies. The further exploration of synthetic biology will exert a profound impact on the biosynthetics of SMs and will continue to promote the discovery process of SMs.

Synthetic biology is the extension and future development direction of biosynthesis research, which is the rebuilding process of the life. It starts from the basic elements to build the parts step by step, hoping to design and synthesize new life processes or organisms based on the human will. It mainly included the construction of chassis cells; the selection and optimization of gene elements; design synthetic pathways, such as gene integration and modular synthesis strategies; and the establishment of cell factories to regulate metabolism, all of which are accompanied by advances in molecular biology techniques and the development of gene editing tools. It is reported that different BGCs of decalin-containing diterpenoid pyrones have been retrieved from five fungal genera, and five natural pathways, one shunt pathway, and four extension pathways were recombined by the selection and integration of genes in *A. oryzae*, producing 15 new compounds, which exhibited the role of synthetic biological methods in enriching the diversity of fungal SMs [85]. Synthetic biology has unique advantages in stabilizing the source and yield of SMs, which is bound to bring new hope for the directional and controllable biosynthesis of SMs.

Author Contributions: Writing—original draft, conceptualization, investigation, methodology, Y.N.; writing—original draft, software, visualization, Y.X.; writing—review and editing, validation, B.J.; writing—review and editing, funding acquisition, validation, X.L. All authors have read and agreed to the published version of the manuscript.

Funding: This research was funded by the National Key R&D Program of China, grant number 2019YFC0312504.

Institutional Review Board Statement: Not applicable.

Informed Consent Statement: Not applicable.

Data Availability Statement: No new data were created or analyzed in this study. Data sharing is not applicable to this article.

Conflicts of Interest: The authors declare no conflict of interest.

References

1. Carroll, A.R.; Copp, B.R.; Davis, R.A.; Keyzers, R.A.; Prinsep, M.R. Marine natural products. *Nat. Prod. Rep.* **2021**, *38*, 362–413. [CrossRef] [PubMed]
2. Aldholmi, M.; Marchand, P.; Ourliac-Garnier, I.; Pape, P.L.; Ganesan, A. A decade of antifungal leads from natural products: 2010–Pharmaceuticals. *Pharmaceuticals* **2019**, *12*, 182. [CrossRef] [PubMed]
3. Butler, M.S. Natural products to drugs: Natural product derived compounds in clinical trials. *Nat. Prod. Rep.* **2005**, *22*, 162–195. [CrossRef] [PubMed]
4. Wang, C.; Tang, S.Y.; Cao, S.G. Antimicrobial compounds from marine fungi. *Phytochem. Rev.* **2021**, *20*, 85–117. [CrossRef]
5. Xu, J.; Yi, M.; Ding, L.; He, S. A review of anti-inflammatory compounds from marine fungi, 2000–2018. *Mar. Drugs.* **2019**, *17*, 636. [CrossRef]
6. Wang, J.Y.; Nielsen, J.; Liu, Z.H. Synthetic biology advanced natural product discovery. *Metabolites* **2021**, *11*, 785. [CrossRef]

7. Alberti, F.; Kaleem, S.; Weaver, J.A. Recent developments of tools for genome and metabolome studies in basidiomycete fungi and their application to natural product research. *Biol. Open.* **2020**, *9*, bio056010. [CrossRef]
8. Vignolle, G.A.; Mach, R.L.; Mach-Aigner, A.R.; Derntl, C. Novel approach in whole genome mining and transcriptome analysis reveal conserved RiPPs in *Trichoderma* spp. *BMC. Genom.* **2020**, *21*, 258. [CrossRef]
9. Alberti, F.; Foster, G.D.; Bailey, A.M. Natural products from filamentous fungi and production by heterologous expression. *Appl. Microbiol. Biot.* **2017**, *101*, 493–500. [CrossRef]
10. Zhang, Y.; Yun, K.Y.; Huang, H.M.; Tu, R.; Hua, E.B.; Wang, M. Antisense RNA interference-enhanced CRISPR/Cas9 base editing method for improving base editing efficiency in *Streptomyces lividans* 66. *ACS. Synth. Biol.* **2021**, *10*, 1053–1063. [CrossRef]
11. Kadooka, C.; Yamaguchi, M.; Okutsu, K.; Yumiko Yoshizaki Takamine, K.; Maruyama, J.; Tamaki, H.; Futagami, T. A CRISPR/Cas9-mediated gene knockout system in *Aspergillus luchuensis* mut. *Kawachii. Biosci. Biotech. Bioch.* **2020**, *84*, 2179–2183. [CrossRef]
12. Oikawa, H. Heterologous production of fungal natural products: Reconstitution of biosynthetic gene clusters in model host *Aspergillus oryzae*. *Proc. Jpn. Academy. Ser. B Phys. Biol. Sci.* **2020**, *96*, 420–430. [CrossRef]
13. Rang, J.; Li, Y.L.; Cao, L.; Shuai, L.; Liu, Y.; He, H.C.; Wan, Q.Q.; Luo, Y.W.; Yu, Z.Q.; Zhang, Y.M.; et al. Deletion of a hybrid NRPS-T1PKS biosynthetic gene cluster via Latour gene knockout system in Saccharopolyspora pogona and its effect on butenyl-spinosyn biosynthesis and growth development. *Microb. Biotechnol.* **2021**, *14*, 2369–2384. [CrossRef]
14. Lebar, M.D.; Cary, J.W.; Majumdar, R.; Carter-Wientjes, C.H.; Mack, B.M.; Wei Qj Uka, V.; Saeger, D.D.; Mavungu, J.D.D. Identification and functional analysis of the aspergillic acid gene cluster in *Aspergillus flavus*. *Fungal. Genet. Biol.* **2018**, *116*, 14–23. [CrossRef]
15. Liu, G.; Qu, Y.B. Engineering of filamentous fungi for efficient conversion of lignocellulose: Tools, recent advances and prospects. *Biotechnol. Adv.* **2019**, *37*, 519–529. [CrossRef]
16. Tomm, H.A.; Ucciferri, L.; Ross, A.C. Advances in microbial culturing conditions to activate silent biosynthetic gene clusters for novel metabolite production. *J. Ind. Microbiol. Biot.* **2019**, *46*, 1381–1400. [CrossRef]
17. Pan, R.; Bai, X.L.; Chen, J.W.; Zhang, H.W.; Wang, H. Exploring Structural Diversity of Microbe Secondary Metabolites Using OSMAC Strategy: A Literature Review. *Front. Microbiol.* **2019**, *108*, 87–94. [CrossRef]
18. Pinedo-Rivilla, C.; Aleu, J.; Durán-Patrón, R. Cryptic metabolites from marine-derived microorganisms using OSMAC and epigenetic approaches. *Mar. Drugs.* **2022**, *20*, 84. [CrossRef]
19. Netzker, T.; Fischer, J.; Weber, J.; Mattern, D.J.; Konig, C.C.; Valiante, V.; Schroeckh, V.; Brakhage, A.A. Microbial communication leading to the activation of silent fungal secondary metabolite gene clusters. *Front. Microbiol.* **2015**, *6*, 299. [CrossRef]
20. Li, C.Y.; Chung, Y.M.; Wu, Y.C.; Hunyadi, A.; Wang, C.C.C.; Chang, F.R. Natural products development under epigenetic modulation in fungi. *Phytochem. Rev.* **2022**, *19*, 1323–1340. [CrossRef]
21. Kjærbølling, I.; Mortensen, U.H.; Vesth, T.; Andersen, M.R. Strategies to establish the link between biosynthetic gene clusters and secondary metabolites. *Fungal. Genet. Biol.* **2019**, *130*, 107–121. [CrossRef] [PubMed]
22. Almeida, H.; Palys, S.; Tsang, A.; Diallo, A.B. TOUCAN: A framework for fungal biosynthetic gene cluster discovery. *NAR Genom. Bioinform.* **2020**, *2*, lqaa098. [CrossRef] [PubMed]
23. Alanjary, M.; Kronmiller, B.; Adamek, M.; Blin, K.; Weber, T.; Huson, D.; Philmus, B.; Ziemert, N. The Antibiotic Resistant Target Seeker (ARTS), an exploration engine for antibiotic cluster prioritization and novel drug target discovery. *Nucleic. Acids. Res.* **2017**, *45*, W42–W48. [CrossRef] [PubMed]
24. Blin, K.; Shaw, S.; Kloosterman, A.M.; Charlop-Powers, Z.; van Wezel, G.P.; Medema, M.H.; Weber, T. antiSMASH 6.0: Improving cluster detection and comparison capabilities. *Nucleic Acids Res.* **2021**, *49*, W29–W35. [CrossRef] [PubMed]
25. Blin, K.; Shaw, S.; Steinke, K.; Villebro, R.; Ziemert, N.; Lee, S.Y.; Medema, M.H.; Weber, T. antiSMASH 5.0: Updates to the secondary metabolite genome mining pipeline. *Nucleic Acids Res.* **2019**, *47*, W81–W87. [CrossRef]
26. Blin, K.; Shaw, S.; Kautsar, S.A.; Medema, M.H.; Weber, T. The antiSMASH database version 3: Increased taxonomic coverage and new query features for modular enzymes. *Nucleic Acids Res.* **2021**, *49*, D639–D643. [CrossRef]
27. Kloosterman, A.M.; Shelton, K.E.; van Wezel, G.P.; Medema, M.H.; Mitchell, D.A. RRE-Finder: A Genome-Mining Tool for Class-Independent RiPP Discovery. *mSystems* **2020**, *5*, e00267-20. [CrossRef]
28. Nguyen, M.; Ekstrom, A.; Li, X.; Yin, Y. HGT-Finder: A New Tool for Horizontal Gene Transfer Finding and Application to *Aspergillus* genomes. *Toxins* **2015**, *7*, 4035–4053. [CrossRef]
29. Yang, Y.; Liu, X.B.; Cai, J.M.; Chen, Y.P.; Li, B.X.; Guo, Z.K.; Huang, G.X. Genomic characteristics and comparative genomics analysis of the endophytic fungus *Sarocladium brachiariae*. *BMC Genom.* **2019**, *20*, 782. [CrossRef]
30. Kumar, A.; Sørensen, J.L.; Hansen, F.T.; Arvas, M.; Syed, M.F.; Hassan, L.; Benz, J.P.; Record, E.; Henrissat, B.; Poggeler, S.; et al. Genome sequencing and analyses of two marine fungi from the North Sea unraveled a plethora of novel biosynthetic gene clusters. *Sci. Rep.* **2018**, *8*, 10187. [CrossRef]
31. Lin, X.J.; Xu, H.; Liu, L.; Li, H.X.; Gao, Z.Z. Draft genome sequence of *Neonectria* sp. DH2 isolated from *Meconopsis grandis* Prain in Tibet. *3 Biotech* **2020**, *10*, 346. [CrossRef]
32. Qi, F.F.; Zhang, W.; Xue, Y.Y.; Geng, C.; Huang, X.N.; Sun, J.; Lu, X.F. Bienzyme-catalytic and dioxygenation-mediated anthraquinone ring opening. *J. Am. Chem. Soc.* **2021**, *143*, 16326–16331. [CrossRef]
33. Won, T.H.; Bok, J.W.; Nadig, N.; Venkatesh, N.; Nickles, G.; Greco, C.; Lim, F.Y.; González, J.B.; Turgeon, B.G.; Keller, N.P.; et al. Copper starvation induces antimicrobial isocyanide integrated into two distinct biosynthetic pathways in fungi. *Nat. Commun.* **2022**, *13*, 4828. [CrossRef]

34. Zheng, L.J.; Wang, H.W.; Fan, A.L.; Li, S.M. Oxepinamide F biosynthesis involves enzymatic D-aminoacyl epimerization, 3H-oxepin formation, and hydroxylation induced double bond migration. *Nat. Commun.* **2020**, *11*, 4914. [CrossRef]
35. Huang, X.N.; Zhang, W.; Tang, S.; Wei, S.H.; Lu, X.F. Collaborative biosynthesis of a class of bioactive azaphilones by two separate gene clusters containing four PKS/NRPSs with transcriptional crosstalk in fungi. *Angew. Chem. Int. Ed.* **2020**, *132*, 4379–4383. [CrossRef]
36. Zheng, L.J.; Yang, Y.L.; Wang, H.W.; Fan, A.L.; Zhang, L.P.; Li, S.M. Ustethylin biosynthesis implies phenethyl derivative formation in *Aspergillus ustus*. *Org. Lett.* **2020**, *22*, 7837–7841. [CrossRef]
37. Ma, K.; Zhang, P.; Tao, Q.Q.; Keller, N.P.; Yang, Y.L.; Yin, W.B.; Liu, H.W. Characterization and biosynthesis of a rare fungal hopane-type triterpenoid glycoside involved in the antistress property of *Aspergillus fumigatus*. *Org. Lett.* **2019**, *21*, 3252–3256. [CrossRef]
38. Wang, C.; Xiao, D.; Dun, B.; Yin, M.; Tsega, A.S.; Xie, L.; Li, W.; Yue, Q.; Wang, S.; Gao, H.; et al. Chemometrics and genome mining reveal an unprecedented family of sugar acid–containing fungal nonribosomal cyclodepsipeptides. *Proc. Natl. Acad. Sci. USA* **2022**, *119*, e2123379119. [CrossRef]
39. Liu, S.H.; Sun, J.L.; Hu, Y.L.; Zhang, L.; Zhang, X.; Yan, Z.Y.; Guo, X.; Guo, Z.K.; Jiao, R.H.; Zhang, B. Biosynthesis of Sordarin Revealing a Diels–Alderase for the Formation of the Norbornene Skeleton. *Angew. Chem. Int. Ed.* **2022**, *61*, e202205577. [CrossRef]
40. Liu, H.; Fan, J.; Zhang, P.; Hu, Y.C.; Liu, X.Z.; Li, S.M.; Yin, W.B. New insights into the disulfide bond formation enzymes in epidithiodiketopiperazine alkaloids. *Chem. Sci.* **2021**, *12*, 4132–4138. [CrossRef]
41. Dai, G.Z.; Han, W.B.; Mei, N.Y.; Xu, K.; Jiao, R.H.; Ge, H.M.; Tan, R.X. Pyridoxal-5′-phosphate–dependent bifunctional enzyme catalyzed biosynthesis of indolizidine alkaloids in fungi. *Proc. Natl. Acad. Sci. USA.* **2019**, *117*, 1174–1180. [CrossRef] [PubMed]
42. Li, X.Y.; Awakawa, T.; Mori, T.; Ling, M.Q.; Hu, D.; Wu, B.; Abe, I. Heterodimeric non-heme iron enzymes in fungal meroterpenoid biosynthesis. *J. Am. Chem. Soc.* **2021**, *143*, 21425–21432. [CrossRef] [PubMed]
43. Lee, C.F.; Chen, L.X.; Chiang, C.Y.; Lai, C.Y.; Lin, H.C. The biosynthesis of norsesquiterpene aculenes requires three cytochrome P450 enzymes to catalyze a stepwise demethylation process. *Angew. Chem. Int. Ed.* **2019**, *58*, 18414–18418. [CrossRef] [PubMed]
44. Yuan, S.W.; Chen, S.H.; Guo, H.; Chen, L.T.; Shen, H.J.; Liu, L.; Gao, Z.Z. Elucidation of the Complete Biosynthetic Pathway of Phomoxanthone A and Identification of a Para–Para Selective Phenol Coupling Dimerase. *Org. Lett.* **2022**, *24*, 3069–3074. [CrossRef]
45. Shu, X.; Wei, G.Z.; Qiao, Y.B.; Zhang, K.X.; Zhang, J.; Ai, G.M.; Tang, M.C.; Zhang, Y.H.; Gao, S.S. TerC Is a Multifunctional and Promiscuous Flavoprotein Monooxygenase That Catalyzes Bimodal Oxidative Transformations. *Org. Lett.* **2021**, *23*, 8947–8951. [CrossRef]
46. Li, B.Q.; Chen, Y.; Zong, Y.Y.; Shang, Y.J.; Zhang, Z.Q.; Xu, X.D.; Wang, X.; Long, M.Y.; Tian, S.P. Dissection of patulin biosynthesis, spatial control and regulation mechanism in *Penicillium expansum*. *Environ. Microbiol.* **2019**, *21*, 1124–1139. [CrossRef]
47. Zhang, T.; Gu, G.; Liu, G.; Su, J.; Zhan, Z.; Zhao, J.; Qian, J.; Cai, G.; Cen, S.; Zhang, D.; et al. Late-stage cascade of oxidation reactions during the biosynthesis of oxalicine B in *Penicillium oxalicum*. *Acta. Pharm. Sin. B* **2022**. [CrossRef]
48. Sun, Y.L.; Chen, B.; Li, X.L.; Yin, Y.; Wang, C.S. Orchestrated Biosynthesis of the Secondary Metabolite Cocktails Enables the Producing Fungus to Combat Diverse Bacteria. *mBio* **2022**, *13*, e0180022. [CrossRef]
49. Wei, Q.; Bai, J.; Yan, D.J.; Bao, X.Q.; Li, W.T.; Liu, B.Y.; Zhang, D.; Qi, X.B.; Yu, D.Q.; Hu, Y.C. Genome mining combined metabolic shunting and OSMAC strategy of an endophytic fungus leads to the production of diverse natural products. *Acta. Pharm. Sin. B.* **2021**, *11*, 572–587. [CrossRef]
50. Qi, F.F.; Zhang, W.; Xue, Y.Y.; Geng, C.; Jin, Z.G.; Li, J.B.; Guo, Q.; Huang, X.N.; Lu, X.F. Microbial production of the plant-derived fungicide physcion. *Metab. Eng.* **2022**, *74*, 130–138. [CrossRef]
51. Xu, D.; Yin, R.Y.; Zhou, Z.Y.; Gu, G.; Zhao, S.J.; Xu, J.R.; Liu, J.F.; Peng, Y.L.; Lai, D.W.; Zhou, L.G. Elucidation of ustilaginoidin biosynthesis reveals a previously unrecognised class of ene-reductases. *Chem. Sci.* **2021**, *12*, 14883–14892. [CrossRef]
52. Tang, Z.J.; Tang, H.Y.; Wang, W.Q.; Xue, Y.F.; Chen, D.D.; Tang, W.H.; Liu, W. Biosynthesis of a New Fusaoctaxin Virulence Factor in *Fusarium graminearum* Relies on a Distinct Path To Form a Guanidinoacetyl Starter Unit Priming Nonribosomal Octapeptidyl Assembly. *J. Am. Chem. Soc.* **2021**, *143*, 19719–19730. [CrossRef]
53. Zeng, H.C.; Yin, G.P.; Wei, Q.; Li, D.H.; Wang, Y.; Hu, Y.C.; Hu, C.H.; Zou, Y. Unprecedented [5.5.5.6] dioxafenestrane ring construction in fungal insecticidal sesquiterpene biosynthesis. *Angew. Chem. Int. Ed.* **2019**, *131*, 6641–6645. [CrossRef]
54. Ning, Y.D.; Hu, B.; Yu, H.B.; Liu, X.Y.; Jiao, B.H.; Lu, X.L. Optimization of Protoplast Preparation and Establishment of Genetic Transformation System of an Arctic-Derived Fungus *Eutypella* sp. *Front. Microbiol.* **2022**, *13*, 769008. [CrossRef]
55. Wei, G.Z.; Shu, X.; Yao, Y.P. Heterologous Production of Unnatural Flavipucine Family Products Provides Insights into Flavipucines Biosynthesis. *Org. Lett.* **2021**, *23*, 7708–7712. [CrossRef]
56. Chen, Q.B.; Yuan, G.Y.; Yuan, T.; Zeng, H.T.; Zou, Z.R.; Tu, Z.C.; Gao, J.; Zou, Y. Set of Cytochrome P450s Cooperatively Catalyzes the Synthesis of a Highly Oxidized and Rearranged Diterpene-Class Sordarinane Architecture. *J. Am. Chem. Soc.* **2022**, *144*, 3580–3589. [CrossRef]
57. Wei, Q.; Zeng, H.C.; Zou, Y. Divergent Biosynthesis of Fungal Dioxafenestrane Sesquiterpenes by the Cooperation of Distinctive Baeyer–Villiger Monooxygenases and α-Ketoglutarate-Dependent Dioxygenases. *ACS. Catal.* **2021**, *11*, 948–957. [CrossRef]
58. Zhong, B.F.; Wan, J.; Shang, C.H.; Wen, J.J.; Wang, Y.J.; Bai, J.; Cen, S.; Hu, Y.C. Biosynthesis of rumbrins and inspiration for discovery of HIV inhibitors. *Acta. Pharm. Sin. B.* **2022**, *12*, 4193–4203. [CrossRef]
59. Yee, D.A.; Kakule, T.B.; Cheng, W.; Chen, M.B.; Chong, C.T.Y.; Hai, Y.; Hang, L.F.; Hung, Y.S.; Liu, N.; Ohashi, M. Genome mining of alkaloidal terpenoids from a hybrid terpene and nonribosomal peptide biosynthetic pathway. *J. Am. Chem. Soc.* **2020**, *142*, 710–714. [CrossRef]

60. Xie, L.N.; Zang, X.; Cheng, W.; Zhang, Z.; Zhou, J.H.; Chen, M.B.; Tang, Y. Harzianic acid from *Trichoderma afroharzianum* is a natural product inhibitor of acetohydroxyacid synthase. *J. Am. Chem. Soc.* **2021**, *143*, 9575–9584. [CrossRef]
61. Liu, L.; Tang, M.C.; Tang, Y. Fungal highly reducing polyketide synthases biosynthesize salicylaldehydes that are precursors to epoxycyclohexenol natural products. *J. Am. Chem. Soc.* **2019**, *141*, 19538–19541. [CrossRef] [PubMed]
62. Zhang, Z.; Qiao, T.Z.; Watanabe, K.; Tang, Y. Concise Biosynthesis of Phenylfuropyridones in Fungi. *Angew. Chem. Int. Ed.* **2020**, *132*, 20061–20065. [CrossRef]
63. Zhang, Z.; Jamieson, C.S.; Zhao, Y.L.; Li, D.H.; Ohashi, M.; Houk, K.N.; Tang, Y. Enzyme-catalyzed inverse-electron demand Diels–Alder reaction in the biosynthesis of antifungal ilicicolin H. *J. Am. Chem. Soc.* **2019**, *141*, 5659–5663. [CrossRef] [PubMed]
64. Gao, D.W.; Jamieson, C.S.; Wang, G.Q.; Yan, Y.; Zhou, J.H.; Houk, K.N.; Tang, Y. A polyketide cyclase that forms medium-ring lactones. *J. Am. Chem. Soc.* **2020**, *143*, 80–84. [CrossRef] [PubMed]
65. Zhao, Q.F.; Zhuang, Z.; Liu, T.; Yang, Q.; He, Q.L.; Liu, W.; Lin, G.Q. Unsymmetrically Regioselective Homodimerization Depends on the Subcellular Colocalization of Laccase/Fasciclin Protein in the Biosynthesis of Phlegmacins. *ACS. Chem. Biol.* **2021**, *17*, 791–796. [CrossRef] [PubMed]
66. Yan, D.; Matsuda, Y. Genome mining-driven discovery of 5-methylorsellinate-derived meroterpenoids from *Aspergillus funiculosus*. *Org. Lett.* **2021**, *23*, 3211–3215. [CrossRef]
67. Chen, L.; Wei, X.X.; Matsuda, Y. Depside Bond Formation by the Starter-Unit Acyltransferase Domain of a Fungal Polyketide Synthase. *J. Am. Chem. Soc.* **2022**, *144*, 19225–19230. [CrossRef]
68. Chen, R.; Feng, T.; Li, M.; Zhang, X.Y.; He, J.; Hu, B.; Deng, Z.X.; Liu, T.G.; Liu, J.; Wang, X.H.; et al. Characterization of Tremulane Sesquiterpene Synthase from the Basidiomycete Irpex lacteus. *Org. Lett.* **2022**, *24*, 5669–5673. [CrossRef]
69. Wang, W.G.; Wang, H.; Du, L.Q.; Li, M.; Chen, L.; Yu, J.; Cheng, G.G.; Zhan, M.T.; Hu, Q.F.; Zhang, L.H.; et al. Molecular basis for the biosynthesis of an unusual chain-fused polyketide, gregatin A. *J. Am. Chem. Soc.* **2020**, *142*, 8464–8472. [CrossRef]
70. Murai, K.; Lauterbach, L.; Teramoto, K.; Quan, Z.Y.; Barra, L.; Yamamoto, T.; Nonaka, K.; Shiomi, K.; Nishiyama, M.; Kuzuyama, T.; et al. An unusual skeletal rearrangement in the biosynthesis of the sesquiterpene trichobrasilenol from *Trichoderma*. *Angew. Chem. Int. Ed.* **2019**, *58*, 15046–15050. [CrossRef]
71. Chen, L.; Tang, J.W.; Liu, Y.Y.; Matsuda, Y. Aspcandine: A Pyrrolobenzazepine Alkaloid Synthesized by a Fungal Nonribosomal Peptide Synthetase-Polyketide Synthase Hybrid. *Org. Lett.* **2022**, *24*, 4816–4819. [CrossRef]
72. Hewage, R.T.; Huang, R.J.; Lai, S.J.; Lien, Y.C.; Weng, S.H.; Li, D.H.; Chen, Y.J.; Wu, S.H.; Chein, R.J.; Lin, H.C. An Enzyme-Mediated Aza-Michael Addition Is Involved in the Biosynthesis of an Imidazoyl Hybrid Product of Conidiogenone B. *Org. Lett.* **2021**, *23*, 1904–1909. [CrossRef]
73. Ye, Y.; Du, L.; Zhang, X.W.; Newmister, S.A.; McCauley, M.; Alegre-Requena, J.V.; Zhang, W.; Mu, S.; Minami, A.; Fraley, A.E.; et al. Fungal-derived brevianamide assembly by a stereoselective semipinacolase. *Nat. Catal.* **2020**, *3*, 497–506. [CrossRef]
74. Yan, D.X.; Matsuda, Y. Biosynthetic Elucidation and Structural Revision of Brevione E: Characterization of the Key Dioxygenase for Pathway Branching from Setosusin Biosynthesis. *Angew. Chem. Int. Ed.* **2022**, e202210938. [CrossRef]
75. Zhang, K.J.; Zhang, G.J.; Hou, X.W.; Ma, C.; Liu, J.Y.; Che, Q.; Zhu, T.J.; Li, D.H. A Fungal Promiscuous UbiA Prenyltransferase Expands the Structural Diversity of Chrodrimanin-Type Meroterpenoids. *Org. Lett.* **2022**, *24*, 2025–2029. [CrossRef]
76. Chen, Y.R.; Naresh, A.; Liang, S.Y.; Lin, C.H.; Chein, R.J.; Lin, H.C. Discovery of a dual function cytochrome P450 that catalyzes enyne formation in cyclohexanoid terpenoid biosynthesis. *Angew. Chem. Int. Ed.* **2020**, *59*, 13537–13541. [CrossRef]
77. Yao, Y.P.; An, C.Y.; Evans, D.; Liu, W.W.; Wang, W.; Wei, G.Z.; Ding, N.; Houk, K.N.; Gao, S.S. Catalase involved in oxidative cyclization of the tetracyclic ergoline of fungal ergot alkaloids. *J. Am. Chem. Soc.* **2019**, *141*, 17517–17521. [CrossRef]
78. Fujii, I.; Hashimoto, M.; Konishi, K.; Unezawa, A.; Sakuraba, H.; Suzuki, K.; Tsushima, H.; Iwasaki, M.; Yoshida, S.; Kudo, A.; et al. Shimalactone Biosynthesis Involves Spontaneous Double Bicyclo-Ring Formation with 8π-6π Electrocyclization. *Angew. Chem. Int. Ed.* **2020**, *59*, 8464–8470. [CrossRef]
79. Feng, C.; Wei, Q.; Hu, C.H.; Zou, Y. Biosynthesis of diphenyl ethers in fungi. *Org. Lett.* **2019**, *21*, 3114–3118. [CrossRef]
80. Li, H.; Gilchrist, C.L.M.; Phan, C.S.; Lacey, H.J.; Vuong, D.; Moggach, S.A.; Lacey, E.; Piggott, A.M.; Chooi, Y.H. Biosynthesis of a New 237 Benzazepine Alkaloid Nanangelenin A from *Aspergillus nanangensis* Involves an Unusual l-238 Kynurenine-Incorporating NRPS Catalyzing Regioselective Lactamization. *J. Am. Chem. Soc.* **2020**, *239*, 15. [CrossRef]
81. Yan, D.J.; Chen, Q.B.; Gao, J.; Bai, J.; Liu, B.Y.; Zhang, Y.L.; Zhang, C.; Zou, Y.; Hu, Y.C. Complexity and diversity generation in the biosynthesis of fumiquinazoline-related peptidyl alkaloids. *Org. Lett.* **2019**, *21*, 1475–1479. [CrossRef] [PubMed]
82. Chen, B.; Sun, Y.L.; Li, S.Q.; Yin, Y.; Wang, C.S. Inductive Production of the Iron-Chelating 2-Pyridones Benefits the Producing Fungus to Compete for Diverse Niches. *Mbio* **2021**, *12*, e0327921. [CrossRef] [PubMed]
83. Guo, J.J.; Cai, Y.S.; Cheng, F.C.; Yang, C.J.; Zhang, W.Q.; Yu, W.L.; Yan, J.J.; Deng, Z.X.; Hong, K. Genome mining reveals a multiproduct sesterterpenoid biosynthetic gene cluster in Aspergillus ustus. *Org. Lett.* **2021**, *23*, 1525–1529. [CrossRef] [PubMed]
84. Li, H.; Shu, S.; Kalaitzis, J.A.; Shang, Z.; Vuong, D.; Crombie, A.; Lacey, E.; Piggott, A.M.; Chooi, Y.H. Genome mining of *Aspergillus hancockii* unearths cryptic polyketide hancockinone A featuring a prenylated 6/6/6/5 carbocyclic skeleton. *Org. Lett.* **2021**, *23*, 8789–8793. [CrossRef] [PubMed]
85. Tsukada, K.; Shinki, S.; Kaneko, A.; Murakami, K.; Irie, K.; Murai, M.; Miyoshi, H.; Dan, S.; Kawaji, K.; Hayashi, H.; et al. Synthetic biology based construction of biological activity-related library of fungal decalin-containing diterpenoid pyrones. *Nat. Commun.* **2020**, *11*, 1830. [CrossRef]

marine drugs

MDPI

Review

Fungal Bergamotane Sesquiterpenoids—Potential Metabolites: Sources, Bioactivities, and Biosynthesis

Maan T. Khayat [1,*], Khadijah A. Mohammad [1], Abdelsattar M. Omar [1,2], Gamal A. Mohamed [3] and Sabrin R. M. Ibrahim [4,5]

1 Department of Pharmaceutical Chemistry, Faculty of Pharmacy, King Abdulaziz University, Jeddah 21589, Saudi Arabia
2 Center for Artificial Intelligence in Precision Medicines, King Abdulaziz University, Jeddah 21589, Saudi Arabia
3 Department of Natural Products and Alternative Medicine, Faculty of Pharmacy, King Abdulaziz University, Jeddah 21589, Saudi Arabia
4 Preparatory Year Program, Department of Chemistry, Batterjee Medical College, Jeddah 21442, Saudi Arabia
5 Department of Pharmacognosy, Faculty of Pharmacy, Assiut University, Assiut 71526, Egypt
* Correspondence: mkhayat@kau.edu.sa; Tel.: +966-555-543-053

Abstract: The marine environment represents the largest ecosystem on the Earth's surface. Marine-derived fungi are of remarkable importance as they are a promising pool of diverse classes of bioactive metabolites. Bergamotane sesquiterpenoids are an uncommon class of terpenoids. They possess diverse biological properties, such as plant growth regulation, phototoxic, antimicrobial, anti-HIV, cytotoxic, pancreatic lipase inhibition, antidiabetic, anti-inflammatory, and immunosuppressive traits. The current work compiles the reported bergamotane sesquiterpenoids from fungal sources in the period ranging from 1958 to June 2022. A total of 97 compounds from various fungal species were included. Among these metabolites, 38 compounds were derived from fungi isolated from different marine sources. Furthermore, the biological activities, structural characterization, and biosynthesis of the compounds are also discussed. The summary in this work provides a detailed overview of the reported knowledge of fungal bergamotane sesquiterpenoids. Moreover, this in-depth and complete review could provide new insights for developing and discovering new valuable pharmaceutical agents from these natural metabolites.

Keywords: bergamotanes; sesquiterpenoids; marine; fungi; biosynthesis; biological activities

Citation: Khayat, M.T.; Mohammad, K.A.; Omar, A.M.; Mohamed, G.A.; Ibrahim, S.R.M. Fungal Bergamotane Sesquiterpenoids—Potential Metabolites: Sources, Bioactivities, and Biosynthesis. *Mar. Drugs* **2022**, *20*, 771. https://doi.org/10.3390/md20120771

Academic Editor: Dehai Li

Received: 5 November 2022
Accepted: 5 December 2022
Published: 8 December 2022

1. Introduction

Nature has substantially participated in the discovery of drugs for human remedial treatments since the beginning of mankind [1]. The marine environment, with more than 70% of the surface of the Earth, represents the largest ecosystem and is characterized by quite variable physicochemical parameters (e.g., limited light access, low temperature, high pressure, and high salinity) [2]. Among the various marine microbes, fungi are a superabundant and ecologically substantial component of marine microbiota [3]. Fungi are one of nature's treasures that inhabit various environments on the earth's surface, including the marine environment [4–7]. They play a growing relevant role in drug development and biomedicine research, either directly as drugs or indirectly as lead structures for bio-inspired drug synthesis [8–12]. In the last decades, natural product chemists and pharmacologists have turned their research interests to marine-derived fungi, which are renowned as a vast unexploited reservoir of metabolic diverseness and found to have the capability to produce structurally unique bio-metabolites [6,7,12–16]. Furthermore, research on fungi-derived metabolites has tremendously increased because of the need for compounds with potential economical values and pharmaceutical applications. Sesquiterpenes belonging to various classes, including hirsutane, alliacane, tremulane, bergamotane, drimane, etc., are reported from fungi [17–19]. The biosynthesis of their C15 skeleton from FPP (farnesyl pyrophosphate) was catalyzed by sesquiterpene synthases [19,20].

Among these metabolites, the bergamotane family represents an uncommon class of natural sesquiterpenes that includes bi-, tri-, or tetracyclic derivatives [19]. Bergamotane sesquiterpenoids having a bridged 6/4 bicyclic skeleton involved in an isopentyl unit are biosynthesized by fungi and plants [21,22]. Interestingly, polyoxygenated derivatives featuring a 6/4/5/5 tetracyclic framework represent a rare class of natural metabolites, and all polycyclic bergamotanes are mainly encountered in fungi [23–26]. Bergamotane sesquiterpenoids have been reported from various marine sources such as sponges, sea mud, deep-sea hydrothermal sulfide deposits, and sea sediments. These metabolites could gain the interest of chemists and biologists because of their unusual structural features and diversified activities, such as phytotoxicity, plant growth regulation, antimicrobial, anti-HIV, cytotoxic, pancreatic lipase inhibition, immunosuppressive, antidiabetic, and anti-inflammatory properties. It is noteworthy that no available work has addressed this class of sesquiterpenes in term of their sources, bioactivities, and biosynthesis. In the current work, the reported fungal bergamotane sesquiterpenoids ranging from 1958 to June 2022 have been listed. They have been classified according to their ring system, i.e., into bi-, tri-, or tetracyclic derivatives (Table 1). Additionally, their fungal sources, structural characterization, biosynthesis, and biological relevance have been provided. Moreover, some of their reported structural characteristics and methods of separation and characterization, as well as their structure–activity relation, are discussed.

Table 1. Naturally occurring fungal bergamotane sesquiterpenoids (name, source, extract/fraction, molecular weights and formulae, and location).

Compound Name	Fungal Source/Host	Extract/Fraction	Mol. Wt.	Mol. Formula	Location	Ref.
Bicyclic Bergamotane Sesquiterpenoids						
α-*trans* Bergamotene (**1**)	*Nectria* sp. HLS206 (Nectriaceae)/*Gelliodas carnosa* (marine sponge Geodiidae)	EtOAc extract	204	$C_{15}H_{24}$	China	[27]
β-*trans* Bergamotene (**2**)	*Aspergillus fumigatus* (Trichocomaceae)/Cultured	Acetone extract	204	$C_{15}H_{24}$	Japan	[28]
β-*trans*-2β,5,15-Trihydroxybergamot-10-ene (**3**)	*Aspergillus fumigatus* YK-7 (Trichocomaceae)/Sea mud	EtOAc extract	254	$C_{15}H_{26}O_3$	Intertidal zone sea mud, Yingkou, China	[29]
E-β-*trans*-5,8,11-Trihydroxybergamot-9-ene (**4**)	*Aspergillus fumigatus* YK-7 (Trichocomaceae)/Sea mud	EtOAc extract	252	$C_{15}H_{24}O_3$	Intertidal zone sea mud, Yingkou, China	[29]
Massarinolin C (**5**)	*Massarina tunicata* (Lophiostomataceae)/Submerged twig	EtOAc extract	266	$C_{15}H_{22}O_4$	Lemonweir River in Adams County, Wisconsin, USA	[23]
	Craterellus odoratus (Cantharellaceae)	EtOAc extract	-	-	Southern part of the Gaoligong-Mountains, Yunnan, China	[30]
	Paraconiothyrium sporulosum YK-03 Verkley (Leptosphaeriaceae)/Sea mud	EtOAc extract	-	-	Intertidal zone of Bohai Bay river in Liaoning, China	[31]
Donacinoic acid B (**6**)	*Montagnula donacina* (Montagnulaceae)/*Craterellus odoratus* (fruiting bodies, Cantharellaceae)	EtOAc extract	266	$C_{15}H_{22}O_4$	Southern part of the Gaoligong Mountains in Yunnan, China	[32]

Table 1. *Cont.*

Compound Name	Fungal Source/Host	Extract/Fraction	Mol. Wt.	Mol. Formula	Location	Ref.
Craterodoratin M (**7**)	*Craterellus odoratus* (Cantharellaceae)	EtOAc extract	252	$C_{15}H_{24}O_3$	Southern part of the Gaoligong-Mountains, Yunnan, China	[30]
Craterodoratin N (**8**)	*Craterellus odoratus* (Cantharellaceae)	EtOAc extract	268	$C_{15}H_{24}O_4$	Southern part of the Gaoligong-Mountains, Yunnan, China	[30]
Craterodoratin O (**9**)	*Craterellus odoratus* (Cantharellaceae)	EtOAc extract	250	$C_{15}H_{22}O_3$	Southern part of the Gaoligong-Mountains, Yunnan, China	[30]
Craterodoratin P (**10**)	*Craterellus odoratus* (Cantharellaceae)	EtOAc extract	250	$C_{15}H_{22}O_3$	Southern part of the Gaoligong-Mountains, Yunnan, China	[30]
Craterodoratin Q (**11**)	*Craterellus odoratus* (Cantharellaceae)	EtOAc extract	308	$C_{17}H_{24}O_5$	Southern part of the Gaoligong-Mountains, Yunnan, China	[30]
Necbergamotenoic acid A (**12**)	*Nectria* sp. HLS206 (Nectriaceae)/*Gelliodas carnosa* (marine sponge, Geodiidae)	EtOAc extract	264	$C_{15}H_{20}O_4$	China	[27]
Necbergamotenoic acid B (**13**)	*Nectria* sp. HLS206 (Nectriaceae)/*Gelliodas carnosa* (marine sponge, Geodiidae)	EtOAc extract	266	$C_{15}H_{22}O_4$	China	[27]
Sporulamide C (**14**)	*Paraconiothyrium sporulosum* YK-03 Verkley (Leptosphaeriaceae)/Sea mud	EtOAc extract	265	$C_{15}H_{23}NO_3$	Intertidal zone of Bohai Bay river in Liaoning, China	[31]
Sporulamide D (**15**)	*Paraconiothyrium sporulosum* YK-03 Verkley (Leptosphaeriaceae)/Sea mud	EtOAc extract	249	$C_{15}H_{23}NO_2$	Intertidal zone of Bohai Bay river in Liaoning, China	[31]
Xylariterpenoid A (**16**)	*Xylariaceae fungus* (No. 63-19-7-3)/*Everniastrum cirrhatum* (Fr.) Haleex Sipman (lichen, Parmeliaceae)	EtOAc extract	252	$C_{15}H_{24}O_3$	Zixi Mountain, Yunnan, China	[33]
	Graphostroma sp. MCCC 3A00421/Deep-sea hydrothermal sulfide deposit	EtOAc extract	-	-	Atlantic Ocean, China	[34]
	Eutypella sp. MCCC 3A00281 (Diatrypaceae)/Deep-sea sediment	EtOAc extract	-	-	South Atlantic Ocean, China	[35]
Xylariterpenoid B (**17**)	*Xylariaceae fungus* (No. 63-19-7-3)/*Everniastrum cirrhatum* (Fr.) Haleex Sipman (lichen, Parmeliaceae)	EtOAc extract	252	$C_{15}H_{24}O_3$	Zixi Mountain, Yunnan, China	[33]
	Graphostroma sp. MCCC 3A00421/Deep-sea hydrothermal sulfide deposit	EtOAc extract	-	-	Atlantic Ocean, China	[34]
	Eutypella sp. MCCC 3A00281 (Diatrypaceae)/Deep-sea sediment	EtOAc extract	-	-	South Atlantic Ocean, China	[35]

Table 1. *Cont.*

Compound Name	Fungal Source/Host	Extract/Fraction	Mol. Wt.	Mol. Formula	Location	Ref.
Eutypeterpene B (**18**)	*Eutypella* sp. MCCC 3A00281 (Diatrypaceae)/Deep-sea sediment	EtOAc extract	268	$C_{15}H_{24}O_4$	South Atlantic Ocean, China	[35]
Eutypeterpene C (**19**)	*Eutypella* sp. MCCC 3A00281 (Diatrypaceae)/Deep-sea sediment	EtOAc extract	266	$C_{15}H_{22}O_4$	South Atlantic Ocean, China	[35]
Eutypeterpene D (**20**)	*Eutypella* sp. MCCC 3A00281 (Diatrypaceae)/Deep-sea sediment	EtOAc extract	250	$C_{15}H_{22}O_3$	South Atlantic Ocean, China	[35]
Eutypeterpene E (**21**)	*Eutypella* sp. MCCC 3A00281 (Diatrypaceae)/Deep-sea sediment	EtOAc extract	250	$C_{15}H_{22}O_3$	South Atlantic Ocean, China	[35]
Eutypeterpene F (**22**)	*Eutypella* sp. MCCC 3A00281 (Diatrypaceae)/Deep-sea sediment	EtOAc extract	252	$C_{15}H_{24}O_3$	South Atlantic Ocean, China	[35]
(10S)-Xylariterpenoid A (**23**)	*Graphostroma* sp. MCCC 3A00421/Deep-sea hydrothermal sulfide deposit	EtOAc extract	252	$C_{15}H_{24}O_3$	Atlantic Ocean. China	[34]
(10R)-Xylariterpenoid B (**24**)	*Graphostroma* sp. MCCC 3A00421/Deep-sea hydrothermal sulfide deposit	EtOAc extract	252	$C_{15}H_{24}O_3$	Atlantic Ocean. China	[34]
Xylariterpenoid E (**25**)	*Graphostroma* sp. MCCC 3A00421/Deep-sea hydrothermal sulfide deposit	EtOAc extract	208	$C_{12}H_{16}O_3$	Atlantic Ocean. China	[34]
Xylariterpenoid F (**26**)	*Graphostroma* sp. MCCC 3A00421/Deep-sea hydrothermal sulfide deposit	EtOAc extract	270	$C_{15}H_{26}O_4$	Atlantic Ocean. China	[34]
Xylariterpenoid G (**27**)	*Graphostroma* sp. MCCC 3A00421/Deep-sea hydrothermal sulfide deposit	EtOAc extract	270	$C_{15}H_{26}O_4$	Atlantic Ocean. China	[34]
Eutypeterpene A (**28**)	*Eutypella* sp. MCCC 3A00281 (Diatrypaceae)/Deep-sea sediment	EtOAc extract	294	$C_{16}H_{22}O_5$	South Atlantic Ocean, China	[35]
Craterodoratin A (**29**)	*Craterellus odoratus* (Cantharellaceae)	EtOAc extract	252	$C_{15}H_{24}O_3$	Southern part of the Gaoligong-Mountains, Yunnan, China	[30]
Craterodoratin C (**30**)	*Craterellus odoratus* (Cantharellaceae)	EtOAc extract	268	$C_{15}H_{24}O_4$	Southern part of the Gaoligong-Mountains, Yunnan, China	[30]
Craterodoratin D (**31**)	*Craterellus odoratus* (Cantharellaceae)	EtOAc extract	268	$C_{15}H_{24}O_4$	Southern part of the Gaoligong-Mountains, Yunnan, China	[30]
Craterodoratin E (**32**)	*Craterellus odoratus* (Cantharellaceae)	EtOAc extract	284	$C_{15}H_{24}O_5$	Southern part of the Gaoligong-Mountains, Yunnan, China	[30]

Table 1. *Cont.*

Compound Name	Fungal Source/Host	Extract/Fraction	Mol. Wt.	Mol. Formula	Location	Ref.
Craterodoratin F (**33**)	*Craterellus odoratus* (Cantharellaceae)	EtOAc extract	284	$C_{15}H_{24}O_5$	Southern part of the Gaoligong-Mountains, Yunnan, China	[30]
Dihydroprehelminthosporol (**34**)	*Bipolaris* sp. No. 36/Johnson grass leaf	EtOAc extract	238	$C_{15}H_{26}O_2$	Wake County, North Carolina, USA	[36,37]
Helminthosporal acid (**35**)	*Bipolaris* sp. No. 36/Johnson grass leaf	EtOAc extract	250	$C_{15}H_{22}O_3$	Wake County, North Carolina, USA	[36]
Helminthosporol (**36**)	*Bipolaris* sp. No. 36/Johnson grass leaf	EtOAc extract	236	$C_{15}H_{24}O_2$	Wake County, North Carolina, USA	[36]
Helminthosporic acid (**37**)	*Bipolaris* sp. No. 36/Johnson grass leaf	EtOAc extract	252	$C_{15}H_{24}O_3$	Wake County, North Carolina, USA	[36]
Tricyclic Bergamotane Sesquiterpenoids						
Prehelminthosporol (**38**)	*Bipolaris* sp. No. 36/Johnson grass leaf	EtOAc extract	236	$C_{15}H_{24}O_2$	Wake County, North Carolina, USA	[36,37]
Prehelminthosporolactone (**39**)	*Bipolaris* sp. No. 36/Johnson grass leaf	EtOAc extract	234	$C_{15}H_{22}O_2$	Wake County, North Carolina, USA	[37]
Victoxinine (**40**)	*Helminthosporium victoriae* (Totiviridae)	Diethyl ether extract	263	$C_{17}H_{29}NO$	USA	[36,38,39]
	Helminthosporium sativum (Totiviridae)	Diethyl ether fraction/$CHCl_3$ extract	-	-	Canada	[40]
Victoxinine-α-glycerophosphate (**41**)	*H. sativum* (Totiviridae)	*n*-BuOH extract	417	$C_{20}H_{36}NO_6P$	USA	[41]
Craterodoratin S (**42**)	*Craterellus odoratus* (Cantharellaceae)	EtOAc extract	277	$C_{17}H_{27}NO_2$	Southern part of the Gaoligong-Mountains, Yunnan, China	[30]
Isosativenediol (**43**)	*Bipolaris* sp. No. 36/Johnson grass leaf	EtOAc extract	236	$C_{15}H_{24}O_2$	Wake County, North Carolina, USA	[36]
Pinthunamide (**44**)	*Ampulliferina* sp. No. 27 (Ampullicephala)/*Pinus thunbergii* (dead tree, Pinaceae)	Acetone extract	277	$C_{15}H_{19}NO_4$	Japan	[42]
	Paraconiothyrium brasiliense Verkley (M3–3341) (Leptosphaeriaceae)/*Acer truncatum* Bunge (branches, Sapindaceae)	Acetone extract	-	-	Dongling Mountain, Beijing, China	[43]
Brasilamide A (**45**)	*Paraconiothyrium brasiliense* Verkley (M3–3341) (Leptosphaeriaceae)/*Acer truncatum* Bunge (branches, Sapindaceae)	Acetone extract	293	$C_{15}H_{19}NO_5$	Dongling Mountain, Beijing, China	[43,44]

Table 1. *Cont.*

Compound Name	Fungal Source/Host	Extract/Fraction	Mol. Wt.	Mol. Formula	Location	Ref.
Brasilamide B (**46**)	*Paraconiothyrium brasiliense* Verkley (M3–3341) (Leptosphaeriaceae)/ *Acer truncatum* Bunge (branches, Sapindaceae)	Acetone extract	265	$C_{15}H_{23}NO_3$	Dongling Mountain, Beijing, China	[43]
Brasilamide C (**47**)	*Paraconiothyrium brasiliense* Verkley (M3–3341) (Leptosphaeriaceae)/ *Acer truncatum* Bunge (branches, Sapindaceae)	Acetone extract	279	$C_{15}H_{21}NO_4$	Dongling Mountain, Beijing, China	[43,44]
Brasilamide D (**48**)	*Paraconiothyrium brasiliense* Verkley (M3–3341) (Leptosphaeriaceae)/ *Acer truncatum* Bunge (branches, Sapindaceae)	Acetone extract	321	$C_{17}H_{23}NO_5$	Dongling Mountain, Beijing, China	[43]
Brasilamide K (**49**)	*Paraconiothyrium brasiliense* Verkley (M3–3341) (Leptosphaeriaceae)/ *Acer truncatum* Bunge (branches, Sapindaceae)	EtOAc extract	279	$C_{15}H_{21}NO_4$	Dongling Mountain, Beijing, China	[44]
Brasilamide L (**50**)	*Paraconiothyrium brasiliense* Verkley (M3–3341) (Leptosphaeriaceae)/ *Acer truncatum* Bunge (branches, Sapindaceae)	EtOAc extract	265	$C_{15}H_{23}NO_3$	Dongling Mountain, Beijing, China	[44]
Brasilamide M (**51**)	*Paraconiothyrium brasiliense* Verkley (M3–3341) (Leptosphaeriaceae)/ *Acer truncatum* Bunge (branches, Sapindaceae)	EtOAc extract	293	$C_{15}H_{19}NO_5$	Dongling Mountain, Beijing, China,	[44]
Brasilamide N (**52**)	*Paraconiothyrium brasiliense* Verkley (M3–3341) (Leptosphaeriaceae)/ *Acer truncatum* Bunge (branches, Sapindaceae)	EtOAc extract	279	$C_{15}H_{21}NO_4$	Dongling Mountain, Beijing, China	[44]
Craterodoratin I (**53**)	*Craterellus odoratus* (Cantharellaceae)	EtOAc extract	250	$C_{15}H_{22}O_3$	Southern part of the Gaoligong-Mountains, Yunnan, China	[30]
Craterodoratin J (**54**)	*Craterellus odoratus* (Cantharellaceae)	EtOAc extract	282	$C_{15}H_{22}O_5$	Southern part of the Gaoligong-Mountains, Yunnan, China	[30]
Craterodoratin K (**55**)	*Craterellus odoratus* (Cantharellaceae)	EtOAc extract	282	$C_{15}H_{22}O_5$	Southern part of the Gaoligong-Mountains, Yunnan, China	[30]
Craterodoratin L (**56**)	*Craterellus odoratus* (Cantharellaceae)	EtOAc extract	278	$C_{15}H_{18}O_5$	Southern part of the Gaoligong-Mountains, Yunnan, China	[30]

Table 1. *Cont.*

Compound Name	Fungal Source/Host	Extract/Fraction	Mol. Wt.	Mol. Formula	Location	Ref.
Sporulosoic acid A (**57**)	*Paraconiothyrium sporulosum* YK-03 Verkley (Leptosphaeriaceae)/Sea mud	EtOAc extract	282	$C_{15}H_{22}O_5$	Intertidal zone of Bohai Bay river in Liaoning, China	[31]
Sporulosoic acid B (**58**)	*Paraconiothyrium sporulosum* YK-03 Verkley (Leptosphaeriaceae)/Sea mud	EtOAc extract	280	$C_{15}H_{20}O_5$	Intertidal zone of Bohai Bay river in Liaoning, China	[31]
Sporulamide A (**59**)	*Paraconiothyrium sporulosum* YK-03 Verkley (Leptosphaeriaceae)/Sea mud	EtOAc extract	265	$C_{15}H_{23}NO_3$	Intertidal zone of Bohai Bay river in Liaoning, China	[31]
Sporulamide B (**60**)	*Paraconiothyrium sporulosum* YK-03 Verkley (Leptosphaeriaceae)/Sea mud	EtOAc extract	249	$C_{15}H_{23}NO_2$	Intertidal zone of Bohai Bay river in Liaoning, China	[31]
Massarinolin B (**61**)	*Massarina tunicata* (Lophiostomataceae)/Submerged twig	EtOAc extract	266	$C_{15}H_{22}O_4$	Lemonweir River in Adams County, Wisconsin, USA	[23]
	Craterellus odoratus (Cantharellaceae)	EtOAc extract	-	-	Southern part of the Gaoligong-Mountains, Yunnan, China	[30]
Massarinolin B methyl ester (**62**)	*Paraconiothyrium sporulosum* YK-03 Verkley (Leptosphaeriaceae)/Sea mud	EtOAc extract	280	$C_{16}H_{24}O_4$	Intertidal zone of Bohai Bay river in Liaoning, China	[31]
Craterodoratin R (**63**)	*Craterellus odoratus* (Cantharellaceae)	EtOAc extract	282	$C_{15}H_{22}O_5$	Southern part of the Gaoligong-Mountains, Yunnan, China	[30]
Craterodoratin G (**64**)	*Craterellus odoratus* (Cantharellaceae)	EtOAc extract	278	$C_{16}H_{22}O_4$	Southern part of the Gaoligong-Mountains, Yunnan, China	[30]
Craterodoratin H (**65**)	*Craterellus odoratus* (Cantharellaceae)	EtOAc extract	278	$C_{16}H_{22}O_4$	Southern part of the Gaoligong-Mountains, Yunnan, China	[30]
Brasilterpene A (**66**)	*Paraconiothyrium brasiliense* HDN15-135 (Leptosphaeriaceae)/Deep-sea sediment	EtOAc extract	294	$C_{16}H_{22}O_5$	Indian Ocean, China	[45]
Brasilterpene B (**67**)	*Paraconiothyrium brasiliense* HDN15-135 (Leptosphaeriaceae)/Deep-sea sediment	EtOAc extract	294	$C_{16}H_{22}O_5$	Indian Ocean, China	[45]
Brasilterpene C (**68**)	*Paraconiothyrium brasiliense* HDN15-135 (Leptosphaeriaceae)/Deep-sea sediment	EtOAc extract	278	$C_{16}H_{22}O_4$	Indian Ocean, China	[45]

Table 1. *Cont.*

Compound Name	Fungal Source/Host	Extract/Fraction	Mol. Wt.	Mol. Formula	Location	Ref.
Brasilterpene D (**69**)	*Paraconiothyrium brasiliense* HDN15-135 (Leptosphaeriaceae)/Deep-sea sediment	EtOAc extract	278	$C_{16}H_{22}O_4$	Indian Ocean, China	[45]
Brasilterpene E (**70**)	*Paraconiothyrium brasiliense* HDN15-135 (Leptosphaeriaceae)/Deep-sea sediment	EtOAc extract	278	$C_{16}H_{22}O_4$	Indian Ocean, China	[45]
Craterodoratin B (**71**)	*Craterellus odoratus* (Cantharellaceae)	EtOAc extract	266	$C_{15}H_{22}O_4$	Southern part of the Gaoligong-Mountains, Yunnan, China	[30]
Tetracyclic Bergamotane Sesquiterpenoids						
Expansolide A (**72**)	*Penicillium expansum* (Trichocomaceae)/Fruit	EtOAc extract	306	$C_{17}H_{22}O_5$	France	[25]
	Aspergillus fumigatus Fresenius (Trichocomaceae)/Leaf litter	EtOAc extract	-	-	Waipoua Forest, New Zealand	[26]
Expansolide C (**73**)	*Penicillium expansum* ACCC37275/Agricultural Culture	Acetone extract	264	$C_{15}H_{20}O_4$	China	[46]
Decipienolide A (**74**)	*Podospora decipiens* Niessl (JS 270) (Podosporaceae)/Sheep dung	EtOAc extract	378	$C_{21}H_{30}O_6$	South Australia	[24]
Decipienolide B (**75**)	*Podospora decipiens* Niessl (JS 270) (Podosporaceae)/Sheep dung	EtOAc extract	378	$C_{21}H_{30}O_6$	South Australia	[24]
Donacinolide B (**76**)	*Montagnula donacina* (Montagnulaceae)/*Craterellus odoratus* (fruiting bodies, Cantharellaceae)	EtOAc extract	246	$C_{15}H_{18}O_3$	Southern part of the Gaoligong Mountains in Yunnan, China	[32]
Massarinolin A (**77**)	*Massarina tunicata* (Lophiostomataceae)/Submerged twig	EtOAc extract	262	$C_{15}H_{18}O_4$	LemonweirRiver in Adams County, Wisconsin, USA	[23]
	Craterellus odoratus (Cantharellaceae)	EtOAc extract	-	-	Southern part of the Gaoligong-Mountains, Yunnan, China	[30]
Sporuloketal A (**78**)	*Paraconiothyrium sporulosum* YK-03 Verkley (Leptosphaeriaceae)/Sea mud	EtOAc extract	262	$C_{15}H_{18}O_4$	Intertidal zone of Bohai Bay river in Liaoning, China	[31]
Sporuloketal B (**79**)	*Paraconiothyrium sporulosum* YK-03 Verkley (Leptosphaeriaceae)/Sea mud	EtOAc extract	262	$C_{15}H_{18}O_4$	Intertidal zone of Bohai Bay river in Liaoning, China	[31]
Expansolide B (**80**)	*Penicillium expansum* (Trichocomaceae)	EtOAc extract	306	$C_{17}H_{22}O_5$	France	[25]
	Aspergillus fumigatus Fresenius (Trichocomaceae)/Leaf litter	EtOAc extract	-	-	Waipoua Forest, New Zealand	[26]

Table 1. *Cont.*

Compound Name	Fungal Source/Host	Extract/Fraction	Mol. Wt.	Mol. Formula	Location	Ref.
Expansolide D (**81**)	*Penicillium expansum* ACCC37275, (Trichocomaceae)/Agricultural Culture	Acetone extract	264	$C_{15}H_{20}O_4$	China	[46]
Donacinolide A (**82**)	*Montagnula donacina* (Montagnulaceae)/*Craterellus odoratus* (fruiting bodies, Cantharellaceae)	EtOAc extract	246	$C_{15}H_{18}O_3$	Southern part of the Gaoligong Mountains in Yunnan, China	[32]
Purpurolide B (**83**)	*Penicillium purpurogenum* IMM003 (Trichocomaceae)/*Edgeworthia Chrysantha* (leaves, Thymelaeaceae)	EtOAc extract	336	$C_{17}H_{20}O_7$	Hangzhou Bay, Hangzhou, Zhejiang, China	[47]
Purpurolide C (**84**)	*Penicillium purpurogenum* IMM003 (Trichocomaceae)/*Edgeworthia Chrysantha* (leaves, Thymelaeaceae)	EtOAc extract	308	$C_{16}H_{20}O_6$	Hangzhou Bay, Hangzhou, Zhejiang, China	[47]
Purpurolide D (**85**)	*Penicillium purpurogenum* IMM003 (Trichocomaceae)/*Edgeworthia Chrysantha* (leaves, Thymelaeaceae)	EtOAc extract	294	$C_{15}H_{18}O_6$	Hangzhou Bay, Hangzhou, Zhejiang, China	[48]
Purpurolide E (**86**)	*Penicillium purpurogenum* IMM003 (Trichocomaceae)/*Edgeworthia Chrysantha* (leaves, Thymelaeaceae)	EtOAc extract	278	$C_{15}H_{18}O_5$	Hangzhou Bay, Hangzhou, Zhejiang, China	[48]
Purpurolide F (**87**)	*Penicillium purpurogenum* IMM003 (Trichocomaceae)/*Edgeworthia Chrysantha* (leaves, Thymelaeaceae)	EtOAc extract	464	$C_{25}H_{36}O_8$	Hangzhou Bay, Hangzhou, Zhejiang, China	[48]
Donacinoic acid A (**88**)	*Montagnula donacina* (Montagnulaceae)/*Craterellus odoratus* (fruiting bodies, Cantharellaceae)	EtOAc extract	264	$C_{15}H_{20}O_4$	Southern part of the Gaoligong Mountains in Yunnan, China	[32]
	Craterellus odoratus (Cantharellaceae)	EtOAc extract	-	-	Southern part of the Gaoligong-Mountains, Yunnan, China	[30]
Sporulaminal A (**89**)	*Paraconiothyrium sporulosum* YK-03 (Leptosphaeriaceae)/Sea mud	EtOAc extract	247	$C_{15}H_{21}NO_2$	Intertidal zone of Bohai river in Liaonign, China	[49]
Sporulaminal B (**90**)	*Paraconiothyrium sporulosum* YK-03 (Leptosphaeriaceae)/Sea mud	EtOAc extract	247	$C_{15}H_{21}NO_2$	Intertidal zone of Bohai river in Liaonign, China	[49]
Ampullicin (**91**)	*Ampulliferina*-like sp. No. 27 (Ampullicephala)/*Pinus thunbergii* (dead tree, Pinaceae)	Acetone extract	259	$C_{15}H_{17}NO_3$	Japan	[50,51]
Isoampullicin (**92**)	*Ampulliferina*-like sp. No. 27 (Ampullicephala)/*Pinus thunbergii* (dead tree, Pinaceae)	Acetone extract	259	$C_{15}H_{17}NO_3$	Japan	[50]

Table 1. *Cont.*

Compound Name	Fungal Source/Host	Extract/Fraction	Mol. Wt.	Mol. Formula	Location	Ref.
Dihydroampullicin (**93**)	*Ampulliferina*-like sp. No. 27 (Ampullicephala)/*Pinus thunbergii* (dead tree, Pinaceae)	Acetone extract	261	$C_{15}H_{19}NO_3$	Japan	[51]
Eutypellacytosporin A (**94**)	*Eutypella* sp. D-1 (Diatrypaceae)/Soil sample	CH_2Cl_2 fraction of EtOAc extract	714	$C_{40}H_{58}O_{11}$	London Island of Kongsfjorden of the Ny-Ålesund District, Arctic, Norway	[52]
Eutypellacytosporin B (**95**)	*Eutypella* sp. D-1 (Diatrypaceae)/Soil sample	CH_2Cl_2 fraction of EtOAc extract	714	$C_{40}H_{58}O_{11}$	London Island of Kongsfjorden of the Ny-Ålesund District, Arctic, Norway	[52]
Eutypellacytosporin C (**96**)	*Eutypella* sp. D-1 (Diatrypaceae)/Soil sample	CH_2Cl_2 fraction of EtOAc extract	714	$C_{40}H_{58}O_{11}$	London Island of Kongsfjorden of the Ny-Ålesund District, Arctic, Norway	[52]
Eutypellacytosporin D (**97**)	*Eutypella* sp. D-1 (Diatrypaceae)/Soil sample	CH_2Cl_2 fraction of EtOAc extract	714	$C_{40}H_{58}O_{11}$	London Island of Kongsfjorden of the Ny-Ålesund District, Arctic, Norway	[52]

Surveying their bioactivities may open a new research area for the synthesis of new agents from these metabolites by synthetic and medicinal chemists. The literature search for the reported data was performed using diverse databases and publishers, including Web of Science, Google Scholar, PubMed, Scopus, SciFinder, Wiley, SpringerLink, and ACS Publications, using specific keywords (bergamotane, marine, fungi, biosynthesis, and biological activities).

2. Structural Assignment and Stereochemistry Determination

A total of 97 metabolites have been separated from various fungal source extracts using different chromatographic techniques and characterized by NMR, MS, and IR spectral analyses as well as chemical derivatization. The relative configuration of these metabolites was established using NOESY or ROESY spectral analyses. Various studies reported the assigning of their absolute stereochemistry using total synthesis [53,54], Mosher's method [26], X-ray diffraction, chemical conversion [34,43,55], and ECD analyses [31]. The reported metabolites have been categorized into bi-, tri-, and tetracyclic derivatives.

3. Biological Activities of Bergamotane Sesquiterpenoids

Various reported studies revealed the assessment of bergamotane sesquiterpenoids for diverse bioactivities, including plant growth regulation, phototoxic, antimicrobial, anti-HIV, cytotoxic, pancreatic lipase inhibition, antidiabetic, anti-inflammatory, and immunosuppressive, which were summarized in this work (Table 2). Additionally, the reported structure–activity relation was included.

Table 2. Biological activities of fungal naturally occurring in bergamotane sesquiterpenoids.

Compound Name	Biological Activity	Assay, Organism, or Cell Line	Biological Results		Ref.
			Compound	Positive Control	
E-β-*trans*-5,8,11-trihydroxybergamot-9-ene **(4)**	Cytotoxicity	MTT/U937	84.9 (IC_{50})	Doxorubicin 0.021 μM (IC_{50})	[29]
Craterodoratin M **(7)**	Immunosuppressive	BALB/c mice T and B lymphocyte/LPS	15.43 μM (IC_{50})	Cyclosporin A 0.47 μM (IC_{50})	[30]
Craterodoratin N **(8)**	Immunosuppressive	BALB/c mice T and B lymphocyte/LPS	13.26 μM (IC_{50})	Cyclosporin A 0.47 μM (IC_{50})	[30]
Craterodoratin O **(9)**	Immunosuppressive	BALB/c mice T and B lymphocyte/LPS	17.12 μM (IC_{50})	Cyclosporin A 0.47 μM (IC_{50})	[30]
Craterodoratin Q **(11)**	Immunosuppressive	BALB/c mice T and B lymphocyte/Concanavalin A	31.50 μM (IC_{50})	Cyclosporin A 0.04 μM (IC_{50})	[30]
Xylariterpenoid A **(16)**	Anti-inflammatory	Spectrophotometrically/LPS	17.5 μM (IC_{50})	Quercetin 17.0 μM (IC_{50}) NG-monomethyl-L-arginine 9.7 μM (IC_{50})	[35]
Xylariterpenoid B **(17)**	Anti-inflammatory	Spectrophotometrically/LPS	21.0 μM (IC_{50})	Quercetin 17.0 μM (IC_{50}) NG-monomethyl-L-arginine 9.7 μM (IC_{50})	[35]
Eutypeterpene B **(18)**	Anti-inflammatory	Spectrophotometrically/LPS	13.4 μM (IC_{50})	Quercetin 17.0 μM (IC_{50}) NG-monomethyl-L-arginine 9.7 μM (IC_{50})	[35]
Eutypeterpene C **(19)**	Anti-inflammatory	Spectrophotometrically/LPS	16.8 μM (IC_{50})	Quercetin 17.0 μM (IC_{50}) NG-monomethyl-L-arginine 9.7 μM (IC_{50})	[35]
Eutypeterpene D **(20)**	Anti-inflammatory	Spectrophotometrically/LPS	21.4 μM (IC_{50})	Quercetin 17.0 μM (IC_{50}) NG-monomethyl-L-arginine 9.7 μM (IC_{50})	[35]
Eutypeterpene E **(21)**	Anti-inflammatory	Spectrophotometrically/LPS	18.7 μM (IC_{50})	Quercetin 17.0 μM (IC_{50}) NG-monomethyl-L-arginine 9.7 μM (IC_{50})	[35]
Eutypeterpene F **(22)**	Anti-inflammatory	Spectrophotometrically/LPS	24.3 μM (IC_{50})	Quercetin 17.0 μM (IC_{50}) NG-monomethyl-L-arginine 9.7 μM (IC_{50})	[35]
(10S)-Xylariterpenoid A **(23)**	Anti-inflammatory	Spectrophotometrically/LPS	86.0 μM (IC_{50})	Aminoguanidine 23.0 μM (IC_{50})	[34]
(10R)-Xylariterpenoid B **(24)**	Anti-inflammatory	Spectrophotometrically/LPS	230.0 μM (IC_{50})	Aminoguanidine 23.0 μM (IC_{50})	[34]
Xylariterpenoid E **(25)**	Anti-inflammatory	Spectrophotometrically/LPS	120.0 μM (IC_{50})	Aminoguanidine 23.0 μM (IC_{50})	[34]
Xylariterpenoid F **(26)**	Anti-inflammatory	Spectrophotometrically/LPS	85.0 μM (IC_{50})	Aminoguanidine 23.0 μM (IC_{50})	[34]

Table 2. *Cont.*

Compound Name	Biological Activity	Assay, Organism, or Cell Line	Biological Results		Ref.
			Compound	Positive Control	
Xylariterpenoid G (**27**)	Anti-inflammatory	Spectrophotometrically/LPS	85.0 μM (IC_{50})	Aminoguanidine 23.0 μM (IC_{50})	[34]
Eutypeterpene A (**28**)	Anti-inflammatory	Spectrophotometrically/LPS	21.0 μM (IC_{50})	Quercetin 17.0 μM (IC_{50}) NG-monomethyl-L-arginine 9.7 μM (IC_{50})	[35]
Craterodoratin C (**30**)	Immunosuppressive	BALB/c mice T and B lymphocyte/LPS	12.62 μM (IC_{50})	Cyclosporin A 0.47 μM (IC_{50})	[30]
Craterodoratin S (**42**)	Immunosuppressive	BALB/c mice T and B lymphocyte/LPS	22.68 μM (IC_{50})	Cyclosporin A 0.47 μM (IC_{50})	[30]
Craterodoratin J (**54**)	Immunosuppressive	BALB/c mice T and B lymphocyte/LPS	19.40 μM (IC_{50})	Cyclosporin A 0.47 μM (IC_{50})	[30]
Craterodoratin L (**56**)	Immunosuppressive	BALB/c mice T and B lymphocyte/LPS	13.71 μM (IC_{50})	Cyclosporin A 0.47 μM (IC_{50})	[30]
Massarinolin B (**61**)	Immunosuppressive	BALB/c mice T and B lymphocyte/Concanavalin A	0.98 μM (IC_{50})	Cyclosporin A 0.04 μM (IC_{50})	[30]
Brasilterpene A (**66**)	Hypoglycemic	Spectrophotometrically/ Diabetic zebrafish model	449.3 pmol/larva (IC_{50})	Rosiglitazone 395.6 pmol/larva (IC_{50})	[45]
Brasilterpene C (**68**)	Hypoglycemic	Spectrophotometrically/ Diabetic zebrafish model	420.4 pmol/larva (IC_{50})	Rosiglitazone 395.6 pmol/larva (IC_{50})	[45]
Expansolide C (**73**)	α-Glucosidase inhibition	Spectrophotometrically/α-glucosidase enzyme	0.50 mM (IC_{50})	Acarbose 1.90 mM (IC_{50})	[46]
Expansolide D (**81**)	α-Glucosidase inhibition	Spectrophotometrically/α-glucosidase enzyme	0.50 mM (IC_{50})	acarbose 1.90 mM (IC_{50})	[46]
Purpurolide B (**83**)	Pancreatic lipase inhibition	Spectrophotometrically/ pancreatic lipase enzyme	5.45 μM (IC_{50})	Kaempferol 1.50 μM (IC_{50})	[47]
Purpurolide C (**84**)	Pancreatic lipase inhibition	Spectrophotometrically/ pancreatic lipase enzyme	6.63 μM (IC_{50})	Kaempferol 1.50 μM (IC_{50})	[47]
Purpurolide D (**85**)	Pancreatic lipase inhibition	Spectrophotometrically/ pancreatic lipase enzyme	1.22 μM (IC_{50})	Kaempferol 1.50 μM (IC_{50})	[48]
Purpurolide E (**86**)	Pancreatic lipase inhibition	Spectrophotometrically/ pancreatic lipase enzyme	6.50 μM (IC_{50})	Kaempferol 1.50 μM (IC_{50})	[48]
Purpurolide F (**87**)	Pancreatic lipase inhibition	Spectrophotometrically/ pancreatic lipase enzyme	7.88 μM (IC_{50})	Kaempferol 1.50 μM (IC_{50})	[48]
Donacinoic acid A (**88**)	Immunosuppressive	BALB/c mice T and B lymphocyte/LPS	13.23 μM (IC_{50})	Cyclosporin A 0.47 μM (IC_{50})	[30]
Eutypellacytosporin A (**94**)	Cytotoxicity	CCK-8/DU145	17.1 μM (IC_{50})	Cisplatin 2.9 μM (IC_{50})	[52]
		CCK-8/SW1990	7.3 μM (IC_{50})	Cisplatin 1.2 μM (IC_{50})	[52]
		CCK-8/Huh7	8.4 μM (IC_{50})	Cisplatin 2.2 μM (IC_{50})	[52]
		CCK-8/PANC-1	9.7 μM (IC_{50})	Cisplatin 4.5 μM (IC_{50})	[52]

Table 2. *Cont.*

Compound Name	Biological Activity	Assay, Organism, or Cell Line	Biological Results		Ref.
			Compound	Positive Control	
Eutypellacytosporin B (**95**)	Cytotoxicity	CCK-8/DU145	11.0 μM (IC_{50})	Cisplatin 2.9 μM (IC_{50})	[52]
		CCK-8/SW1990	4.9 μM (IC_{50})	Cisplatin 1.2 μM (IC_{50})	[52]
		CCK-8/Huh7	4.9 μM (IC_{50})	Cisplatin 2.2 μM (IC_{50})	[52]
		CCK-8/PANC-1	7.9 μM (IC_{50})	Cisplatin 4.5 μM (IC_{50})	[52]
Eutypellacytosporin C (**96**)	Cytotoxicity	CCK-8/DU145	13.5 μM (IC_{50})	Cisplatin 2.9 μM (IC_{50})	[52]
		CCK-8/SW1990	9.6 μM (IC_{50})	Cisplatin 1.2 μM (IC_{50})	[52]
		CCK-8/Huh7	11.2 μM (IC_{50})	Cisplatin 2.2 μM (IC_{50})	[52]
		CCK-8/PANC-1	10.2 μM (IC_{50})	Cisplatin 4.5 μM (IC_{50})	[52]
Eutypellacytosporin D (**97**)	Cytotoxicity	CCK-8/DU145	13.4 μM (IC_{50})	Cisplatin 2.9 μM (IC_{50})	[52]
		CCK-8/SW1990	8.2 μM (IC_{50})	Cisplatin 1.2 μM (IC_{50})	[52]
		CCK-8/Huh7	9.6 μM (IC_{50})	Cisplatin 2.2 μM (IC_{50})	[52]
		CCK-8/PANC-1	7.5 μM (IC_{50})	Cisplatin 4.5 μM (IC_{50})	[52]

3.1. *Anti-Inflammatory Activity*

NO (nitric oxide) is a substantial pro-inflammatory mediator, and its excessive production is accompanied with various inflammatory illnesses; therefore, it possesses a remarkable role for regulating immune responses and inflammation [56]. NO production inhibitors may represent the potential capacity for treating various inflammatory disorders. Thus, further research for fungal metabolites must be conducted to discover novel anti-inflammation agents.

The epigenetic chemical manipulation of *Eutypella* sp. derived from deep-sea hydrothermal sulfide deposit by co-treatment with SBHA (histonedeacetylase inhibitor, suberohydroxamic acid) and 5-Aza (DNA methyltransferase inhibitor, 5-azacytidine) was shown to activate a biosynthetic sesquiterpene-linked gene cluster [35]. From elicitor-treated cultures EtOAc extract, eutypeterpenes A–F (**18**–**22** and **28**) along with xylariterpenoids A (**16**) and B (**17**) were purified using SiO_2/RP-18/HPLC that were identified by spectral analyses, as well as by using chemical conversion, X-ray diffraction, ECD, and calculated NMR for configuration assignments.

Eutypeterpene A (**28**) is the first bergamotene sesquiterpene incorporating a dioxolanone moiety. These metabolites were assessed for their NO production inhibitory capacity induced by LPS-(lipopolysaccharide) in RAW 264.7 macrophages [35]. The results indicated thatcompound **18** and **19** (IC_{50} 13.4 and 16.8 μM, respectively) displayed more effectiveness than quercetin (IC_{50} of 17.0 μM), whereas other metabolites had noticeable potentials (IC_{50} values ranged from 18.7 to 24.3 μM) with weak cytotoxic capacities (IC_{50} > 100 μM). A structure–activity study revealed that the analog with a triol unit (**18**) at the side chain was more effective than compound **16**, **17**, and **19** with a diol unit, which were more potent than compound **20**, **21**, and **28** with one hydroxy group. Furthermore, the α,β-unsaturated ketone unit (as in compound **21** and **22**) and the OH-linked carbon configuration also affected the activities (**16** versus **17**) [35] (Figure 1).

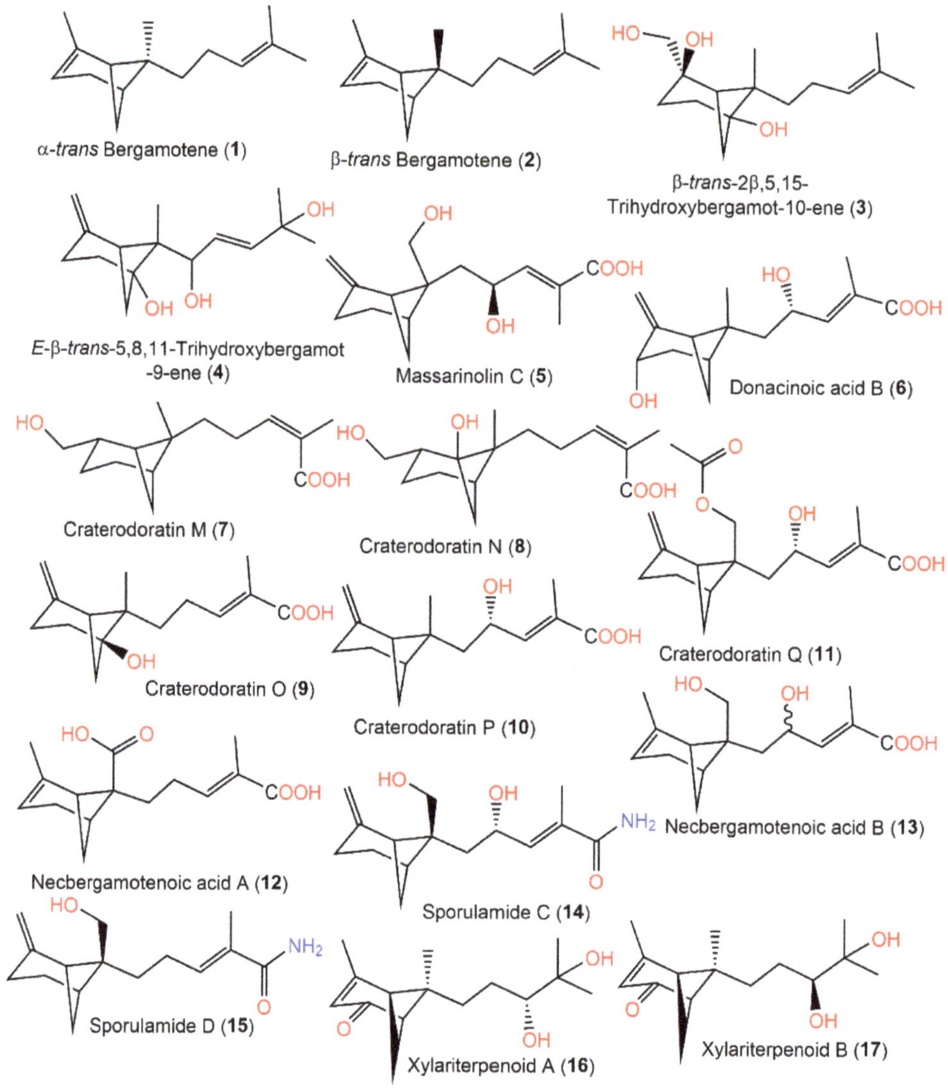

Figure 1. Structures of bicyclic bergamotane sesquiterpenoids (**1–17**).

Biogenetically, compounds **18–22** are derived from FPP that performs a 1,6-cyclization to produce bisabolane (**A**). The 4,7-cyclization of **A** generates bergamotane (**B**), which further generates **18–22** via diverse oxidation and reduction processes. Additionally, compound **28** is formed from **18** by carbonate incorporation [35] (Scheme 1).

The deep-sea-isolated *Graphostroma* sp. MCCC3A00421 associated with the Atlantic Ocean hydrothermal sulfide deposits biosynthesized new bergamotane sesquiterpenoids: (10S)-xylariterpenoid A (**23**), (10R)-xylariterpenoid B (**24**), xylariterpenoid E (**25**), xylariterpenoid F (**26**), and xylariterpenoid G (**27**), which were purified using SiO_2/OSD/Sephadex LH-20/RP-18 CC and preparative TLC. They were characterized by extensive spectral data, and their absolute configuration was established by ECD, Cu-Ka-single-crystal X-ray diffraction, and modified Mosher's method analyses. Compound **25** is trinor-bergamotane. Compounds **23**, **26**, and **27** revealed moderate inhibition potentials (IC_{50}s of 86, 85, and

85 μM, respectively) of NO production in LPS-stimulated RAW264.7 macrophages compared with aminoguanidine (IC_{50} of 23 μM). It was noted that bergamotane moiety's 10S configuration obviously boosted the activity as in compound **23** (10S, IC_{50} of 85 μM) versus compound **24** (10R, of IC_{50} 230 μM) (Figure 2) [34].

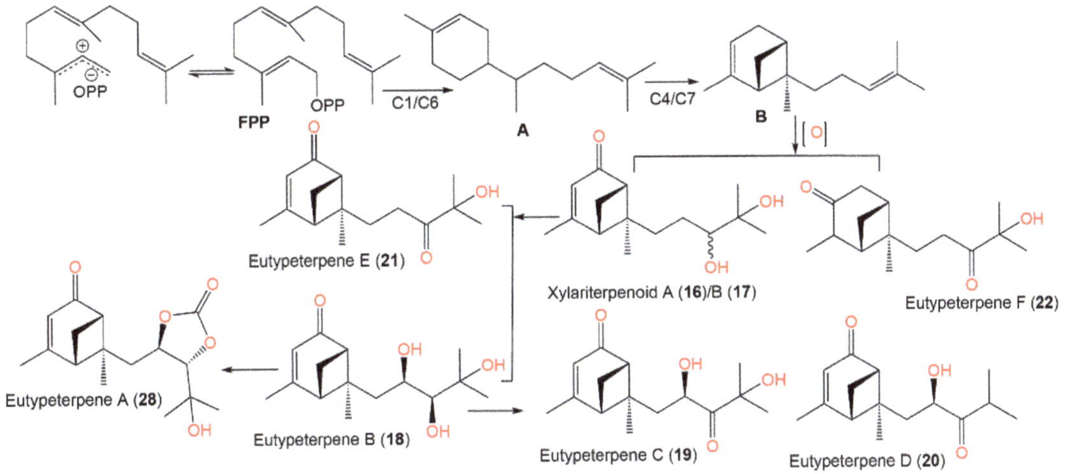

Scheme 1. Biosynthetic pathway of eutypeterpenes A–F (compounds **18–22** and **28**) [35].

3.2. Phytotoxic Activity

Prehelminthosporol (**38**) and dihydroprehelminthosporol (**34**) along with compounds **35–37**, **39**, **40**, and **43** were separated by SiO_2, flash CC, and preparative TLC from the EtOAc extract of the *Bipolaris* species, which is a *Sorghum halepnse* (Johnson grass) pathogen (Figure 3). These metabolites were assessed for their phytotoxic potential towards *Sorghum bicolor* (Sorghum) and *Sorghum balepense* (Johnson grass) in leaf spot assays [36,37]. Compounds **34**, **38**, and **39** produced similar lesions to those caused by the fungus in the field. The lesions appeared as a reddish-brown area (0.3–0.5 cm diameter) surrounded by a black circle with an outer chlorotic zone. Compounds **34** and **38** (concentration of 25 μg/5 μL) had comparable toxic effectiveness, while compound **38** maintained its effect at a lower concentration of 2.5 μg/5 μL; meanwhile, the other compounds were non-toxic [36,37]. Victoxinine was also toxic to cereals in the order of oats > rye and barley > wheat > sorghum in a root inhibition assay [37]. The phytotoxic influence of compounds **34** and **38–40** versus sorghum, corn, bent-grass, sickle-pod, and morning glory was also assessed in leaf spot assays. Moreover, victoxinine caused a water-soaked translucent appearance with defined irregular necrotic edges. It is worth mentioning that 3-deoxyanthocyanidins are sorghum stress response metabolites (phytoalexins), which were accountable for the red wound response. Compounds **34**, **38**, and **39** were elicitors of a very strong reddening compared with the wounding-produced reddening, but compound **40** did not elicit a sorghum phytoalexin response. In bent grass and corn, compounds **34** and **38–40** produced a light-brown area limited by a chlorotic region, whereas in sickle pod and morning glory, they showed necrotic lesions that extended at high concentrations beyond the under-drop area. It is noteworthy that compound **38** was the most toxic compound versus all tested plants except for the morning glory [37].

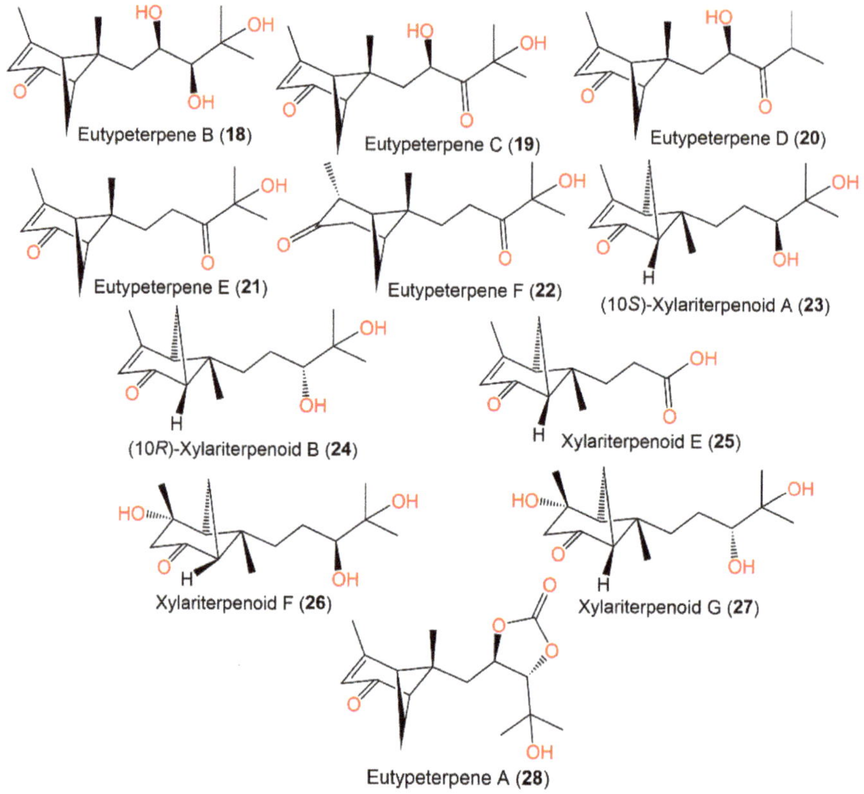

Figure 2. Structures of bicyclic bergamotane sesquiterpenoids (compounds **18**–**28**).

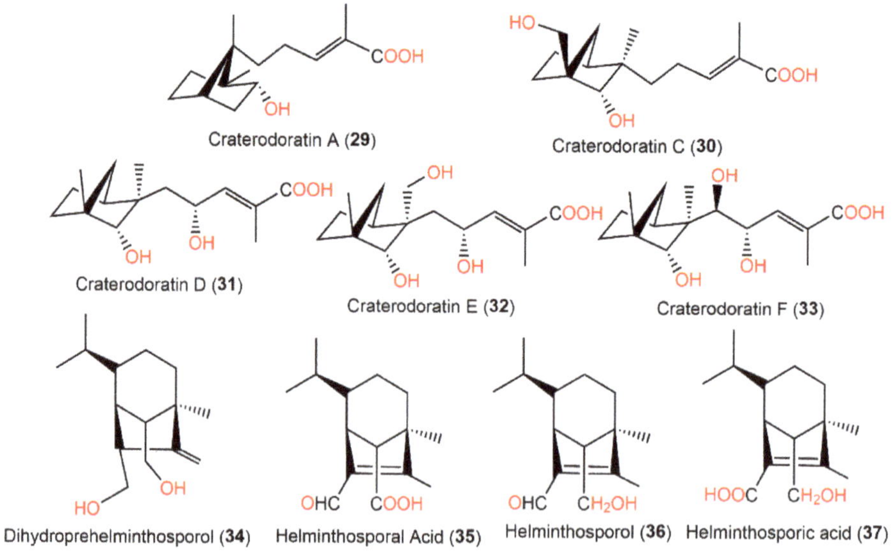

Figure 3. Structures of the bicyclic bergamotane sesquiterpenoids (compounds **29**–**37**).

Helminthosporium victoriae, the causative agent of oats Victoria blight disease yielded phytotoxins, victoxinine (**40**) and victoxinine α-glycerophosphate (**41**), which were separated from its diethyl ether extract using Sephadex LH-20 and SiO_2 CC and detected on the TLC plate by giving a blue color with 5% vanillin:H_2SO_4 [41] (Figure 4). The existence of α-glycerophosphate moiety was established by coupling between the phosphorous and carbon. Compound **40** completely prohibited the root growth of toxin-susceptible and toxin-resistant oats (concentration of 2.5×10^{-4} M); it was ≈ 7500 times more toxic for susceptible plants on a weight basis, while its toxicity for resistant plants was nearly similar, suggesting a role of the victoxinine moiety on the toxicity [38,39,41]. On the other side, compound **41** (concentration of 100 μg/mL) demonstrated little or no growth inhibition effectiveness on either susceptible or resistant oats [41].

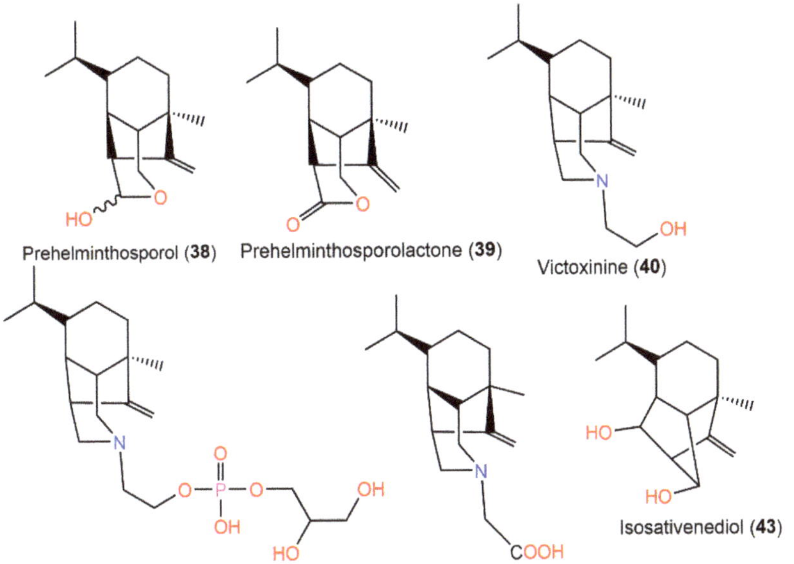

Figure 4. Structures of tricyclic bergamotane sesquiterpenoids (compounds **38–43**).

3.3. *Anti-HIV Activity*

From *Paraconiothyrium brasiliense*, new tricyclic sesquiterpenoids, brasilamides A–D (**45–48**) and the formerly reported pinthunamide (**44**), were separated from the culture's EtOAc extract utilizing SiO_2/Sephadex LH-20 CC and HPLC. Their structures were established using NMR and X-ray analyses (Figure 5). Compounds **45** and **46** are rare metabolites having a 4-oxatricyclo[3.3.1.0^{2,7}]nonane moiety with a tetrahydro-2*H*-pyrone or a tetrahydro-2*H*-pyran linked with bicyclo[3.1.1]heptane ring at C-5 and C-2, whereas compounds **47** and **48** are analogs of **44**, possessing an unprecedented 9-oxatricyclo[4.3.0.0^{4,7}]-nonane core.

The differences of the above-mentioned compounds from **44** were the existence of a tetrahydrofuran moiety connected to the bicycle[3.1.1]heptane unit instead of γ-lactone ring, as well as different C-10 substituents. Compounds **45–48** demonstrated inhibitory effectiveness (EC_{50}s of 108.8, 57.4, and 48.3 μM, respectively) versus HIV-1 replication in C8166 cells compared with indinavir sulfate (EC_{50} of 8.2 nM) [43]. Biogenetically, they were derived from the mevalonate/*trans-cis*-farnesol/bisabolene/bergamotane pathway (Scheme 2).

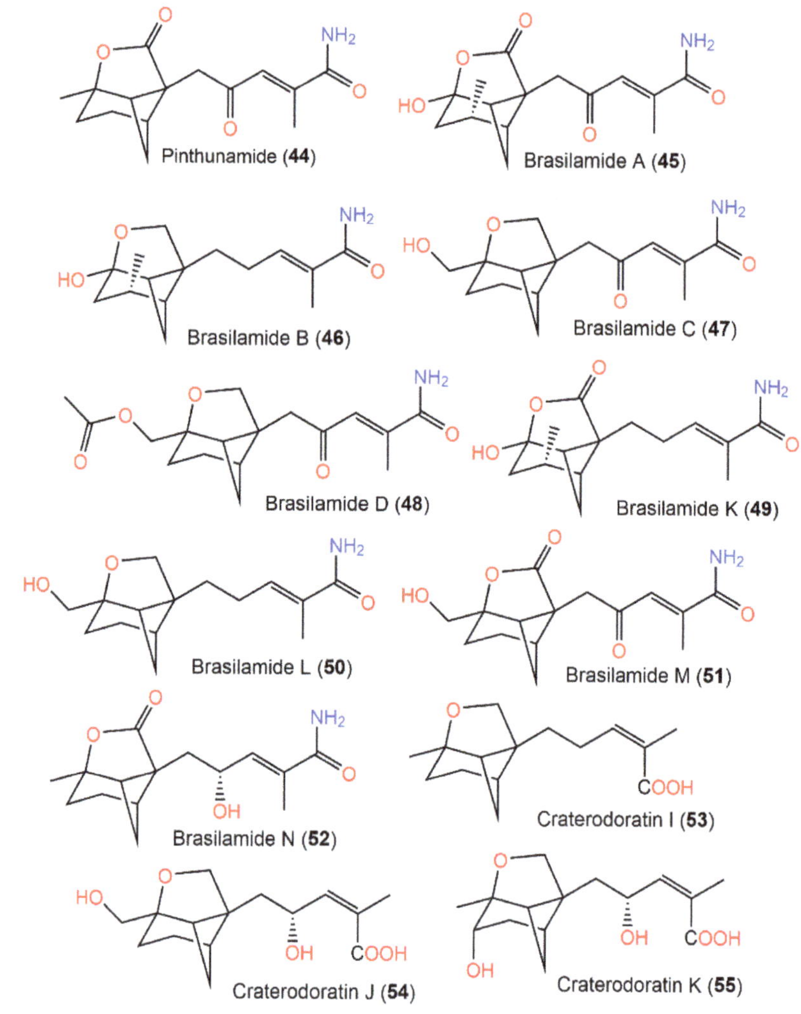

Figure 5. Structures of tricyclic bergamotane sesquiterpenoids (**44–55**).

3.4. Immunosuppressive Activity

Immunosuppressants are drugs that prohibit body immunity and are principally utilized in organ transplantation to overcome rejection and in auto-immune illnesses [57]. Currently, many immunosuppressive agents act by prohibiting T-cell proliferation; however, there is no new, safe, and efficient immune-suppressive agent that prohibits B-cell proliferation [58].

Dai et al. separated eighteen bergamotane sesquiterpenoids from the EtOAc extract of *Craterellus ordoratus*: craterodoratins A–R (**7–11**, **29–33**, **53**, **55**, **56**, **63–65**, and **71**) and a new victoxinine derivative, craterodoratin S (**42**), along with the previously isolated **5**, **61**, **77**, and **88** by SiO_2/RP-18/Sephadex LH-20/preparative HPLC (Figure 6) [30].

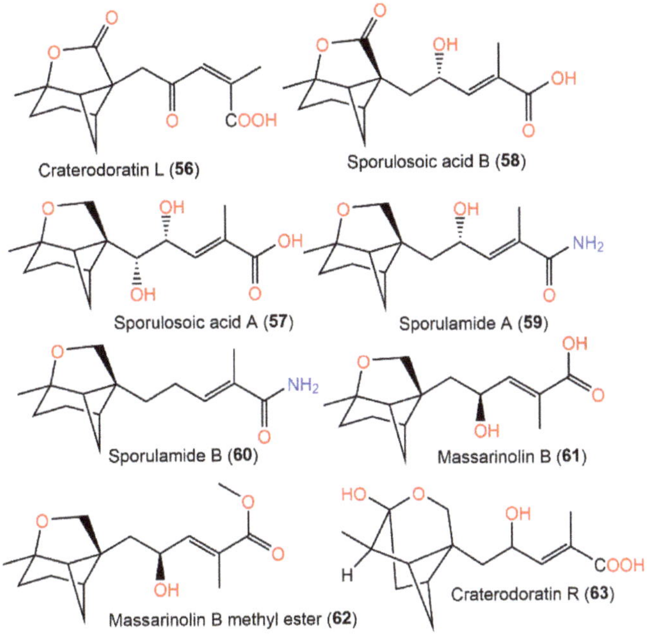

Figure 6. Structures of tricyclic bergamotane sesquiterpenoids (**56**–**63**).

Their structures with absolute configurations were established by spectral, X-ray diffraction, and ECD analyses and NMR calculations. Compounds **29** and **71** possess a rare skeleton, where the C-14methyl in **71** showed a further 1,2-migration. On the other hand, compounds **7**–**11**, **53**, **55**, **56**, and **63**–**65** belong to β-pinene derivatives that produced **30**–**33** through an alkyl migration (Figure 7). Compounds **7**–**10**, **30**, **42**, **55**, **61**, and **88** demonstrated potent inhibitory potential versus LPS-caused B lymphocyte cell proliferation (IC_{50}s ranged from 0.67 to 22.68 μM) in BALB/c mice compared with cyclosporin A (IC_{50} of 0.47 μM), where compound **61** (IC_{50} 0.67 μM) had the most potent effectiveness. Moreover, compounds **11** and **61** possessed inhibition (IC_{50}s of 31.50 and 0.98 μM, respectively) on T lymphocyte cells proliferation induced by ConA (concanavalin A) compared with cyclosporin A (IC_{50} 0.04 μM). Structurally, it was noted that the *α*,*β*-unsaturated-carboxylic acid unit could be the key functional group for the immunosuppressive potential of these metabolites. Furthermore, compounds **61** and **7**–**10** with a *β*-pinene main core had a wider range of bioactivities [30].

3.5. Antimicrobial Activity

From *Podospora decipiens*, two new tetracyclic sesquiterpenoids, decipienolides A (**74**) and B (**75**), were separated from the EtOAc extract by SiO_2 CC and HPLC analyses. They were obtained as a mixture of inseparable epimers, having a 3-hydroxy-2,2-dimethylbutyric acid sidechain as elucidated by an NMR analysis (Figure 8). The **74**/**75** mixture had an antibacterial influence versus *B. subtilis* (inhibition zone diameter of 9–10 mm, concentration of 200 μg/disk). Neither of them demonstrated capacity versus *Ascobolus furfuraceus* NRRL6460, *Sordaria fimicola* NRRL6459, and *C. albicans* ATCC90029 [24]. Donacinolides A (**82**) and B (**76**) (concentration of 50 μg/mL) revealed weak inhibition versus *Salmonella enterica* subsp. *enterica* (inhibition rates of 24.3, 23.9, and 26.2%) in the microdilution assay [32]. Furthermore, there were no observed antibacterial activity for purpurolides B (**83**) and C (**84**) (concentration of 50 μM) versus *E. coli* ATCC25922, *M. smegmatis* mc2155 ATCC70084, *S. aureus* ATCC25923, and *S. epidermidis* ATCC12228 [47].

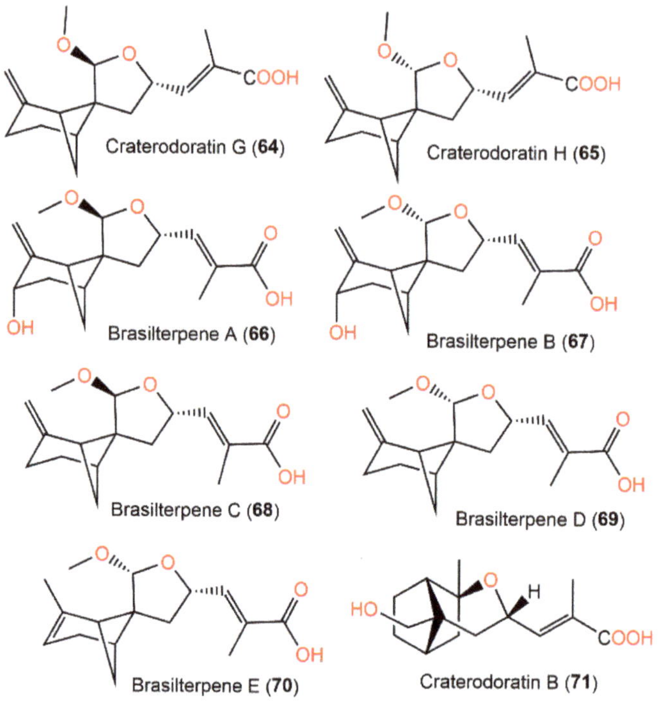

Figure 7. Structures of tricyclic bergamotane sesquiterpenoids (**64**–**71**).

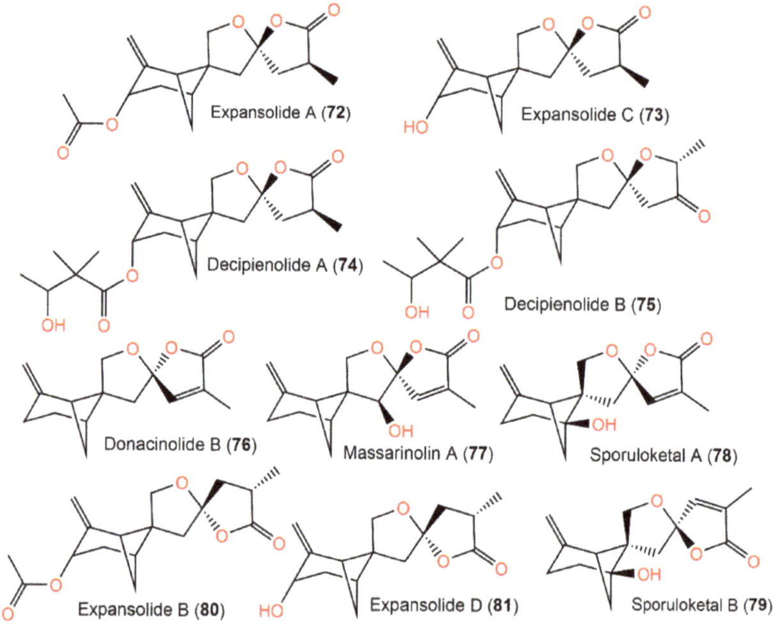

Figure 8. Structures of tetracyclic bergamotane sesquiterpenoids (**72**–**79**).

3.6. Pancreatic Lipase Inhibition

Purpurolides B (**83**) and C (**84**) are new 6/4/5/5-tetracyclic sesquiterpenoids that were separated from *Penicillium purpurogenum* IMM003 cultures by SiO_2/RP-18/preparative HPLC analysis. The structures and configurations of compounds **83** and **84** were established using spectral and X-ray analyses as well as ECD and GIAO NMR data calculations (Figure 9).

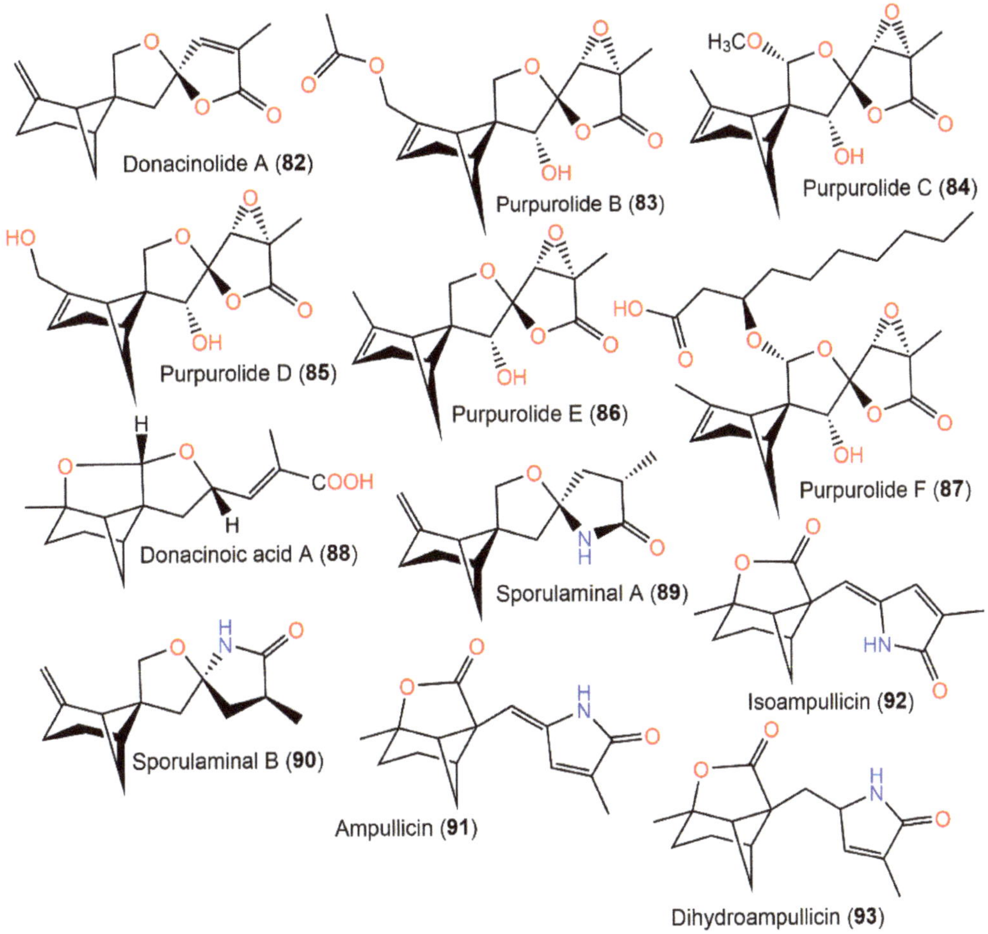

Figure 9. Structures of tetracyclic bergamotane sesquiterpenoids (**80–93**).

Compounds **83** and **84** demonstrated potent pancreatic lipase inhibition (IC_{50}s of 5.45 and 6.63 μM, respectively), compared with kaempferol (IC_{50} of 1.50 μM) [47]. These compounds were possibly biosynthesized via numerous the cyclization and enzyme-catalyzed oxidation of FPP (farnesyl pyrophosphate), leading to four- and six-membered rings and the formation of two five-membered heterocyclic rings (Scheme 3) [47].

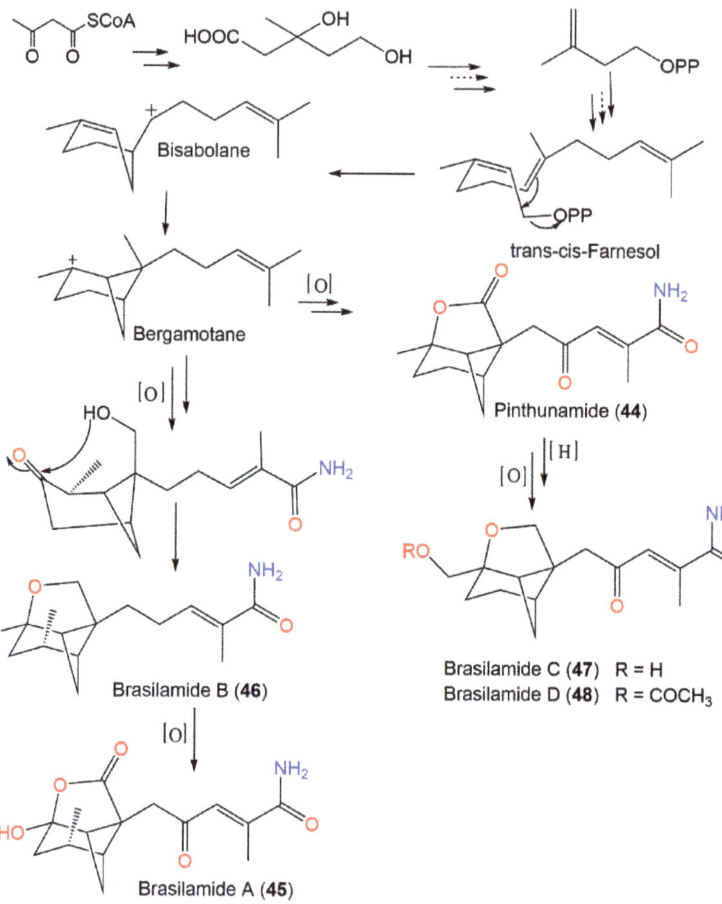

Scheme 2. Biosynthetic pathways of brasilamides A–D (**45–48**) [43].

Xia et al. separated from *Penicillium purpurogenum* IMM003 purpurolides D–F (**85–87**), which are new polyoxygenated 6/4/5/5-tetracyclic bergamotanes, using SiO_2/Sephadex-LH-20/RP-18 CC and preparative HPLC processing [48]. Their elucidation was accomplished using spectral ^{13}C NMR calculations coupled with DP^{4+} probability and ECD analyses. Compound **87** had potent pancreatic lipase inhibition potential (IC_{50} of 1.22 μM) compared with kaempferol (IC_{50} of 1.50 μM) and orlistat (IC_{50} of 0.75 μM), whereas compounds **85** and **86** (IC_{50}s of 6.50 and 7.88 μM, respectively) were five or six-fold less powerful than **87**, revealing that the C-14 hydroxylated decanoic acid moiety increased the potency [48]. Therefore, polyoxygenated bergamotanes could be viable candidates as pancreatic lipase inhibitors for further clinical development [48].

3.7. Antidiabetic Activity

From the deep sea-derived *Paraconiothyrium brasiliense* HDN15-135 EtOAc extract, new bergamotane sesquiterpenoids, brasilterpenes A-E (**66–70**), featuring an uncommon 6/4/5-tricyclic ring system, were separated by SiO_2/RP-18/Sephadex LH-20/HPLC and assigned by diverse NMR analyses and X-ray diffraction, ECD, and DFT-NMR (density functional theory calculations of nuclear magnetic resonance) data [45]. Their hypoglycemic potential was estimated utilizing β-cell-ablated zebrafish larvae. Compounds **66** and **68** (concentration of 10 μM) remarkably lessened the glucose level down to 449.3 and

420.4 pmol/larva respectively, compared with the β-cell-ablated group (Teton+) (glucose level of 502.8 pmol/larva) and rosiglitazone (glucose level 395.6 pmol/larva) with no toxic influence on zebrafish larvae up to 200 μM. It was found that compounds **66** and **68** notably minimized free blood glucose in vivo in hyperglycemic zebrafish by suppressing gluconeogenesis and improving insulin sensitivity, which revealed that compound **68** had promising antidiabetic potential [45]. The structure–activity study revealed that the activity may be linked to the C-14 *S*-configuration of compounds **66** and **68**, which represent the main structural difference from **67** and **69.** The existence of C-3-OH may weaken the influence in **68** versus **66**; however, the Δ^2 endocyclic double bond may enhance the potential in **70** versus **69** [45]. Therefore, compound **68** may provide a scaffold for hypoglycemic drug development. Compounds **66–70** are also biosynthesized by the FPP pathway (Scheme 4). The cyclization of FPP via NPP (nerolidyl diphosphate) followed by a bisabol intermediate yields the bergamotane skeleton. These compounds are created by further oxidation, 9-OH-nucleophilic attack, and methylation processes. Because of the nucleophilic attack direction flexibility during the furan ring formation, compounds **66–69** appeared as C14-epimers in pairs [45] (Scheme 4).

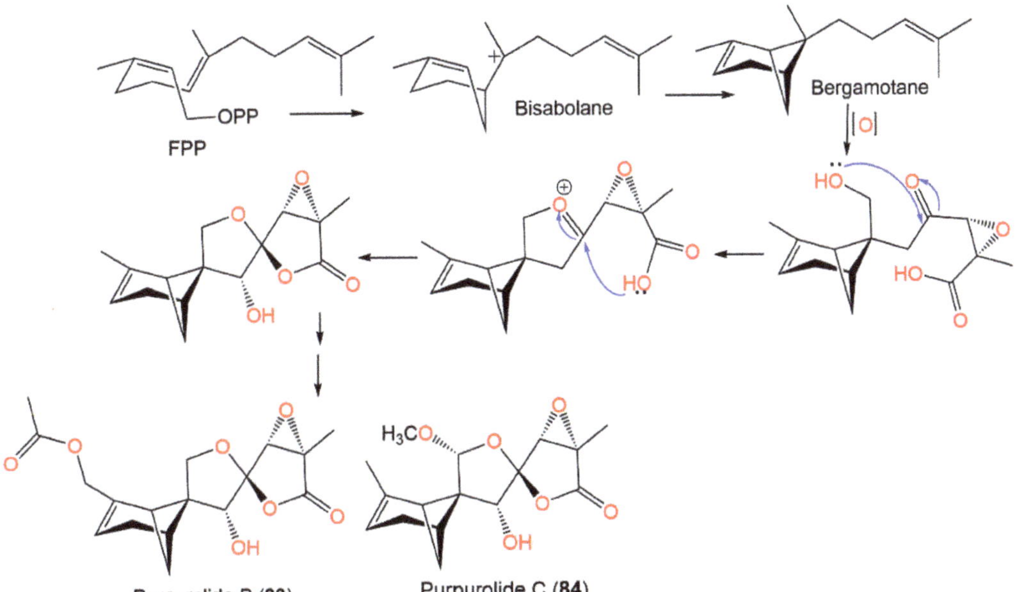

Scheme 3. Biosynthetic pathway of purpurolides B and C (**83** and **84**) [47].

Ying et al. isolated two new derivatives, expansolides C (**73**) and D (**81**), in addition to **72** and **80** from the plant pathogen *Penicillium expansum* ACCC37275 [46]. In an α-glucosidase inhibition assay; the **73/81** epimeric mixture (ratio 2:1) possessed a more powerful effectiveness (IC_{50} of 0.50 mM) compared with acarbose (IC_{50} 1.90 mM), while the **72/80** epimeric mixture possessed no apparent potential. It was assumed that the acetyl group in compounds **72** and **80** impeded their binding with the α-glucosidase, resulting in loss of activity [46].

3.8. Plant Growth Regulation

Kimura et al. purified the tricyclic amide sesquiterpenoid pinthunamide (**44**) from the acetone extract of *Ampulliferina* sp. at pH 2.0 utilizing SiO_2 and sephadex LH20 CC processing as well as crystallization from EtOAc extract, which gave positive NH_2OH-HCl-$FeCl_3$ and $KMnO_4$ reactions [42]. The compound was assigned by X-ray diffraction and NMR methods. Its plant growth regulation effectiveness was evaluated using a lettuce seedling assay, where it (dose 300 mg/L) produced a 150% root growth acceleration over the control seedlings (100%) while scarcely influencing the hypocotyl elongation at the tested concentrations [42]. Its structure combined a unique configuration of six-, five-, and four-membered rings that was proposed to be biosynthesized via the mevalonate/*trans-cis*-farnesol/bisabolene/bergamotane pathway (Scheme 5) [42].

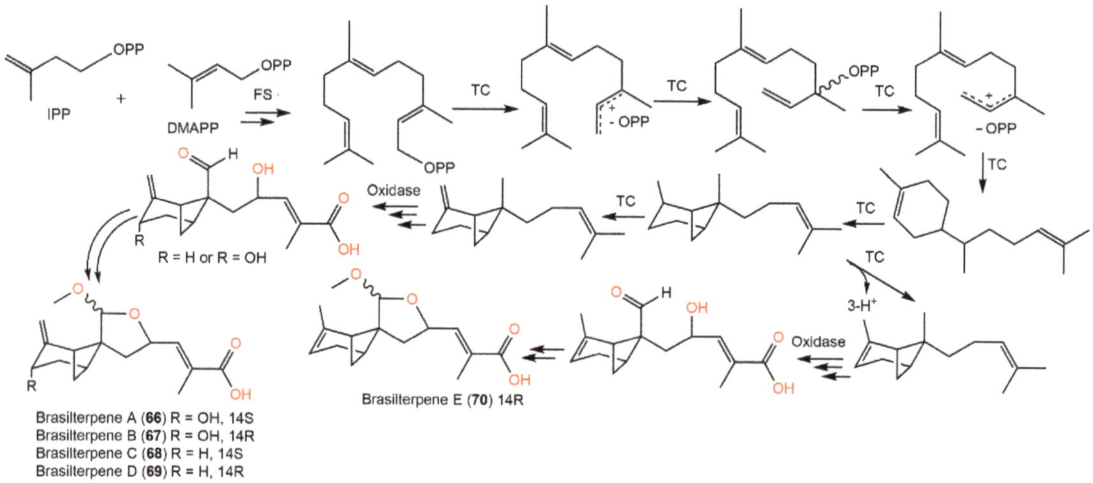

Scheme 4. Biosynthetic pathway of brasilterpenes A-E (**66–70**) [45]. IPP: isopentenyl diphosphate; FS: farnesyl synthase; NPP: nerolidyl diphosphate; TC: terpenyl cyclase; DMAPP: dimethylallyl diphosphate; FPP: farnesyl diphosphate.

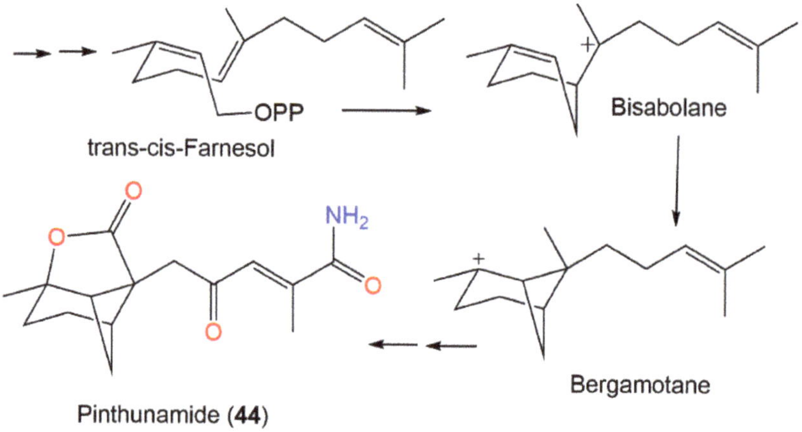

Scheme 5. Biosynthesis pathway of pinthunamide (**44**) [42].

Furthermore, in 1990, Kimura et al. purified another two new plant growth regulators, ampullicin (**91**) and isoampullicin (**92**) from *Ampulliferina* sp. No. 27 associated with *Pinus thunbergii* dead tree by SiO_2 CC utilizing benzene:acetone as an eluent [50] (Figure 10). They were stereoisomers that had γ-lactam rings. Additionally, they (doses of 300 and 30 mg/L) were shown to promote lettuce seedling root growth by 200% over the control lettuce seedlings (100%) [50]. In 1993, the same group separated a minor metabolite, dihydroampullicin (**93**), characterized by the absence of the C8-C9 double bond. The compound promoted a 160% growth rate in lettuce seedling roots (dose of 300 mg/L) compared with the control; however, it had no influence on the hypocotyl growth, indicating that the C8-C9-double bond (C8-C9) was substantial in lettuce seedlings' root growth [51]. Bermejo et al. reported the synthesis of (+)−**91** and **92** from (*R*)-(-)-carvone with a 4.5% overall yield using a stereo-selective 18-step sequence application [59]. The EtOAc extract of *Aspergillus fumigatus* Fresenius separated from leaf litter yielded expansolides A (**72**) and B (**80**). They had 2*S*/4*S*/6*S*/7*R*/9*R*/11*S* and 2*S*/4*R*/6*S*/7*R*/9*R*/11*S*, respectively, based on modified Mosher's method. The compounds noticeably prohibited etiolated wheat coleoptiles growth by 100% and 59% at 10^{-3} M and 10^{-4} M solution compared with LOGRAN (commercial herbicide) (%inhibition of 80 and 42%) at the same concentrations [26].

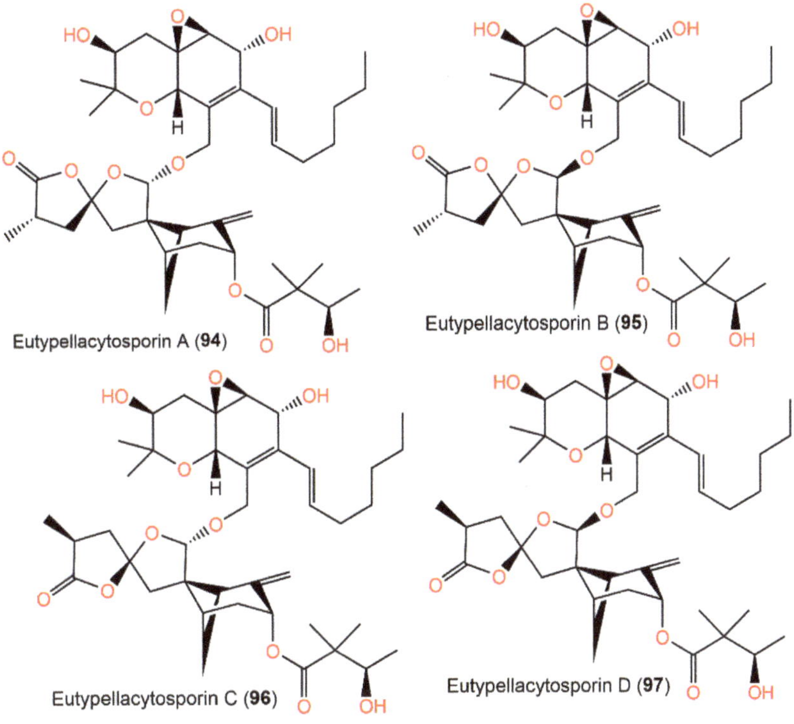

Figure 10. Structures of tetracyclic bergamotane sesquiterpenoids (**94**–**97**).

3.9. Cytotoxic Activity

Compounds **3** and **4**, which were new β-bergamotane sesquiterpenoids, were separated by SiO_2/RP-18/HPLC from the marine-associated *Aspergillus fumigatus*-YK-7 EtOAc extract. Their antiproliferative effects on the U937 and PC-3 cell lines were measured in vitro in an MTT assay. Compound **4** revealed a weak growth inhibition capacity (IC_{50} of 84.9 μM) versus the U937 cell line, while **3** had no activity (IC_{50} > 100 μM) compared with doxorubicin hydrochloride (IC_{50} of 0.021 μM). On the other sides, both had no effect versus PC-3 cells [29]. Wu et al. reported the separation of two new derivatives, xylariterpenoids

A and B (**16** and **17**), from the EtOAc extract of *Xylariaceae* fungus by Sephadex LH-20/ODS CC and reversed-phase HPLC processing [33]. Their structures and stereo-configuration were proved utilizing NMR and CD methods. They are C-10 epimers having 2S/6S/7S/10R and 2S/6S/7S/10S configurations, respectively. Unfortunately, they (IC_{50} > 40 μM) exhibited no cytotoxic potential versus HL-60, MCF-7, SMMC-7721, A-549, and SW480 in an MTT assay [33].

From *Paraconiothynium brasiliense* Verkley, new bergamotane sesquiterpenoids brasilamides K-N (**49–52**), featuring 4-oxatricyclo-(3.3.1.0^{2,7})-nonane (as in **49**) and 9-oxatricyclo-(4.3.0.0^{4,7})-nonane (as in **50–52**) skeletons in addition to the formerly reported brasilamides A and C (**45** and **46**), were purified from the fungus scale-up fermentation cultures using SiO_2/Sephadex LH-20/HPLC processing. They were elucidated via NMR analyses and compound **52**'s configuration was assured using modified Mosher's method. Compound **49** is a **45**-hydrogenated analog that has a tetrahydro-2H-pyrone unit linked at C-2 and C-5 to the bicyclo(3.1.1)heptane framework, forming a 4-oxatricyclo-(3.3.1.0 ^{2,7})-nonane skeleton, whereas compounds **50–52** displayed unusual 9-oxatricyclo-(4.3.0.0 ^{4,7})-nonane skeletons. Compounds **50** and **51** are hydrogenated and oxygenated derivatives of **46**, respectively, while **52** differed from **46** by having a C-8-carbonyl, C-1-methyl, and C-12 hydroxyl group instead of methylene, oxy-methylene, and ketone carbonyl, respectively. These metabolites (concentration of 50 μM) possessed no potential versus A549, A375, MCF-7, CNE1-LMP1, EC109, MGC, PANC-1, and Hep3B-2 in the MTS assay [44].

Montagnula donacina (edible mushroom) biosynthesized rare tetracyclic bergamotane sesquiterpenoids, donacinolides A (**82**) and B (**76**) and donacinoic acids A (**88**) and B (**6**), which were separated using SiO_2 CC/Sephadex LH-20 CC/HPLC processing and were characterized using spectroscopic data, X-ray diffraction analysis, and computational methods. Compounds **76** and **82** are C9 epimers with a spiroketal moiety having 1S/5S/6S/9R and 1S/5S/6S/9S configurations, respectively, whereas **88** and **6** exhibited α,β-unsaturated carboxylic acid moiety and had 1R/2R/5S/6S/9S/14S and 1R/3S/5R/6R/9S configurations, respectively. These metabolites lacked a marked cytotoxic potential (IC_{50} > 40 μM) versus HL-60, SW480, A549, SMMC-7721, and MCF-7 [32].

In addition, purpurolides B (**83**) and C (**84**) had no cytotoxicity versus M14, HCT-116, U87, A2780, BGC-823, Bel-7402, and A549 [47], whereas compounds **85–87** (concentration of 50 μM) were inactive versus HCT-116, BGC-823, and Bel-7402 cell lines [48].

The chemical investigation of Arctic fungus *Eutypella* sp. D-1's EtOAc extract yielded new derivatives, eutypellacytosporins A–D (**94–97**), which were established by spectroscopic analysis and modified Mosher's method. Structurally, these metabolites are related to decipienolides and cytosporins. They exhibited (IC_{50}s ranging from 4.9 to 17.1 μM) weak-to-moderate cytotoxic influence versus DU145, SW1990, Huh7, and PANC-1 in the CCK-8 assay, whereas Huh7 and SW1990 cell lines had more sensitivity to **94–97** (IC_{50}s ranging from 4.9 to 8.4 μM). On the other hand, compounds **95** and **97** possessed noticeable potential versus PANC-1 (IC_{50}s of 7.9 and 7.5 μM, respectively) compared with cisplatin (IC_{50} 4.5 μM). The results revealed that the decipienolide moiety was substantial for activity; however, the C-33 configuration did not affect the activity [52]. It was proposed that compounds **94–97** are created from gentisaldehyde precursor with subsequent isoprenyl unit addition, double bond epoxidation, keto group hydrogenation, and an aliphatic chain addition (Scheme 6). The other precursor, the 14-OH of decipienolide A **74** or B **75**, is produced from hydroxylation, allylic oxidation, and cyclization of farnesyl diphosphate to give **I** with a bicycle[3.1.1]heptane. Additionally, (14S)-14-OH-expansolide C, (14R)-14-OH-expansolide C, (14S)-14-OH-expansolide D, and (14R)-14-OH-expansolide D are formed via two steps of reface- and si-face attacks of the OH groups on the ketone and aldehyde groups, respectively. After these steps, compounds **94–97** were produced from the two groups of 14-OH-expansolides C and D through condensation reactions with (S)-3-hydroxy-2,2-dimethylbutanoic acid and cytosporin D, respectively [52].

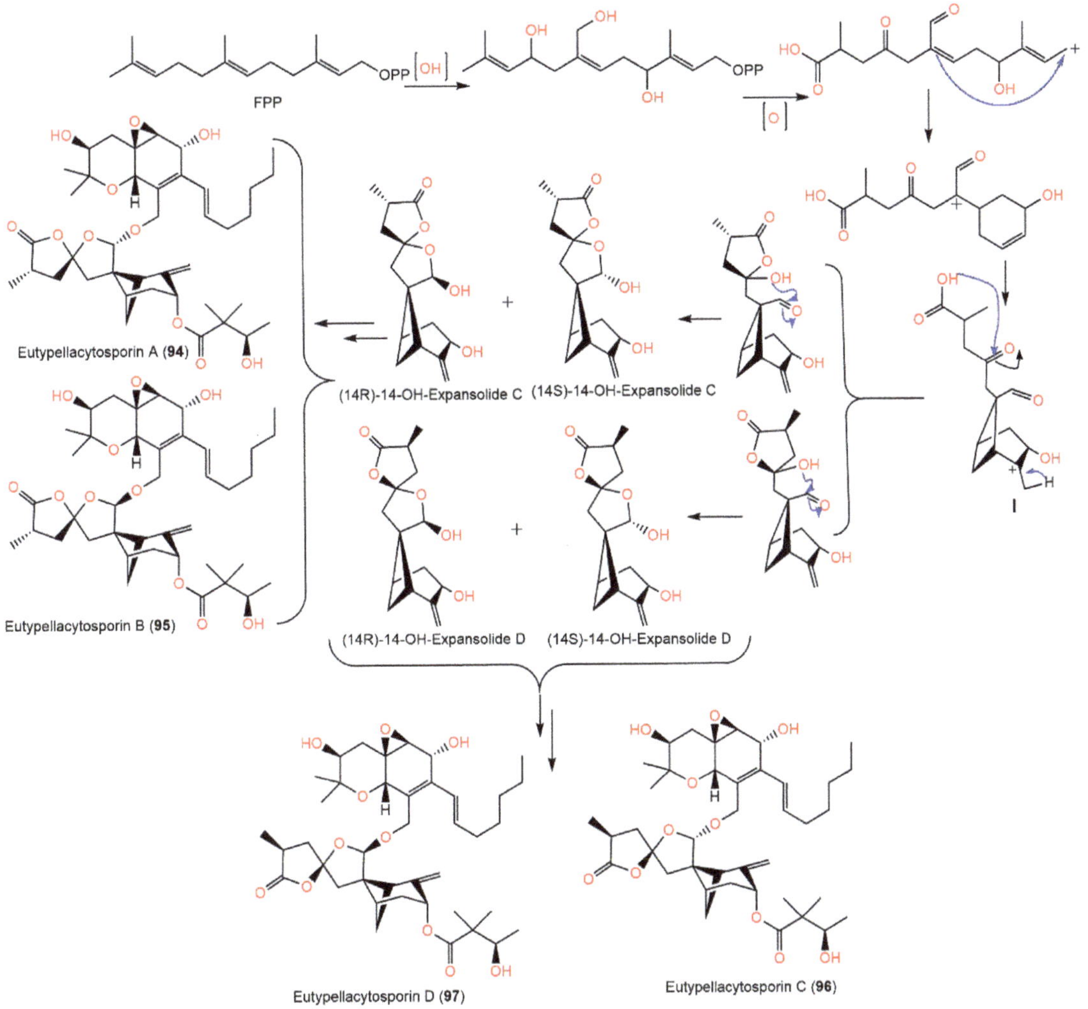

Scheme 6. Biosynthetic pathway of eutypellacytosporins A–D (**94**–**97**) [52].

4. Conclusions

Fungal metabolites are an unparalleled pool for pharmaceutical lead discovery. Sesquiterpenoids involving the bergamotane skeleton have been separated from various sources, including fungi. In the current work, 97 bergamotane sesquiterpenoids were reported from various fungal species derived from different sources, including endophytic (24 compounds), mushroom (21 compounds), sea mud (14 compounds), sea sediment (13 compounds), deep-sea deposit (8 compounds), and sponges (3 compounds) (Figure 11).

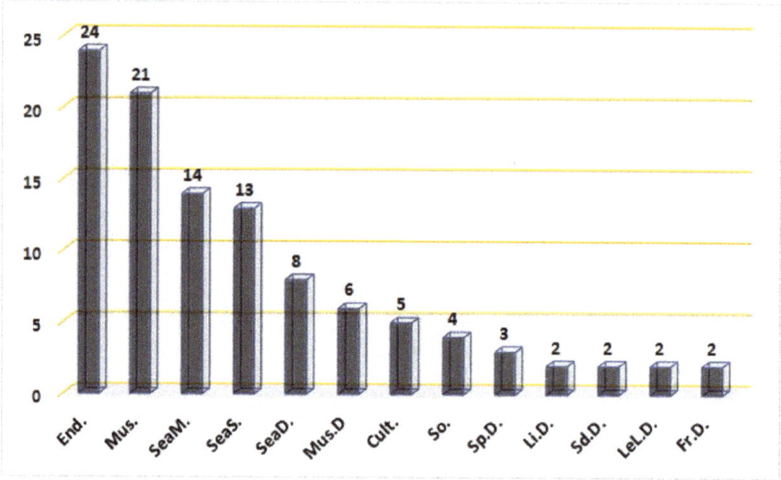

Figure 11. Number of bergamotane sesquiterpenoids reported from fungi derived from various sources. End.: endophytic; Mus.: mushroom; SeaM.: sea mud-derived; SeaS.: sea sediment-derived; SeaD.: sea deposit-derived; Mus.D: mushroom-derived; Cult.: cultured; So.: soil-derived; Sp.D.: sponge-derived; Li.D.: lichen-derived; Sd.D.: sheep dung-derived; LeL.D.: leaf litter-derived; Fr.D.: fruit-derived.

The majority of compounds have been reported from *Paraconiothyrium* (25 compounds), *Craterellus* (23 compounds), and *Eutypella* (12 compounds) species (Figure 12). Interestingly, many of these metabolites normally occurred as inseparable mixtures. These metabolites were assessed for diverse bio-activities. It is obvious that cytotoxic evaluation accounts for the largest proportion of biological assessments, where they had weak or no effectiveness on the tested cell lines. On the other hand, there are limited reports on their phytotoxic, plant growth regulation, antimicrobial, anti-HIV, cytotoxic, anti-inflammatory, pancreatic lipase inhibition, immunosuppressive, and antidiabetic activities. Therefore, this suggested more potential for trying other types of pharmacological effectiveness. Victoxinine (**40**) and prehelminthosporolactone (**39**) displayed potential phytotoxic capacities; therefore, they could be utilized as bioherbicides or as lead metabolites for synthesizing more efficacious phytotoxic compounds against various weeds. Pinthunamide (**44**), ampullicin (**91**), isoampullicin (**92**), and dihydroampullicin (**93**) were found to selectively promote the root growth. However, the phytotoxic and plant growth promotion potential should be transferred from laboratory experiments into field settings for assessing the environmental influences on these activities. Purpurolide F (**87**) had potent pancreatic lipase inhibition potential that could be a viable candidate as a pancreatic lipase inhibitor for further clinical development. Massarinolin B (**61**) had prominent immunosuppressive potential, suggesting further in vivo and mechanistic investigations for the development of this metabolite as an immunosuppressant. In silico studies for the reported metabolites that have not been tested or have had no noticeable effectiveness in the estimated activities could be a possible area of future research. Moreover, synthesis and structural modifications of these metabolites may produce more potential and useful tags of these metabolites through click chemistry, which is a new approach for synthesizing drug-like molecules that can boost the drug discovery process.

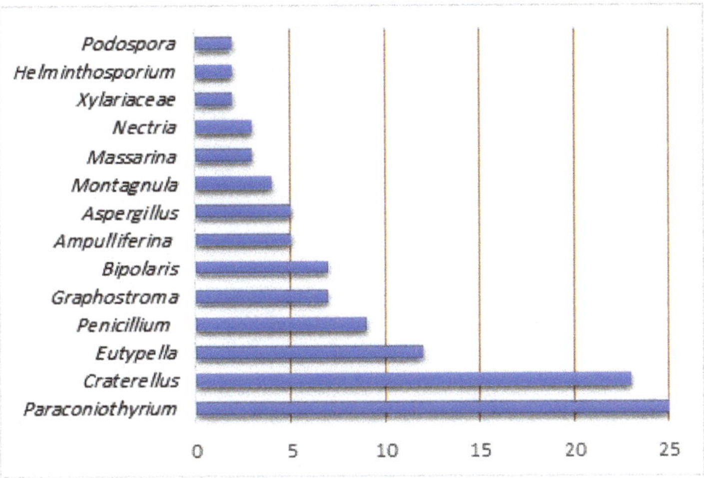

Figure 12. Bergamotane sesquiterpenoids from various fungal species.

Biogenetically, these metabolites are generated from acyclic farnesyl-diphosphate, which undertakes various condensation and rearrangement reactions. This work could be a beneficial reference for researchers studying this class of fungal metabolites. Several strategies, including co-culture, molecular and epigenetic manipulations, OSMAC (one strain many compounds), heterologous gene expression, and inter-species cross-talk approaches could be successively employed to access undescribed natural metabolites from silent biosynthetic pathways. It was found that the selective epigenetic target manipulation utilizing small molecule inhibitors toward DNA methyltransferase and histone deacetylase activities resulted in the enhancement of biosynthetic pathway expression for new secondary metabolite production. Highlighting the biosynthesis of these metabolites in this review could draw the attention of molecular biologists and genetics-interested researchers for isolating genes accountable for the biosynthesis of these interesting metabolites; this could allow for the discovery of the detailed mechanisms of their formation by various enzymes, which could allow for the preparation these metabolites and their analogs by engineering their biosynthetic pathways.

Author Contributions: Conceptualization, S.R.M.I. and G.A.M.; resources, M.T.K., K.A.M. and A.M.O.; discussion of the contents, A.M.O., G.A.M. and S.R.M.I.; writing—original draft, M.T.K., K.A.M., A.M.O., G.A.M. and S.R.M.I.; writing—review and editing, G.A.M. and S.R.M.I. All authors have read and agreed to the published version of the manuscript.

Funding: This research work was funded by Institutional Fund Projects under grant no. (IFPDP-254-22). Therefore, the authors gratefully acknowledge technical and financial support from Ministry of Education and Deanship of Scientific Research (DSR), King Abdulaziz University (KAU), Jeddah, Saudi Arabia.

Institutional Review Board Statement: Not applicable.

Informed Consent Statement: Not applicable.

Data Availability Statement: Not applicable.

Acknowledgments: This research work was funded by Institutional Fund Projects under grant no. (IFPDP-254-22). Therefore, the authors gratefully acknowledge technical and financial support from Ministry of Education and Deanship of Scientific Research (DSR), King Abdulaziz University (KAU), Jeddah, Saudi Arabia.

Conflicts of Interest: The authors declare no conflict of interest.

Abbreviations

5-Aza	5-azacytidine
A2780	Human ovarian cancer cell line
4-MUO	4-methylumbelliferyl oleate
A549	Lung adenocarcinoma epithelial cell line
Bel-7402	Human hepatoma cell line
BGC-823	Human stomach cancer cell line
BuOH	*n*-Butanol
C8166	Human T-cell leukaemia
CC	Column chromatography
CC_{50}	The 50% cytotoxic concentration
CCK-8	Cell Counting Kit-8
$CHCl_3$	Chloroform
CH_2Cl_2	Dichloromethane
CNE1-LMP1	Stable oncoprotein LMP1 integrated nasopharyngeal carcinoma cell line
DU145	Human prostate carcinoma cell line
EC109	Human esophageal cancer cell line
ED_{50}	Half-maximal effective concentration
H_2SO_4	Sulfuric acid
Hep3B-2	Human hepatoma carcinoma cell line
HCT-116	Human colon cancer cell line
HIV	Human immunodeficiency virus
HPLC	High-performance liquid chromatography
Huh7	Human hepatoma adenocarcinoma cell line
IR	Infrared
HL-60	Human myeloid leukemia cell line
LPS	Lipopolysaccharide
M14	Human melanoma cell line
MCF-7	Human breast cancer cell line
MGC	Human gastric cancer cell line
MTS	(3-(4,5-Dimethylthiazol-2-yl)-5-(3-carboxymethoxyphenyl)-2-(4-sulfophenyl)-2H-tetrazoliuminner salt)
MTT	3-(4,5-Dimethylthiazol-2-yl)-2,5-diphenyltetrazolium bromide
NMR	Nuclear magnetic resonance
NO	Nitric oxide
RP-18	Reversed phase-18
SBHA	Histonedeacetylase inhibitor, suberohydroxamic acid
SiO_2	Silica gel
SMMC-7721	Hepatocellular carcinoma cell line
SW480	Colon cancer cell line
SW1990	Human pancreatic adenocarcinoma cell line
PANC-1	Human pancreatic carcinoma cell line
PC-3	Human prostate cancer cell line
TLC	Thin-layer chromatography
U937	Human leukemic monocyte lymphoma cell line

References

1. Shen, B. A New Golden Age of Natural Products Drug Discovery. *Cell* **2015**, *163*, 1297–1300. [CrossRef] [PubMed]
2. Krabberød, A.K.; Deutschmann, I.M.; Bjorbækmo, M.F.; Balagué, V.; Giner, C.R.; Ferrera, I.; Garcés, E.; Massana, R.; Gasol, J.M.; Logares, R. Long-Term Patterns of an Interconnected Core Marine Microbiota. *Environ. Microbiome* **2022**, *17*, 1–24. [CrossRef] [PubMed]
3. Imhoff, J.F. Natural Products from Marine fungi—Still an Underrepresented Resource. *Mar. Drugs* **2016**, *14*, 19. [CrossRef] [PubMed]
4. Alves, A.; Sousa, E.; Kijjoa, A.; Pinto, M. Marine-Derived Compounds with Potential use as Cosmeceuticals and Nutricosmetics. *Molecules* **2020**, *25*, 2536. [CrossRef]

5. Ibrahim, S.R.; Fadil, S.A.; Fadil, H.A.; Hareeri, R.H.; Alolayan, S.O.; Abdallah, H.M.; Mohamed, G.A. *Dactylospongia Elegans*—A Promising Drug Source: Metabolites, Bioactivities, Biosynthesis, Synthesis, and Structural-Activity Relationship. *Mar. Drugs* **2022**, *20*, 221. [CrossRef]
6. Ibrahim, S.R.; Fadil, S.A.; Fadil, H.A.; Hareeri, R.H.; Abdallah, H.M.; Mohamed, G.A. Genus Smenospongia: Untapped Treasure of Biometabolites—Biosynthesis, Synthesis, and Bioactivities. *Molecules* **2022**, *27*, 5969. [CrossRef]
7. Mohamed, G.A.; Ibrahim, S.R. Untapped Potential of Marine-Associated *Cladosporium* Species: An Overview on Secondary Metabolites, Biotechnological Relevance, and Biological Activities. *Mar. Drugs* **2021**, *19*, 645. [CrossRef]
8. Ibrahim, S.R.M.; Mohamed, G.A.; Al Haidari, R.A.; El-Kholy, A.A.; Zayed, M.F.; Khayat, M.T. Biologically Active Fungal Depsidones: Chemistry, Biosynthesis, Structural Characterization, and Bioactivities. *Fitoterapia* **2018**, *129*, 317–365. [CrossRef]
9. Omar, A.M.; Mohamed, G.A.; Ibrahim, S.R.M. Chaetomugilins and Chaetoviridins-Promising Natural Metabolites: Structures, Separation, Characterization, Biosynthesis, Bioactivities, Molecular Docking, and Molecular Dynamics. *J. Fungi* **2022**, *8*, 127. [CrossRef]
10. Ibrahim, S.R.M.; Sirwi, A.; Eid, B.G.; Mohamed, S.G.A.; Mohamed, G.A. Fungal Depsides-Naturally Inspiring Molecules: Biosynthesis, Structural Characterization, and Biological Activities. *Metabolites* **2021**, *11*, 683. [CrossRef]
11. Ibrahim, S.R.M.; Fadil, S.A.; Fadil, H.A.; Eshmawi, B.A.; Mohamed, S.G.A.; Mohamed, G.A. Fungal Naphthalenones; Promising Metabolites for Drug Discovery: Structures, Biosynthesis, Sources, and Pharmacological Potential. *Toxins* **2022**, *14*, 154. [CrossRef] [PubMed]
12. Ameen, F.; AlNadhari, S.; Al-Homaidan, A.A. Marine Microorganisms as an Untapped Source of Bioactive Compounds. *Saudi J. Biol. Sci.* **2021**, *28*, 224–231. [CrossRef] [PubMed]
13. Hasan, S.; Ansari, M.I.; Ahmad, A.; Mishra, M. Major Bioactive Metabolites from Marine Fungi: A Review. *Bioinformation* **2015**, *11*, 176. [CrossRef] [PubMed]
14. Wibowo, J.T.; Ahmadi, P.; Rahmawati, S.I.; Bayu, A.; Putra, M.Y.; Kijjoa, A. Marine-Derived Indole Alkaloids and their Biological and Pharmacological Activities. *Mar. Drugs* **2021**, *20*, 3. [CrossRef]
15. Giddings, L.; Newman, D.J. Extremophilic Fungi from Marine Environments: Underexplored Sources of Antitumor, Anti-Infective and Other Biologically Active Agents. *Mar. Drugs* **2022**, *20*, 62. [CrossRef]
16. Hafez Ghoran, S.; Taktaz, F.; Ayatollahi, S.A.; Kijjoa, A. Anthraquinones and their Analogues from Marine-Derived Fungi: Chemistry and Biological Activities. *Mar. Drugs* **2022**, *20*, 474. [CrossRef]
17. Kramer, R.; Abraham, W. Volatile Sesquiterpenes from Fungi: What are they Good for? *Phytochem. Rev.* **2012**, *11*, 15–37. [CrossRef]
18. Gui, P.; Fan, J.; Zhu, T.; Fu, P.; Hong, K.; Zhu, W. Sesquiterpenoids from the Mangrove-Derived Aspergillus Ustus 094102. *Mar. Drugs* **2022**, *20*, 408. [CrossRef]
19. Dai, Q.; Zhang, F.; Feng, T. Sesquiterpenoids Specially Produced by Fungi: Structures, Biological Activities, Chemical and Biosynthesis (2015–2020). *J. Fungi* **2021**, *7*, 1026. [CrossRef]
20. Minami, A.; Ozaki, T.; Liu, C.; Oikawa, H. Cyclopentane-Forming Di/Sesterterpene Synthases: Widely Distributed Enzymes in Bacteria, Fungi, and Plants. *Nat. Prod. Rep.* **2018**, *35*, 1330–1346. [CrossRef]
21. Fraga, B.M. Natural Sesquiterpenoids. *Nat. Prod. Rep.* **2013**, *30*, 1226–1264. [CrossRef] [PubMed]
22. Cane, D.E. Enzymic Formation of Sesquiterpenes. *Chem. Rev.* **1990**, *90*, 1089–1103. [CrossRef]
23. Oh, H.; Gloer, J.B.; Shearer, C.A. Massarinolins A–C: New Bioactive Sesquiterpenoids from the Aquatic Fungus *Massarina Tunicata*. *J. Nat. Prod.* **1999**, *62*, 497–501. [CrossRef] [PubMed]
24. Che, Y.; Gloer, J.B.; Koster, B.; Malloch, D. Decipinin A and Decipienolides A and B: New Bioactive Metabolites from the Coprophilous Fungus *Podospora Decipiens*. *J. Nat. Prod.* **2002**, *65*, 916–919. [CrossRef]
25. Massias, M.; Rebuffat, S.; Molho, L.; Chiaroni, A.; Riche, C.; Bodo, B. Expansolides A and B: Tetracyclic Sesquiterpene Lactones from *Penicillium Expansum*. *J. Am. Chem. Soc.* **1990**, *112*, 8112–8115. [CrossRef]
26. Macías, F.A.; Varela, R.M.; Simonet, A.M.; Cutler, H.G.; Cutler, S.J.; Hill, R.A. Absolute Configuration of Bioactive Expansolides A and B from *Aspergillus Fumigatus* Fresenius. *Tetrahedron Lett.* **2003**, *44*, 941–943. [CrossRef]
27. Wen, Y.; Chen, T.; Jiang, L.; Li, L.; Guo, M.; Peng, Y.; Chen, J.; Pei, F.; Yang, J.; Wang, R. Unusual (2R, 6R)-Bicyclo [3.1. 1] Heptane Ring Construction in Fungal A-*Trans*-Bergamotene Biosynthesis. *Iscience* **2022**, *25*, 104030. [CrossRef]
28. Nozoe, S.; Kobayashi, H.; Morisaki, N. Isolation of Beta Trans Bergamotene from *Aspergillus Fumigatus*, a Fumagillin Producing Fungi. *Tetrahedron Lett.* **1976**, *50*, 4625–4626. [CrossRef]
29. Wang, Y.; Li, D.; Li, Z.; Sun, Y.; Hua, H.; Liu, T.; Bai, J. Terpenoids from the Marine-Derived Fungus *Aspergillus Fumigatus* YK-7. *Molecules* **2016**, *21*, 31. [CrossRef]
30. Dai, Q.; Zhang, F.; Li, Z.; He, J.; Feng, T. Immunosuppressive Sesquiterpenoids from the Edible Mushroom *Craterellus Odoratus*. *J. Fungi* **2021**, *7*, 1052. [CrossRef]
31. Pei, Y.; Zhang, L.; Wu, X.; Wu, H.; Wang, H.; Wang, Y.; Chen, G. Polyhydroxylated Bergamotane-Type Sesquiterpenoids from Cultures of *Paraconiothyrium Sporulosum* YK-03 and their Absolute Configurations. *Phytochemistry* **2022**, *194*, 113000. [CrossRef] [PubMed]
32. Zhao, Z.; Zhao, K.; Chen, H.; Bai, X.; Zhang, L.; Liu, J. Terpenoids from the Mushroom-Associated Fungus *Montagnula Donacina*. *Phytochemistry* **2018**, *147*, 21–29. [CrossRef] [PubMed]
33. Wu, Z.; Wu, Y.; Chen, G.; Hu, D.; Li, X.; Sun, X.; Guo, L.; Li, Y.; Yao, X.; Gao, H. Xylariterpenoids A–D, Four New Sesquiterpenoids from the Xylariaceae Fungus. *RSC Adv.* **2014**, *4*, 54144–54148. [CrossRef]

34. Niu, S.; Xie, C.; Zhong, T.; Xu, W.; Luo, Z.; Shao, Z.; Yang, X. Sesquiterpenes from a Deep-Sea-Derived Fungus *Graphostroma* Sp. MCCC 3A00421. *Tetrahedron* **2017**, *73*, 7267–7273. [CrossRef]
35. Niu, S.; Liu, D.; Shao, Z.; Liu, J.; Fan, A.; Lin, W. Chemical Epigenetic Manipulation Triggers the Production of Sesquiterpenes from the Deep-Sea Derived *Eutypella* Fungus. *Phytochemistry* **2021**, *192*, 112978. [CrossRef]
36. Pena-Rodriguez, L.M.; Armingeon, N.A.; Chilton, W.S. Toxins from Weed Pathogens, I. Phytotoxins from a *Bipolaris* Pathogen of Johnson Grass. *J. Nat. Prod.* **1988**, *51*, 821–828. [CrossRef]
37. Pena-Rodriguez, L.M.; Chilton, W.S. Victoxinine and Prehelminthosporolactone, Two Minor Phytotoxic Metabolites Produced by *Bipolaris* Sp., a Pathogen of Johnson Grass. *J. Nat. Prod.* **1989**, *52*, 899–901. [CrossRef]
38. Pringle, R.B.; Braun, A.C. Constitution of the Toxin of *Helminthosporium Victoriae*. *Nature* **1958**, *181*, 1205–1206. [CrossRef]
39. Dorn, F.; Arigoni, D. Structure of Victoxinine. *J. Chem. Soc. Chem. Commun.* **1972**, *24*, 1342–1343. [CrossRef]
40. Pringle, R. Comparative Biochemistry of the Phytopathogenic Fungus Helminthosporium. XVI. the Production of Victoxinine by *H. Sativum* and *H. Victoriae*. *Can. J. Biochem.* **1976**, *54*, 783–787. [CrossRef]
41. Kono, Y.; Takeuchi, S.; Daly, J.M. Isolation and Structure of a New Victoxinine Derivative Produced by *Helminthosporium Victoriae*. *Agric. Biol. Chem.* **1983**, *47*, 2701–2702. [CrossRef]
42. Kimura, Y.; Nakajima, H.; Hamasaki, T.; Sugawara, F.; Parkanyi, L.; Clardy, J. Pinthunamide, a New Tricyclic Sesquiterpene Amide Produced by a Fungus, *Ampullifernia* Sp. *Tetrahedron Lett.* **1989**, *30*, 1267–1270. [CrossRef]
43. Liu, L.; Gao, H.; Chen, X.; Cai, X.; Yang, L.; Guo, L.; Yao, X.; Che, Y. Brasilamides A–D: Sesquiterpenoids from the Plant Endophytic Fungus *Paraconiothyrium Brasiliense*. *Eur. J. Org. Chem.* **2010**, 3302–3306. [CrossRef]
44. Guo, Z.; Ren, F.; Che, Y.; Liu, G.; Liu, L. New Bergamotane Sesquiterpenoids from the Plant Endophytic Fungus *Paraconiothyrium Brasiliense*. *Molecules* **2015**, *20*, 14611–14620. [CrossRef]
45. Wang, W.; Shi, Y.; Liu, Y.; Zhang, Y.; Wu, J.; Zhang, G.; Che, Q.; Zhu, T.; Li, M.; Li, D. Brasilterpenes A-E, Bergamotane Sesquiterpenoid Derivatives with Hypoglycemic Activity from the Deep Sea-Derived Fungus *Paraconiothyrium Brasiliense* HDN15-135. *Mar Drugs* **2022**, *20*, 514. [CrossRef]
46. Ying, Y.; Fang, C.; Yao, F.; Yu, Y.; Shen, Y.; Hou, Z.; Wang, Z.; Zhang, W.; Shan, W.; Zhan, Z. Bergamotane Sesquiterpenes with Alpha-glucosidase Inhibitory Activity from the Plant Pathogenic Fungus *Penicillium Expansum*. *Chem. Biodivers.* **2017**, *14*, e1600184. [CrossRef]
47. Wang, Y.; Xia, G.; Wang, L.; Ge, G.; Zhang, H.; Zhang, J.; Wu, Y.; Lin, S. Purpurolide A, 5/5/5 Spirocyclic Sesquiterpene Lactone in Nature from the Endophytic Fungus *Penicillium Purpurogenum*. *Org. Lett.* **2018**, *20*, 7341–7344. [CrossRef]
48. Xia, G.; Wang, L.; Zhang, J.; Wu, Y.; Ge, G.; Wang, Y.; Lin, P.; Lin, S. Three New Polyoxygenated Bergamotanes from the Endophytic Fungus *Penicillium Purpurogenum* IMM 003 and their Inhibitory Activity Against Pancreatic Lipase. *Chin. J. Nat. Med.* **2020**, *18*, 75–80. [CrossRef]
49. Zhang, L.; Feng, B.; Chen, G.; Li, S.; Sun, Y.; Wu, H.; Bai, J.; Hua, H.; Wang, H.; Pei, Y. Sporulaminals A and B: A Pair of Unusual Epimeric Spiroaminal Derivatives from a Marine-Derived Fungus *Paraconiothyrium Sporulosum* YK-03. *RSC Adv.* **2016**, *6*, 42361–42366. [CrossRef]
50. Kimura, Y.; Nakajima, H.; Hamasaki, T.; Matsumoto, T.; Matsuda, Y.; Tsuneda, A. Ampullicin and Isoampullicin, New Metabolites from an *Ampulliferina*-Like Fungus Sp. no. 27. *Agric. Biol. Chem.* **1990**, *54*, 813–814. [CrossRef]
51. Kimura, Y.; Matsumoto, T.; Nakajima, H.; Hamasaki, T.; Matsuda, Y. Dihydroampullicin, a New Plant Growth Regulators Produced by the *Ampulliferina*-Like Fungus Sp. no. 27. *Biosci. Biotechnol. Biochem.* **1993**, *57*, 687–688. [CrossRef]
52. Zhang, Y.; Yu, H.; Xu, W.; Hu, B.; Guild, A.; Zhang, J.; Lu, X.; Liu, X.; Jiao, B. Eutypellacytosporins A–D, Meroterpenoids from the Arctic Fungus *Eutypella* Sp. D-1. *J. Nat. Prod.* **2019**, *82*, 3089–3095. [CrossRef] [PubMed]
53. Bermejo, F.A.; Mateos, A.F.; Escribano, A.M.; Lago, R.M.; Burón, L.M.; López, M.R.; González, R.R. Ti (III)-Promoted Cyclizations. Application to the Synthesis of (E)-Endo-Bergamoten-12-Oic Acids. Moth Oviposition Stimulants Isolated from Lycopersicon Hirsutum. *Tetrahedron* **2006**, *62*, 8933–8942. [CrossRef]
54. López, M.R.; Bermejo, F.A. Total Synthesis of ( )-Massarinolin B and ( )-4-Epi-Massarinolin B, Fungal Metabolites from Massarina Tunicata. *Tetrahedron* **2006**, *62*, 8095–8102. [CrossRef]
55. Zhao, J.; Feng, J.; Tan, Z.; Liu, J.; Zhang, M.; Chen, R.; Xie, K.; Chen, D.; Li, Y.; Chen, X. Bistachybotrysins A-C, Three Phenylspirodrimane Dimers with Cytotoxicity from *Stachybotrys Chartarum*. *Bioorg. Med. Chem. Lett.* **2018**, *28*, 355–359. [CrossRef]
56. Zamora, R.; Vodovotz, Y.; Billiar, T.R. Inducible Nitric Oxide Synthase and Inflammatory Diseases. *Mol. Med.* **2000**, *6*, 347–373. [CrossRef]
57. Kovarik, J.M.; Burtin, P. Immunosuppressants in Advanced Clinical Development for Organ Transplantation and Selected Autoimmune Diseases. Expert Opin. *Emerg. Drugs* **2003**, *8*, 47–62. [CrossRef]
58. Klinker, M.W.; Lundy, S.K. Multiple Mechanisms of Immune Suppression by B Lymphocytes. *Mol. Med.* **2012**, *18*, 123–137. [CrossRef]
59. Bermejo, F.A.; Rico-Ferreira, R.; Bamidele-Sanni, S.; García-Granda, S. Total Synthesis of (+)-Ampullicin and (+)-Isoampullicin: Two Fungal Metabolites with Growth Regulatory Activity Isolated from Ampulliferina Sp. 27. *J. Org. Chem.* **2001**, *66*, 8. [CrossRef]

marine drugs

Article

Chemomodulatory Effect of the Marine-Derived Metabolite "Terrein" on the Anticancer Properties of Gemcitabine in Colorectal Cancer Cells

Reham Khaled Abuhijjleh [1], Dalia Yousef Al Saeedy [1], Naglaa S. Ashmawy [2,3], Ahmed E. Gouda [4], Sameh S. Elhady [5] and Ahmed Mohamed Al-Abd [4,6,*]

1 Department of Pharmaceutical Sciences, College of Pharmacy, Gulf Medical University, Ajman 4184, United Arab Emirates; 2020mdd01@mygmu.ac.ae (R.K.A.); 2020mdd03@mygmu.ac.ae (D.Y.A.S.)
2 Department of Pharmacognosy, Faculty of Pharmacy, Ain Shams University, Abbassia, Cairo 11591, Egypt; naglaa.saad@pharma.asu.edu.eg
3 Research Institute for Medical and Health Sciences, University of Sharjah, Sharjah 27272, United Arab Emirates
4 Life Science Unit, Biomedical Research Division, Nawah Scientific, Al-Mokkatam, Cairo 11571, Egypt; ahmed.gouda@nawah-scientific.com
5 Department of Natural Products, Faculty of Pharmacy, King Abdulaziz University, Jeddah 21589, Saudi Arabia; ssahmed@kau.edu.sa
6 National Research Centre, Department of Pharmacology, Medical and Clinical Research Institute, Cairo 12622, Egypt
* Correspondence: ahmedmalabd@pharma.asu.edu.eg; Tel.: +971-564642929

Abstract: Background: Terrein (Terr) is a bioactive marine secondary metabolite that possesses antiproliferative/cytotoxic properties by interrupting various molecular pathways. Gemcitabine (GCB) is an anticancer drug used to treat several types of tumors such as colorectal cancer; however, it suffers from tumor cell resistance, and therefore, treatment failure. Methods: The potential anticancer properties of terrein, its antiproliferative effects, and its chemomodulatory effects on GCB were assessed against various colorectal cancer cell lines (HCT-116, HT-29, and SW620) under normoxic and hypoxic ($pO_2 \leq 1\%$) conditions. Further analysis via flow cytometry was carried out in addition to quantitative gene expression and ^{1}HNMR metabolomic analysis. Results: In normoxia, the effect of the combination treatment (GCB + Terr) was synergistic in HCT-116 and SW620 cell lines. In HT-29, the effect was antagonistic when the cells were treated with (GCB + Terr) under both normoxic and hypoxic conditions. The combination treatment was found to induce apoptosis in HCT-116 and SW620. Metabolomic analysis revealed that the change in oxygen levels significantly affected extracellular amino acid metabolite profiling. Conclusions: Terrein influenced GCB's anti-colorectal cancer properties which are reflected in different aspects such as cytotoxicity, cell cycle progression, apoptosis, autophagy, and intra-tumoral metabolism under normoxic and hypoxic conditions.

Keywords: terrein; gemcitabine; combination analysis; colorectal cancer; cell cycle; apoptosis; autophagy; metabolomics; qPCR

Citation: Abuhijjleh, R.K.; Al Saeedy, D.Y.; Ashmawy, N.S.; Gouda, A.E.; Elhady, S.S.; Al-Abd, A.M. Chemomodulatory Effect of the Marine-Derived Metabolite "Terrein" on the Anticancer Properties of Gemcitabine in Colorectal Cancer Cells. *Mar. Drugs* **2023**, *21*, 271. https://doi.org/10.3390/md21050271

Academic Editor: Dehai Li

Received: 25 March 2023
Revised: 19 April 2023
Accepted: 22 April 2023
Published: 26 April 2023

1. Introduction

Cancer is a major health problem worldwide, and the global burden of cancer has caused 10 million deaths in the past year [1]. While recent advances in the development of antitumor agents have contributed to cancer therapy, resistance to chemotherapy has led to recurrence and relapse [2]. In addition, the use of current antitumor agents has been limited due to their toxic and deleterious effects [3]. This calls for the development of novel anticancer agents with high efficacy to combat these issues.

Solid tumors suffer from a harsh microenvironment that has unique features and characteristics such as having areas with compromised endothelium, poor, or avascularized areas hindering drugs from reaching their targets, hypoxia in certain areas within the solid tumor, and a high level of acidosis with a noticeable pH gradient along the tumor tissue [4]. All the above features, nonetheless hypoxia, can play a role as drug targets due to the challenges faced with traditional chemotherapies. Oxygen levels in addition to the availability of nutrients differ drastically during the development of tumor cells through angiogenesis, and the recruitment of leukocytes and fibroblasts. Therefore, hypoxia can affect gene expressions, signaling pathways, many metabolic reactions, and the response to stress as well as the response to cytotoxic drugs [5]. Anticancer agents target various survival/death mechanisms or pathways in cancer cells, including angiogenesis, cell cycle regulation, apoptosis, and autophagy [6–9].

The secondary marine metabolite (+)-terrein was first isolated from *Aspergillus terreus* in 1935 and subsequently drew significant attention due to its various bioactivities, including anticancer properties [10]. Promising studies showed that terrein inhibits angiogenin production and secretion [11,12], induces cell cycle arrest [13,14] and apoptosis [15], and inhibits cell proliferation [16]. There is also evidence that terrein has anti-inflammatory activity that is mediated via inactivating the nuclear factor kappa B (NF-kB) signaling pathway. This occurs because of various mechanisms, most notably the inhibition of p60/p50 heterodimer translocation into the nucleus and the DNA-binding activity of the p65 subunit [12,17]. The NF-kB pathway has been implicated in various types of cancer due to its role in regulating apoptosis [18], and its abnormal activation can lead to malignant tumors and oncogenesis [19]. Yet, one of the major obstacles to using terrein on a large scale is the very low yield of this compound from different marine sources [20]. In addition, terrein and other marine-derived secondary metabolites are known for their abundance in diverse geographical distribution and their unique chemical structure [21].

Gemcitabine is the standard drug of choice for locally advanced and metastatic pancreatic cancer [22]. However, it is frequently associated with treatment failure due to intrinsic or acquired resistance. The failure of achieving good clinical outcomes in terms of survival could partially be associated with the hypo-vascularized and dense tumor stroma, and therefore, poor drug penetration and hypoxia [23]. Most patients acquire resistance after weeks of treatment, resulting in poor survival. Gemcitabine resistance can be either intrinsic or acquired and can result from molecular and cellular changes, such as nucleotide metabolism, apoptosis pathway suppression [24], ABC transporter protein overactivation/over-expression [25], activation of the cancer stem cells CSCs [26], activation of the epithelial-to-mesenchymal transition pathway (EMT) [27], and extracellular signal-regulated protein kinase (ERK) 1/2 overactivity [28]. Gemcitabine resistance is also associated with multiple genetic and epigenetic abnormalities. Changes in one or a few genes remain crucial for maintaining drug resistance, cell survival, and malignant phenotype [22]. There is evidence to indicate that NF-κB [29–32], AKT [29,33,34], MAPK [34,35], and HIF-1α pathways [36] are directly related to the resistance of gemcitabine in vitro and in vivo models.

Metabolomics is considered one of the best approaches to studying the effectiveness of drugs towards cancer [37] as well as assessing the reasons behind developing resistance toward drugs [38]. In this study, a metabolomic analysis has been conducted to compare the effect of terrein, gemcitabine, and a combination of both drugs on colorectal cells under normoxic as well as hypoxic conditions. ^{1}HNMR spectroscopy was utilized in combination with chemometric statistical methods to compare the extracellular metabolites after each treatment.

Herein, we evaluated the interaction between terrein and gemcitabine in colorectal cancer cells under normal and hypoxic conditions in terms of a potential chemomodulatory effect and mutual intra-tumoral metabolic influence.

2. Results

2.1. Cytotoxicity Assessment

To study the effect of Terr on the cytotoxic profile of GCB in colorectal cancer cell lines (HCT-116, HT-29, and SW620) under normoxic and hypoxic conditions, the viability dose–response curves of both agents, alone and in combination, were assessed using the E_{max} model as described in the Materials and Methods section. The IC_{50}s of either agent, alone or in combination with their CI indices, are summarized in Table 1.

Table 1. Combination analysis for GCB and Terr against HCT-116, HT-29, and SW620 colorectal cancer cell lines. (N) indicates normoxia, and (H) indicates hypoxia.

	HCT-116		HT-29		SW620	
	IC_{50} (μM)	R-Fraction (%)	IC_{50} (μM)	R-Fraction (%)	IC_{50} (μM)	R-Fraction (%)
GCB (N)	0.19 ± 0.028	38.14 ± 1.40	0.01 ± 0.006	19.87 ± 12.0	0.21 ± 0.0003	32.61 ± 1.78
Terr (N)	75.22 ± 0.97	N/A	56.24 ± 11.39	N/A	72.28 ± 1.35	8.34 ± 1.37
GCB + Terr (N)	0.023 ± 0.005	24.40 ± 1.22	0.027 ± 0.005	46.45 ± 2.76	0.018 ± 0.04	32.30 ± 3.56
CI value	0.13		2.32		0.09	
GCB (H)	0.01 ± 0.002	0.0	0.04 ± 0.004	42.64 ± 2.10	0.20 ± 0.008	32.59 ± 0.28
Terr (H)	20.26 ± 2.89	8.13 ± 0.98	83.30 ± 4.41	3.02 ± 3.39	59.82 ± 8.50	2.52 ± 3.39
GCB + Terr (H)	0.024 ± 0.03	41.02 ± 2.33	0.324 ± 7.95	66.74 ± 6.84	0.027 ± 0.007	48.51 ± 2.97
CI value	1.78		7.28		0.14	

In the HCT-116 cells, GCB exerted potent cytotoxic activity despite a resistant fraction of 38.14 ± 1.4% in the normoxia condition; viability started to drop significantly ($p < 0.05$) from the control value at 0.03 μM and 0.01 μM in normoxia and hypoxia, respectively. The cellular log kill was gradual in profile with IC_{50} of 0.19 ± 0.028 μM and 0.01 ± 0.002 μM in normoxia and hypoxia, respectively. Terr exerted abrupt cytotoxic activity with increasing concentration; viability started to drop significantly at 100 μM in both conditions. The cellular log kill showed IC_{50} of (75.22 ± 0.97 μM and 20.26 ± 2.89 μM) in normoxia and hypoxia, respectively. An equitoxic combination of Terr with GCB improved the cytotoxic profile of GCB in the HCT-116 cell line in normoxia, decreasing the resistant fraction to 24.40 ± 1.22%; however, it did not improve the cytotoxic profile of GCB in hypoxia with an increased resistant fraction to 41.02 ± 2.33%. More so, the IC_{50} of GCB after combination with Terr decreased significantly compared to the single GCB treatment in normoxia and increased in hypoxia (Supplementary Figure S1A,B). The calculated CI values for GCB with Terr were 0.129 and 1.779 in normoxia and hypoxia, respectively. These CI values are indicative of synergistic interaction characteristics in the HCT-116 cell line under normoxic conditions and an antagonistic interaction under hypoxic conditions (Table 1).

For the HT-29 cells, GCB had a resistant fraction of 19.87 ± 12.0% and 42.64 ± 2.10% in normoxia and hypoxia, respectively; viability dropped significantly ($p < 0.05$) compared to the control cells at 0.01 μM and 0.03μM in normoxia and hypoxia, respectively (Supplementary Figure S2A,B). The IC_{50}s of GCB were 0.01 ± 0.006 μM and 0.04 ± 0.004 μM in normoxia and hypoxia, respectively. The viability of the cells treated with Terr started to drop significantly ($p < 0.05$) from the concentration of 30 μM in normoxia and 100 μM in hypoxia. The IC_{50}s of Terr were 56.24 ± 11.39 μM and 83.30 ± 4.41 μM in normoxia and hypoxia, respectively. An equitoxic combination of Terr with GCB did not improve the cytotoxic profile of GCB; however, it increased the resistant fractions to 46.45 ± 2.76% and 66.74 ± 6.84% in normoxia and hypoxia, respectively. The IC_{50} of GCB after the combination with Terr significantly increased compared to the single GCB treatment in normoxia and hypoxia. Yet, the calculated CI values for GCB with Terr were 2.318 and 7.277 in normoxia and hypoxia, respectively. These CI values are indicative of antagonistic interaction characteristics in the HT-29 cell line under both oxygen conditions.

With respect to SW620 cells, the resistant fraction after treatment with GCB was 32.62 ± 1.78% and 32.59 ± 0.28% in normoxia and hypoxia, respectively; viability dropped significantly ($p < 0.05$) at 0.3 µM in both normoxia and hypoxia. The IC_{50} of GCB was 0.21 ± 0.0003 µM, and 0.20 ± 0.008 µM in normoxia and hypoxia, respectively. The viability of the cells treated with Terr dropped significantly ($p < 0.05$) from the control value at 100 µM in both normoxia and hypoxia. The IC_{50} of Terr was 72.28 ± 1.35 µM, and 59.82 ± 8.50 µM in normoxia and hypoxia, respectively. An equitoxic combination of Terr with GCB improved the cytotoxic profile of GCB, keeping the resistant fraction at 32.30 ± 3.56% and 48.51 ± 2.97% in normoxia and hypoxia, respectively. IC_{50} of GCB after combination with Terr decreased compared to single GCB treatment in normoxia and hypoxia (Supplementary Figure S3A,B). The calculated CI values for GCB with Terr were 0.092 and 0.142 in normoxia and hypoxia, respectively. These CI values are an indicator of synergistic interaction characteristics in the SW620 cell line under both oxygen conditions.

2.2. *The Influence of Terr on GCB-Induced Apoptotic Cell Death in Colorectal Cell Lines (HCT-116, HT-29, and SW620)*

The tested colorectal cancer cells were exposed to the predetermined IC_{50} for 24 h and 48 h and stained with annexin V-FITC/PI. In HCT-116, only the combination treatment (GCB + Terr) significantly induced apoptosis after 24 h and 48 h of exposure (0.51 ± 0.22% and 9.59 ± 0.85%, respectively) compared to the control untreated cells (0.133 ± 0.05% and 3.99 ± 0.37%, respectively) (Figure 1A,D). Single treatments showed no significant difference in terms of apoptosis compared to the control untreated cells. Alternatively, GCB alone induced significant necrosis after 24 h of exposure (14.77 ± 1.06%) compared to the control untreated cells.

In the HT-29 cells, the GCB treatment significantly induced apoptosis after 24 h and 48 h of exposure (8.29 ± 0.92% and 6.06 ± 0.069%, respectively) compared to the control untreated cells (1.24 ± 0.21% and 1.26 ± 0.11%, respectively). The combination treatment (GCB + Terr) was found to significantly decrease apoptosis after 24 h and 48 h of exposure when compared to GCB alone (10.5 ± 0.45% and 10.2 ± 0.09%, respectively). On the other hand, there was no significant difference between GCB and Terr in terms of apoptosis at both time points (Figure 1B,E,H).

In the SW620 cells, treatment with GCB significantly induced apoptosis after 24 h and 48 h of exposure (6.84 ± 0.47% and 2.1 ± 0.13%, respectively) compared to the control untreated cells (1.61 ± 0.28% and 2.24 ± 0.01%, respectively) (Figure 1C,F,I). The combination treatment (GCB + Terr) also increased apoptosis significantly after 24 h and 48 h of exposure (2.97 ± 0.91% and 2.84 ± 0.16%, respectively) when compared to the control untreated cells or GCB treatment alone. The combination treatment (GCB + Terr) induced significant necrosis after 24 h of exposure compared to the control untreated cells (5.17 ± 0.53% and 1.77 ± 0.17%, respectively). This effect was carried forward and influenced the total cell death, where the combination treatment (GCB + Terr) showed a significant increase in total cell death (14.14 ± 1.4%) compared to the control untreated cells (5.38 ± 0.19%) as well as single treatments (GCB or Terr).

To further confirm the flowcytometric apoptosis-driven results, we examined apoptotic regulator genes using the RT-qPCR technique and calculated their fold changes after treatment under normoxic and hypoxic conditions. In HCT-116, the antiapoptotic gene, BCL2, was over-expressed by 25-fold when the cells were treated with terrein alone under normoxic conditions. The same gene showed no significant change in expression after the combination treatment (GCB + Terr). The rest of the treatments in both normoxic and hypoxic conditions resulted in the under-expression of BCL2. Similarly, the apoptosis inhibitor gene, BIRC5, was under-expressed in all treatments in both oxygen conditions. On the other hand, the expression of the tumor suppressor gene, TP53, was not affected after treatment with terrein in normoxia but was under-expressed in all other treatments in both conditions. Similarly, the apoptotic FOXO3 gene showed significant over-expression by 11-fold in normoxia when the cells were treated with terrein. FOXO3 showed no change in expression when the cells were treated with the combination treatment (GCB + Terr) under hypoxic conditions. The rest of the treatments in both conditions resulted in the under-expression of FOXO3 (Figure 2A,B). In HT-29, all genes tested for apoptosis were under-expressed for all treatments under both normoxia and hypoxia (Figure 2C,D). Similarly, in the SW620 cells, all genes that were tested for apoptosis were under-expressed with all treatments under both normoxia and hypoxia (Figure 2E,F).

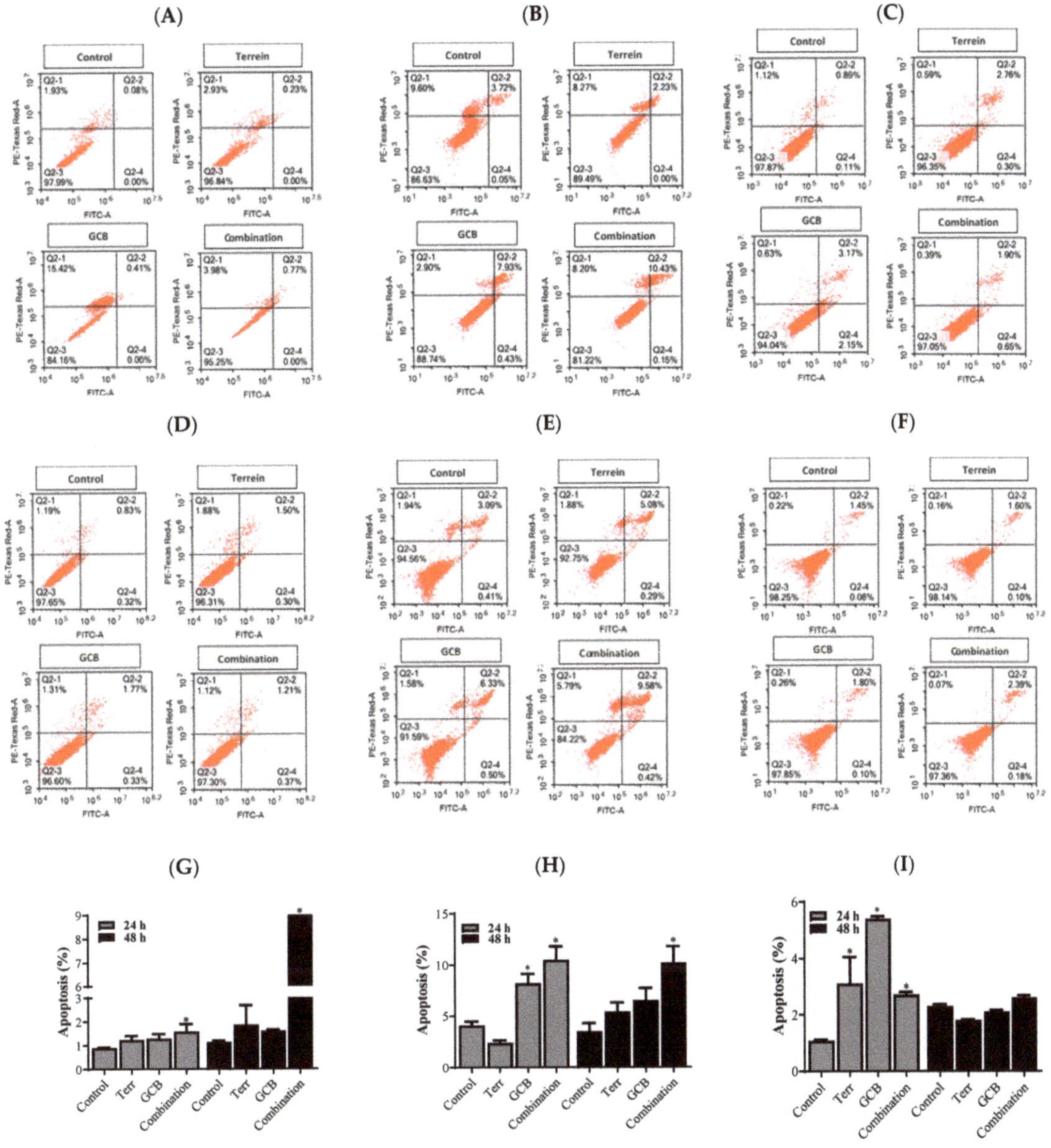

Figure 1. Programmed cell death (apoptosis) after treatment with Terr, GCB, and their combination for 24 h and 48 h. The cells were stained with annexin V-FITC/PI and different cell populations were plotted as a percentage of total events. HCT-116 under normoxia at 24 h and 48 h (**A**,**D**,**G**), HT-29 under normoxia at 24 h and 48 h (**B**,**E**,**H**), and SW620 under normoxia at 24 h and 48 h (**C**,**F**,**I**). Data are presented as mean ± SD; $n = 3$. * Significantly different from control group.

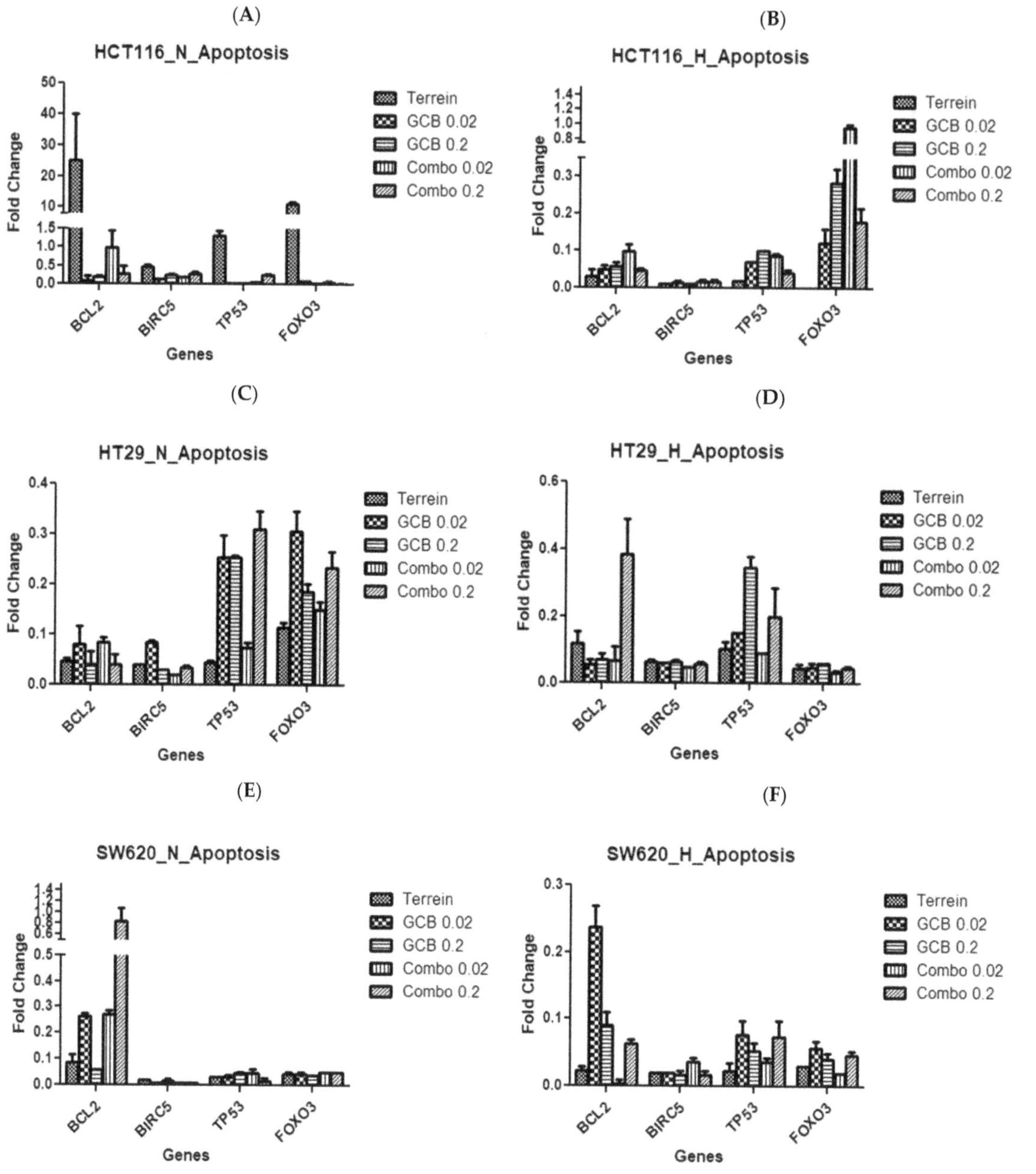

Figure 2. Fold change of apoptosis regulator genes after treatment with Terr, GCB, and their combination under normoxic and hypoxic conditions. A fold change value below 0.5 indicates under-expression, above 2.0 indicates over-expression, and between 0.5 and 2.0 indicates no change in expression. HCT-116 at 24 h under normoxia and hypoxia (**A**,**B**), HT-29 at 24 h under normoxia and hypoxia (**C**,**D**), and SW620 at 24 h under normoxia and hypoxia (**E**,**F**). The data are presented as mean fold change ± SD.

Caspase-3 is crucial in the apoptosis process and is considered the executioner active caspase family member; its concentration indicates the actual progression of apoptosis.

Herein, active caspase-3 was increased in HCT-116 in response to all single and combined treatments after 24 h and 48 h as well (Figure 3A). However, it was significantly increased in HT-29 when treated with the terrein and GCB combination for 24 and 48 h (Figure 3B). Similarly, the combination of terrein and GCB activated caspase-3 after 48 h only. Surprisingly, GCB induced the activation of caspase-3 after 24 h and 48 h as well (Figure 3C).

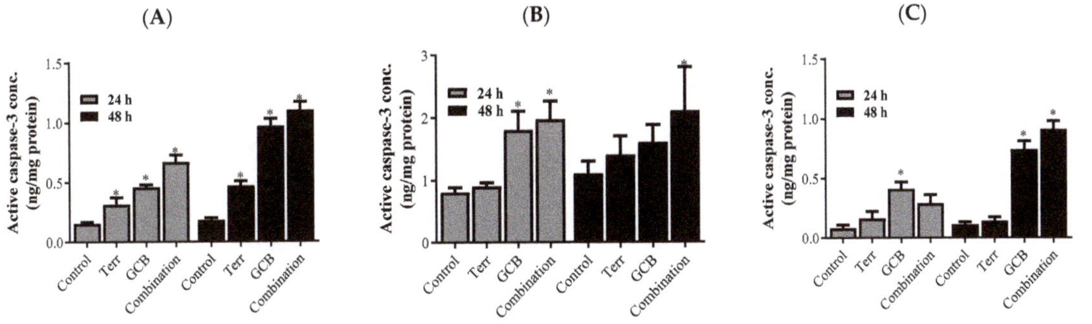

Figure 3. Caspase-3 level. The active caspase-3 concentration after treatment with Terr, GCB, and their combination in HCT-116 cells (**A**), HT-29 cells (**B**), and SW-620 cells (**C**) for 24 h. The data are presented as mean fold change ± SD. * Significantly different from control untreated cells.

2.3. The Effect of Terr on the Autophagic Cell Death of Colorectal Cell Lines Treated with GCB

In the HCT-116 cells, GCB induced a significant increase in autophagic cell death by 54.63% compared to the control untreated cells after 24 h under normoxic conditions. Surprisingly, not only Terr but also the combination of GCB + Terr had no significant autophagic effect (Figure 4A,D,G).

In the HT-29 cells, only a combination of GCB + Terr for 48 h induced significant autophagic cell death by 38.18% increase in acridine orange-fluorescent signal compared to the control untreated cells (Figure 4B,D,H).

In the SW620 cells, after 24 h of exposure to the treatment, GCB induced a significant increase in autophagic cell death by 87.62% when compared to the control untreated cells. There was no significant difference between GCB and Terr as single treatments. However, combination treatment (GCB + Terr) induced a significant decrease in autophagic cell death when compared to GCB alone by 35.58%. After 48 h of treatment, the combination treatment induced a significant increase in autophagic cell death by 36.79% when compared to the control untreated cells. However, the combination treatment induced no significant difference when compared with GCB alone (Figure 4C,F,I).

To further confirm flowcytometric-driven results, we examined autophagy regulator genes (ATG5 and Beclin-1) using the RT-qPCR technique and calculated their fold changes after treatment under normoxic and hypoxic conditions. The autophagic-forming vesicle regulator gene, ATG5, was over-expressed when HCT-116 cells were treated with terrein under normoxic conditions by 12-fold. However, it was significantly under-expressed with the other treatments in both oxygen and hypoxia conditions. The golden autophagy standard gene, Beclin-1, was under-expressed in response to all treatments in both oxygen and hypoxia conditions (Figure 5A,B). In HT-29 and SW620, both genes tested for autophagy were under-expressed for all treatments under both normoxia and hypoxia (Figure 5C–F).

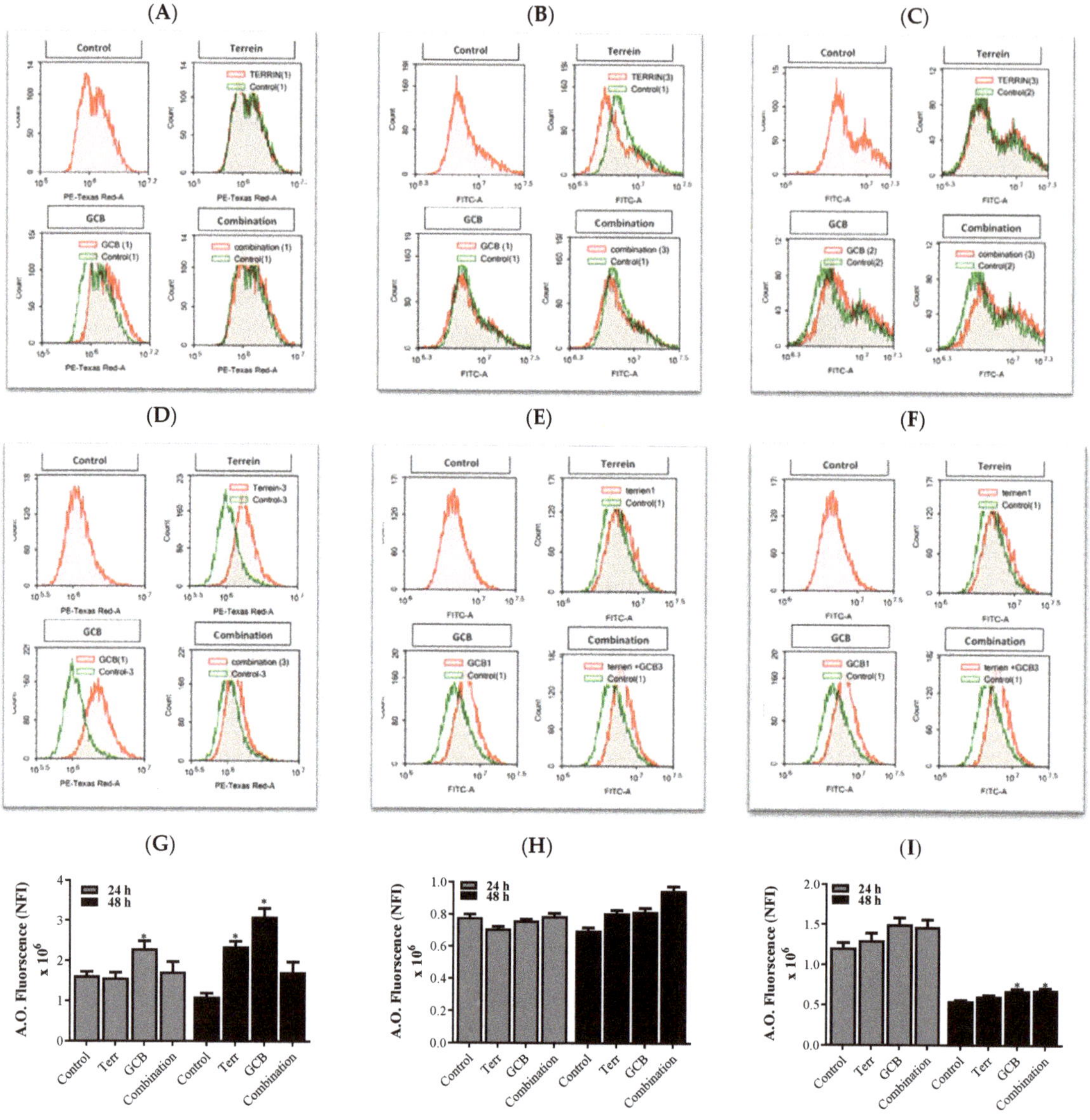

Figure 4. Induction of programmed cell death (autophagy) by Terr, GCB, and their combination for 24 h and 48 h. The cells were stained with acridine orange. The Average Net Fluorescent Intensity (NFI) values were plotted and compared to control cells. HCT-116 after 24 h and 48 h (**A**,**D**,**G**), HT-29 after 24 h and 48 h (**B**,**E**,**H**), and SW620 after 24 h and 48 h (**C**,**F**,**I**). Data are presented as mean ± SD; $n = 3$. * Significantly different from control untreated cells.

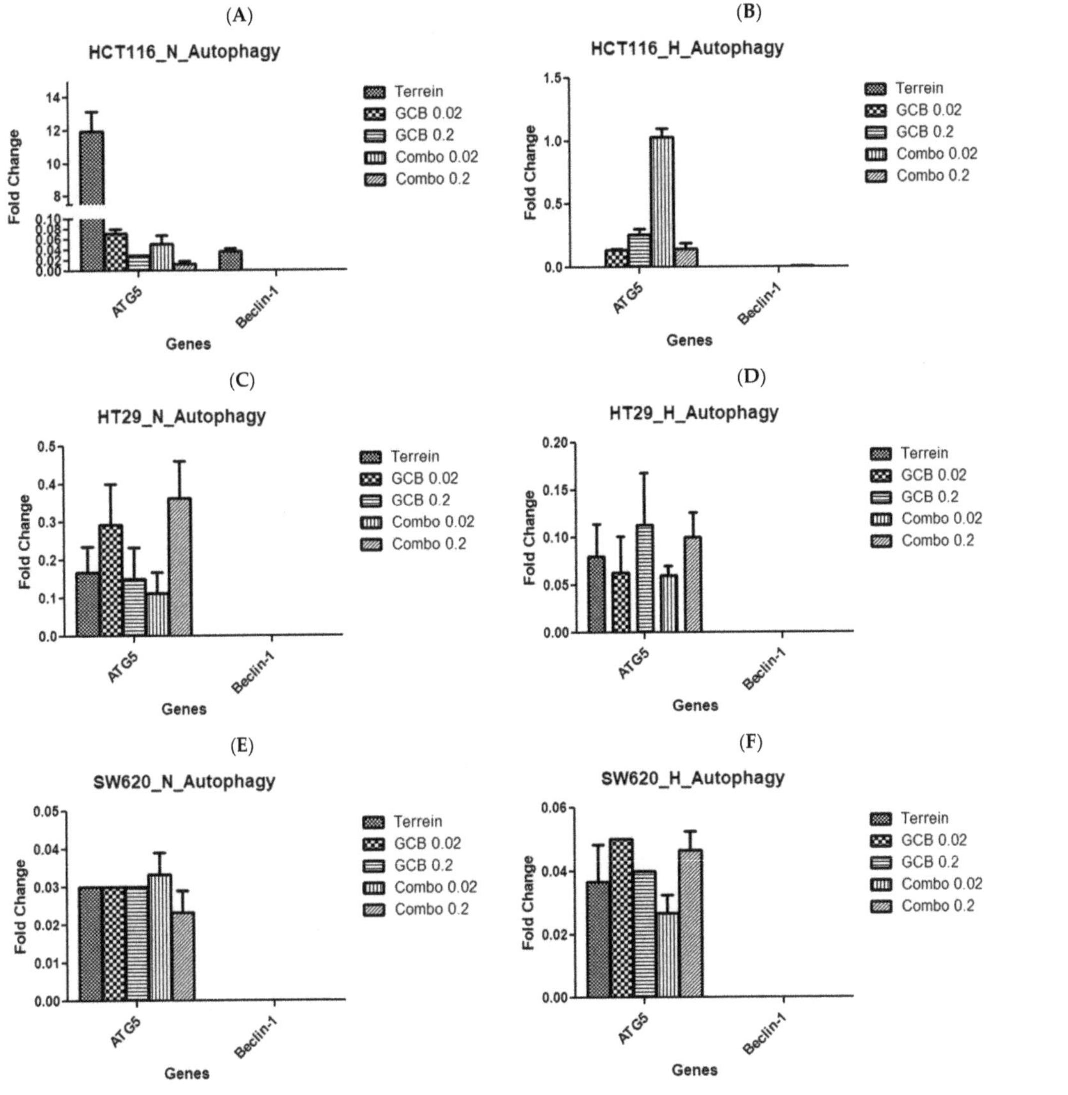

Figure 5. Fold change of autophagy regulator genes by Terr, GCB, and their combination under normoxic and hypoxic conditions. A fold change value below 0.5 indicates under-expression, above 2.0 indicates over-expression, and between 0.5 and 2.0 indicates no change in expression. HCT-116 cells under normoxia and hypoxia (**A**,**B**), HT-29 cells under normoxia and hypoxia (**C**,**D**), and SW620 cells under normoxia and hypoxia (**E**,**F**). The data are presented as mean ± SD.

2.4. The Effect of Terr, GCB, and Their Combination on the Cell Cycle Distribution of Colorectal Cell Lines

In the HCT-116 cells, Terr induced a significant G_2/M phase arrest and increased the cell population after 24 h from 27.8 ± 1.1% to 38 ± 1.15%. This effect was reversed by GCB where the cell population in the S phase dropped significantly to 32.61 ± 0.5%. As a result, there was no significant difference between the combination treatment (GCB + Terr) and GCB and Terr alone. On the other hand, after 48 h of exposure, there was no significant difference observed between all treatments in terms of the S phase population (Figure 6A,B).

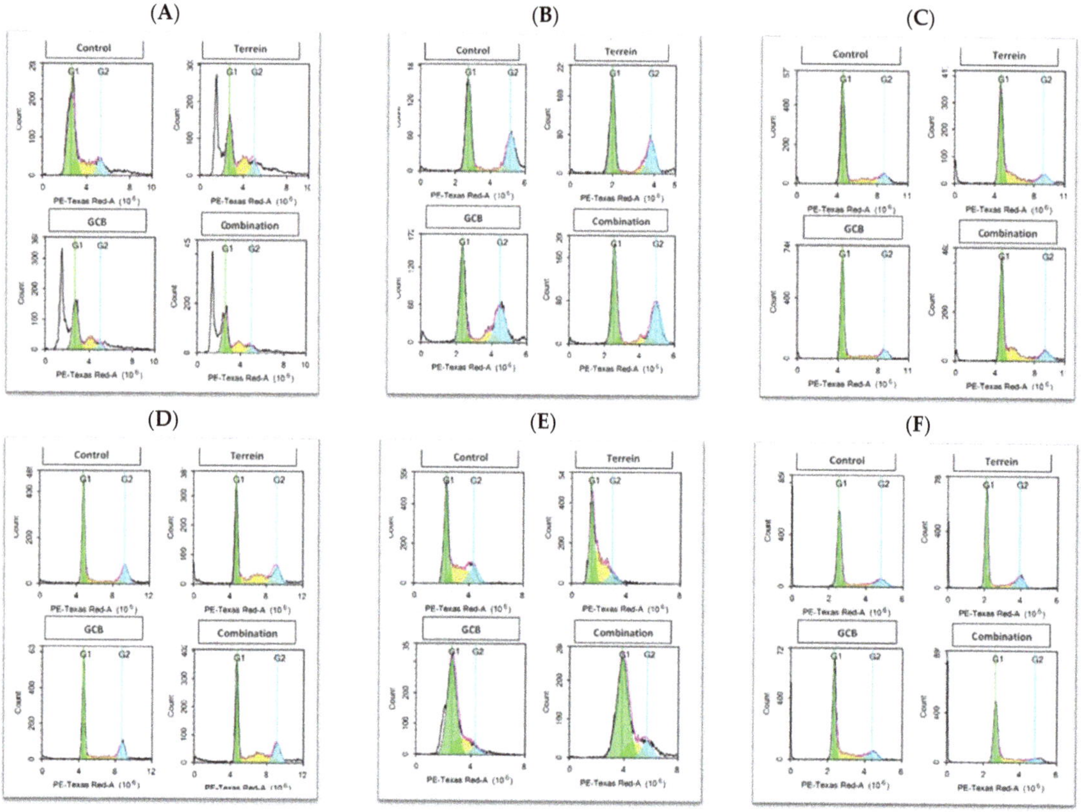

Figure 6. Effect of Terr, GCB, and their combination on the cell cycle distribution after 24 h and 48 h. The cell cycle distribution was determined using DNA cytometry analysis and different cell phases were plotted as the percentage of total events. HCT-116 after 24 h and 48 h (**A**,**B**), HT-29 after 24 h and 48 h (**C**,**D**), and SW620 after 24 h and 48 h (**E**,**F**).

In the HT-29 cells, Terr did not induce significant changes at the G_0/G_1 phase after 24 h of exposure. However, after further exposure (48 h), Terr induced a significant reduction in the cell population at the G_0/G_1 phase compared to the control untreated cells from 54.81 ± 0.7% to 42.78 ± 1.61%. Terr induced significant G_2/M phase arrest after 24 h (from 22.84 ± 0.57% to 26.5 ± 1.99%) and after 48 h (from 22.42 ± 0.99% to 34.42 ± 0.84%). After 24 h of exposure, the combination treatment (GCB + Terr) induced a decrease in the G_0/G_1cell population when compared to GCB alone from 67.3 ± 1.81% to 54.32 ± 1.49%, where the combination treatment (GCB + Terr) increased the S phase cell population compared to GCB (from 18.48 ± 0.09% to 31.39 ± 0.92%). Similar results were observed after 48 h of exposure (Figure 6C,D).

In the SW620 cells, GCB increases in the G_0/G_1 phase cell population after 24 h of exposure from 48.04 ± 1.28% to 77.43 ± 2.57%. This effect was also seen for the combination treatment (GCB + Terr), where the cell population in the G_0/G_1 phase increased significantly to 52.77% when compared to the control untreated cells. After 48 h of exposure, the combination treatment induced significant G_2/M phase arrest and increased the S phase cell population from 18.96 ± 0.86% to 27.88 ± 1.14% when compared to the control untreated cells. It also increased the G_0/G_1 phase cell population from 53.01 ± 2.72% to 60.16 ± 1.08% compared to GCB alone (Figure 6E,F).

In addition, we examined cell cycle regulatory genes by RT-qPCR (CCND1, CDK4, and MCM7). In HCT-116, only CDK4 was over-expressed due to treatment with terrein by 6.7-fold under normoxic conditions. CCND1 was under-expressed in response to all treatments under both oxygen and hypoxia conditions. Similar results were observed with the MCM7 gene (Figure 7A,B). In HT-29 and SW620, all genes tested for cell cycle regulation were under-expressed in response to all treatments under both normoxia and hypoxia (Figure 7C–E).

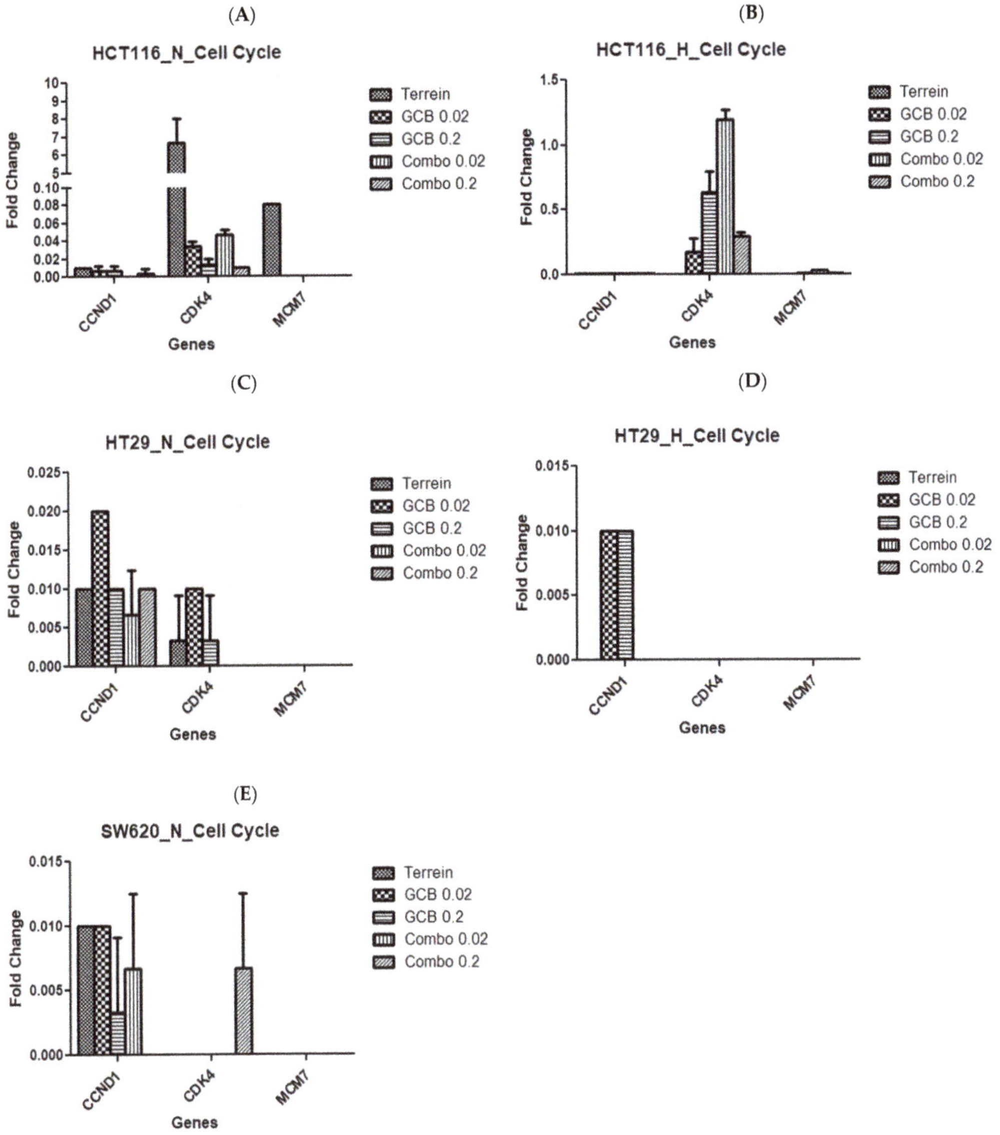

Figure 7. Fold change of cell cycle regulatory genes due to treatment with Terr, GCB, and their combination under normoxic and hypoxic conditions. A fold change value below 0.5 indicates under-expression, above 2.0 indicates over-expression, and between 0.5 and 2.0 indicates no change in expression. HCT-116 under normoxia and hypoxia (**A**,**B**), HT-29 under normoxia and hypoxia (**C**,**D**), and SW620 under normoxia and hypoxia (**E**). The data are presented as mean fold change ± SD.

2.5. The Effect of GCB, Terr, and Their Combination on the Colorectal Cell Lines' Growth and Proliferation

The combination of Terr and GCB was found to show a synergistic effect in normoxia and an antagonistic effect in hypoxia. This was evident when checking the cell growth regulators and cell proliferation regulators in both normoxia and hypoxia conditions. We examined and quantified the fold changes of several genes responsible for cellular growth (AKT1, TGF-B1, HIF1-α, and PRKDC) and cellular proliferation (PCNA and RAD18) using the RT-qPCR technique.

In HCT-116, the AKT1 gene was over-expressed when the cells were treated with terrein in normoxia by 45.8-fold and were under-expressed for other treatments in the same condition. The contrary was evident in hypoxia. TGF-1β showed no change in expression for cells treated with terrein but was under-expressed when the cells were treated with other treatment conditions under normoxia. HIF1-α showed similar results to the ones seen by AKT1 under normoxic conditions. PRKDC showed no change in expression for all treatments except the combination treatment (GCB + Terr) under normoxic conditions. At hypoxia, TGF-1β, HIF1-α, and PRKDC were under-expressed in all treatments (Figures 8A,B and 9A,B).

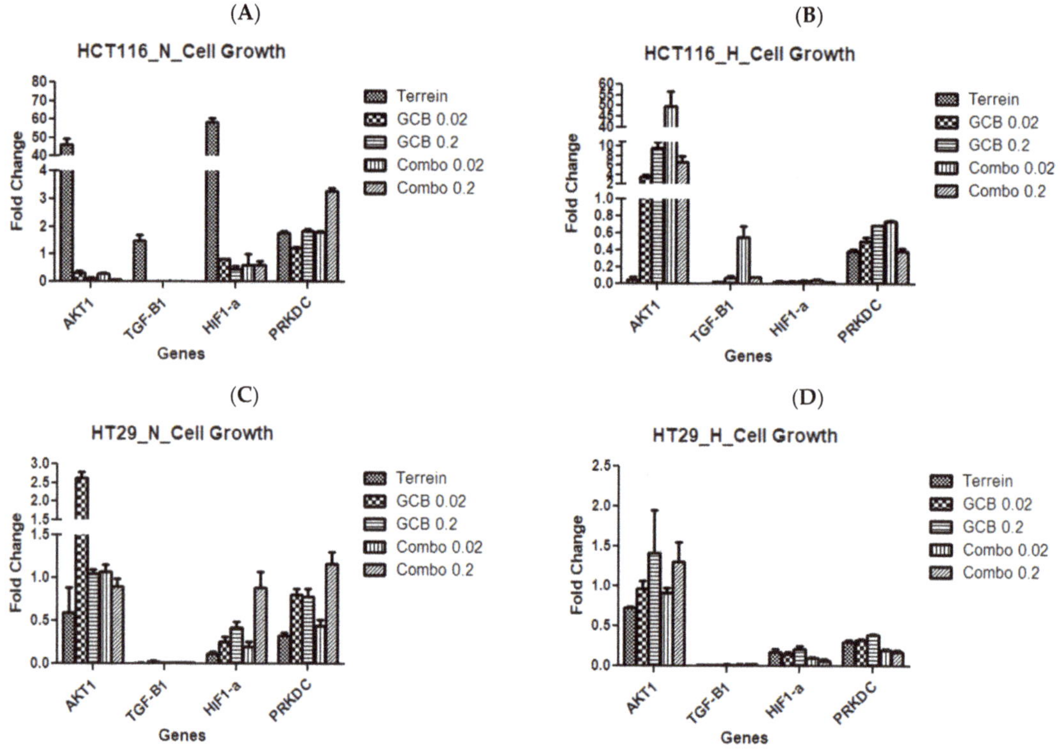

Figure 8. *Cont.*

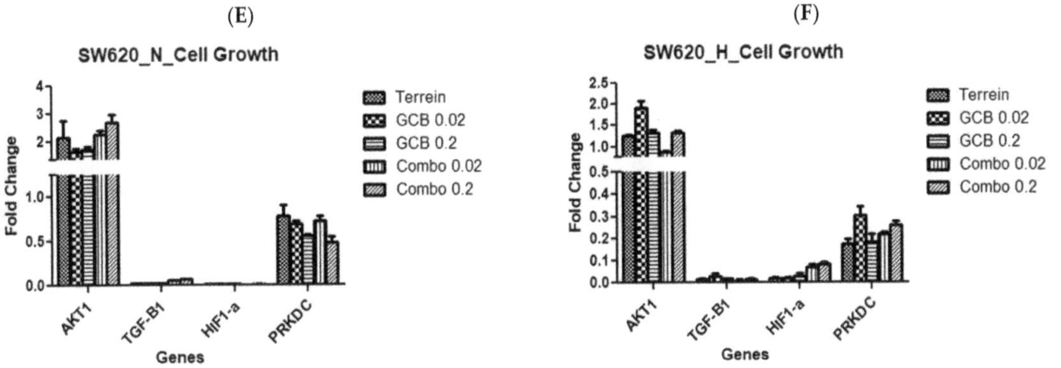

Figure 8. Fold change of genes by Terr, GCB, and their combination under normoxic and hypoxic conditions. A fold change value below 0.5 indicates under-expression, above 2.0 indicates over-expression, and between 0.5 and 2.0 indicates no change in expression. HCT-116 under normoxia and hypoxia (**A**,**B**), HT-29 under normoxia and hypoxia (**C**,**D**), and SW620 under normoxia and hypoxia (**E**,**F**). The data are presented as mean fold change ± SD.

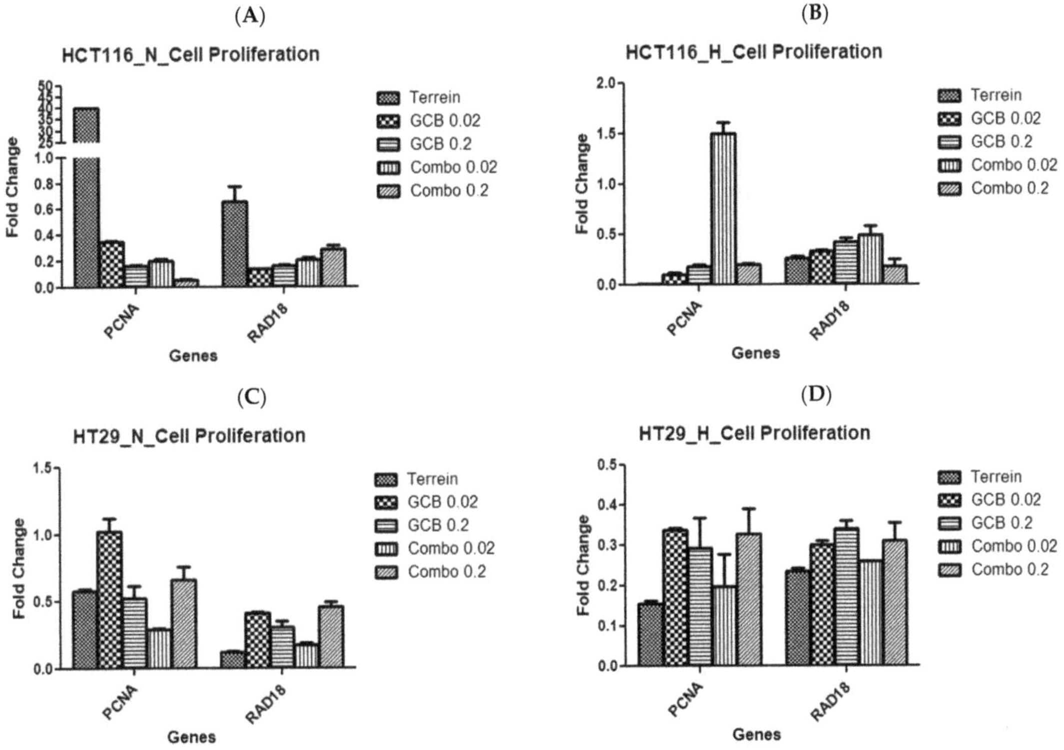

Figure 9. *Cont.*

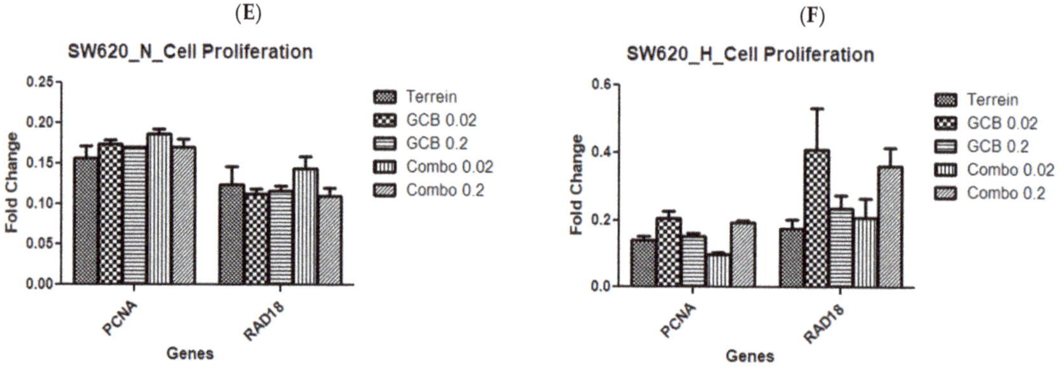

Figure 9. Fold change of genes by Terr, GCB, and their combination under normoxic and hypoxic conditions. A fold change value below 0.5 indicates under-expression, above 2.0 indicates over-expression, and between 0.5 and 2.0 indicates no change in expression. HCT-116 at 24 h under normoxia and hypoxia (**A**,**B**), HT-29 at 24 h under normoxia and hypoxia (**C**,**D**), and SW620 at 24 h under normoxia and hypoxia (**E**,**F**). The data are presented as mean fold change ± SD.

On the other hand, the cellular proliferation gene, PCNA, was over-expressed when the cells were treated with terrein in normoxic conditions by 40-fold. However, it was under-expressed for the rest of the treatments. At hypoxia, the combination treatment (GCB + Terr), resulted in no change in the expression of PCNA. Yet, the remaining treatments resulted in the under-expression of the gene. The RAD18 gene was under-expressed in all treatments in both oxygen conditions except for terrein in normoxia where the gene did not change in expression (Figure 8A,B).

In HT-29, AKT1 was only over-expressed 2.62-fold after treatment with GCB. TGF-1β was under-expressed in all treatments under both oxygenation conditions. HIF1-α was under-expressed in all treatments and oxygen conditions, except after treatment with (GCB + Terr) where it showed no change in expression. PRKDC showed no change in expression in normoxia in response to all treatment conditions (Figure 8C,D). PCNA was under-expressed due to hypoxic conditions and no change in expression was observed under normoxia (Figure 9C,D).

In SW620, AKT1 showed no change in gene expression for all treatments under normoxia and hypoxia. However, TGF-1β and HIF1-α were under-expressed in all treatments under normoxia and hypoxia. The PRKDC gene was only expressed under hypoxia conditions for all treatments (Figure 8C,D). PCNA and RAD18 were under-expressed in this cell line for all treatment groups under both oxygenation conditions (Figure 9E,F).

2.6. The Effect of Terr, GCB, and Their Combination on the Extracellular Metabolites within Colorectal Cell Llines

^{1}H-NMR comparative analysis was carried out between the metabolites released from HCT-116 cells after three different drug treatments: Terr only, GCB only, and a combination of both Terr and GCB under normoxic conditions. Three metabolites, namely, ethyl malonate, tyrosine, and methylhistidine were detected extracellularly from HCT-116 cells treated with GCB only. While other metabolites such as hypoxanthine and imidazole were identified in the HCT-116 cells treated with Terr and GCB + Terr but not in the extracellular fluid of the HCT-116 cells treated with GCB only. On the other hand, methionine was detected extracellularly in the HCT-116 cells treated with Terr only. Metabolites such as formate and pipocolate appeared extracellularly in both cell lines treated with Terr and GCB only but were not detected in cell lines treated with (GCB + Terr).

Another similar comparative analysis was performed between the metabolites released from HCT-116 cells after the same three different drug treatments but under hypoxic conditions. Three metabolites, namely, 4-hydroxyphenyl acetate, leucine, and pyruvate were detected only in the extracellular fluid of HCT-116 cells treated with both drugs (GCB + Terr) and were not detected in the cells treated only with single drugs. Furthermore, methionine and phenylalanine were detected in all the treatments except the GCB-treated cell lines.

Then, a third ^{1}H-NMR metabolomic analysis was conducted to compare the extracellular metabolites released by HCT-116 cells in both hypoxic and normoxic conditions. A total of 25 metabolites were detected from HCT-116 cells treated with the three drugs under normoxic and hypoxic conditions. In total, 19 of them were common between the two conditions (Table 2). Three metabolites, namely, 2-hydroxy valerate and 2-phosphoglycerate, were completely absent from all normoxic treatments and detected in all hypoxic ones, while pyruvate was exceptionally detected in the hypoxic cells treated with the combined drugs (GCB + Terr). On the other hand, methylhistidine was uniquely detected in the normoxic cells treated only with GCB drugs.

Table 2. Extracellular metabolites identified by ^{1}H NMR-based profiling for HCT-116 cells under both hypoxic and normoxic conditions. The quantification of metabolites was achieved via fitting with its reference spectrum from the library of the Chenomx NMR suite. The mean concentration ± standard error is shown for metabolites from a set of three biological replicates.

Name	NMR Chemical Shift	Concentration (mM)							
		Normoxia				Hypoxia			
		Control	Terr	GCB	Terr + GCB	Control	Terr	GCB	Terr + GCB
4-hydroxyphenyl acetate	7.16 (d), 6.68 (t), 3.44 (s)	0	3.2866	2.342	2.4149	0	0	0	1.332
2-hydroxyvalerate		0	0	0	0	0	7.351	26.6519	7.033
2-phosphoglycerate	4.86 (dt), 3.76 (dd), 3.62 (dd)	0	0	0	0	0	92.55	165.354	10.4
Acetate	1.9 (s)	0	4.203	6.2544	2.8006	0	3.2218	6.9808	2.04
Alanine	1.46 (d)	8.7211	18.9587	20.264	16.0552	48.545	25.8158	19.2287	1.1259
Dimethylamine	2.5 (s)	0	20.6557	6.4366	14.3796	0	18.5201	4.2958	13.026
Ethylmalonate		0	0	11.0424	0	0	10.0587	9.8134	4.29
Formate	8.46 (s)	1.3573	3.5965	4.6718	0	0	2.0378	4.6042	1.2927
Glucose	3.23, 3.39, 3.45, 3.50, 3.71, 3.81, 3.88, 4.63, 5.22	71.7887	143.9592	193.9061	102.7318	257.7746	71.1131	108.3805	36.567
Glutamate	2.34 (m)	7.9295	17.862	8.3554	31.9252	31.8823	5.9122	29.2106	7.0509
Histamine	7.99 (s), 7.14 (s), 3.29 (t), 3.03 (m)	0	2.2367	2.0502	2.993	0	1.2942	1.0424	1.9017
Hypoxanthine	8.19 (s), 8.21 (s)	0	1.1172	0	1.3887	0	0.7355	0	1.53
Imidazole	7.28 (s), 8.18 (s)	0	3.6693	0	1.4086	0	0.8819	0	0.99
Lactate	1.31, 4.09	62.6687	144.9777	201.7872	73.772	400.2995	103.39	166.539	65.067
Leucine	0.96 (t),1.70 (m)	3.7907	8.944	9.7745	4.7783	0	0	0	3.048
Methionine	3.86 (dd), 2.65 (t), 2.23 (m)	0	6.5637	0	0	0	0.4214	0	0.84
Phenylalanine	7.37 (m), 3.98 (m), 3.27 (m), 3.11 (m)	1.4766	3.7121	1.9949	3.3628	4.721	1.3442	0	1.1259
Pyruvate	2.36 (s)	0	0	0	0	0	0	0	4.7677
Pipecolate	3.58 (dd), 3.42 (m), 3.02 (td), 2.22 (m), 1.89 (m), 1.69 (m)	0	31.4662	12.192	0	0	0	0	0
Succinate	2.39 (s)	1.4136	2.4953	0.6909	2.3995	6.888	0.4962	0.1574	0.1986
Tyrosine	7.17 (m), 6.8 (m), 3.9 (m)	1.0549	0	2.8125	0	5.7714	1.2658	1.7629	1.28
Tyramine	2.92 (t), 3.23 (t), 6.9 (m), 7.2 (m)	0	3.4448	2.7521	2.6895	0	0	0	0
Valine	3.6 (d), 2.29 (m), 1.04 (d), 0.98 (d)	2.1199	3.6865	5.2247	2.535	7.3598	2.204	2.6834	2.037
Xanthine	7.89 (s)	0	1.7549	1.6411	1.6685	0	0.9064	0	0.6265
Methylhistidine	7.67 (s), 7.0 (s), 3.97 (dd), 3.68 (s), 3.18 (dd), 3.09 (dd)	0.5439	0	1.535	0	0	0	0	0

Two multivariate statistical analyses, hierarchical cluster analysis (HCA) and partial least squares discriminant analysis (PLS-DA), were utilized to study the overall difference in the metabolites released because of the three different drug treatments under each of the normoxic and hypoxic conditions individually and then were thirdly conducted to compare between the two conditions. The first HCA classified the released metabolites under normoxic conditions after the three drug treatments into clusters based on their abundance (Figure 10). Furthermore, supervised multivariate analysis, PLS-DA, showed that metabolites from different treatments under normoxic conditions are separately clustered (Figure 11A). VIP scores were selected as a criterion for choosing the most important variables of the PLS-DA model (Figure 11B). For normoxic conditions, the significantly different metabolites with a VIP score of greater than 1 are glutamate, histamine, hypoxanthine, phenylalanine, xanthine, dimethylamine, tyramine, 4-hydroxyphenyl alanine, and succinate.

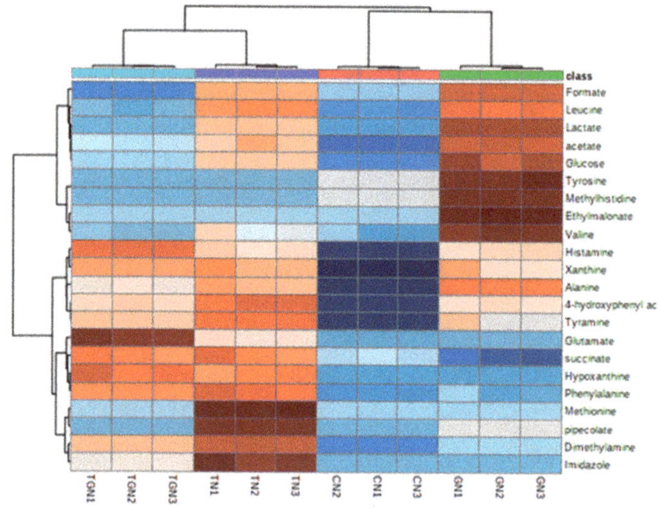

Figure 10. Unsupervised hierarchical clustering and heat map analysis of extracellular metabolites released from HCT-116 under normoxic conditions.

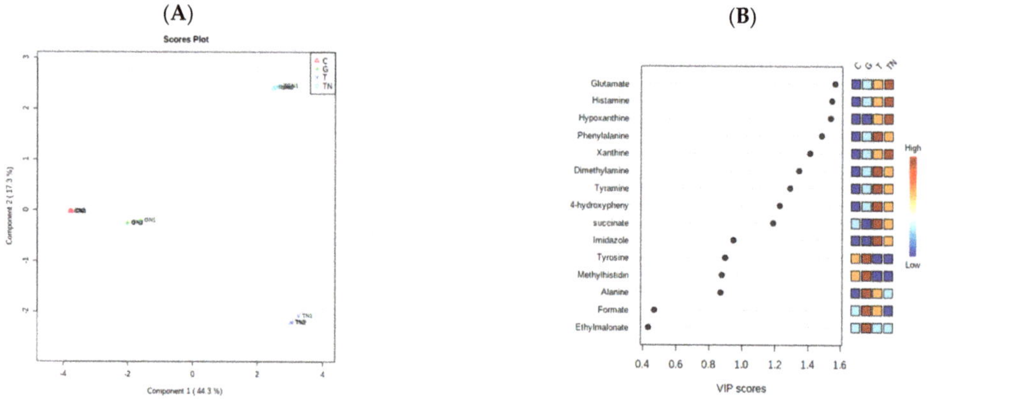

Figure 11. Supervised partial least squares discriminant analysis (PLS-DA) of the metabolite profiling of HCT-166 cells treated with Terr (T), gemcitabine (G), and Terr + GCB (TG) under normoxic conditions. (**A**) Two-dimensional score plot for HCT-116 cells treatments. (**B**) VIP score plot for treatments of HCT-116 cells.

Similarly, hierarchical cluster analysis (HCA) and partial least squares discriminant analysis (PLS-DA) were conducted to investigate the difference in the metabolites released because of different kinds of drug treatments under hypoxic conditions. HCA clustered the released metabolites based on their abundance (Figure 12). PLS-DA revealed that metabolites from the three treatments are separately clustered (Figure 13A). VIP scores were also selected as criteria for choosing the most important variables of the PLS-DA model. The most significantly different metabolites with a VIP score of more than 1 as shown in (Figure 13B) are histamine, methionine, hypoxanthine, glucose, lactate, imidazole, glutamate, alanine, valine, tyrosine, dimethylamine, pyruvate, 4- hydroxyphenyl alanine, leucine, and succinate.

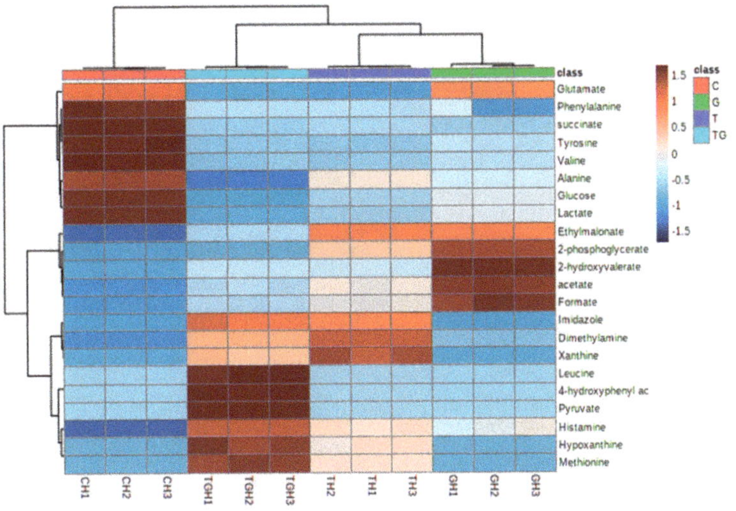

Figure 12. Unsupervised hierarchical clustering and heat map analysis of extracellular metabolites released from HCT-116 under hypoxic conditions.

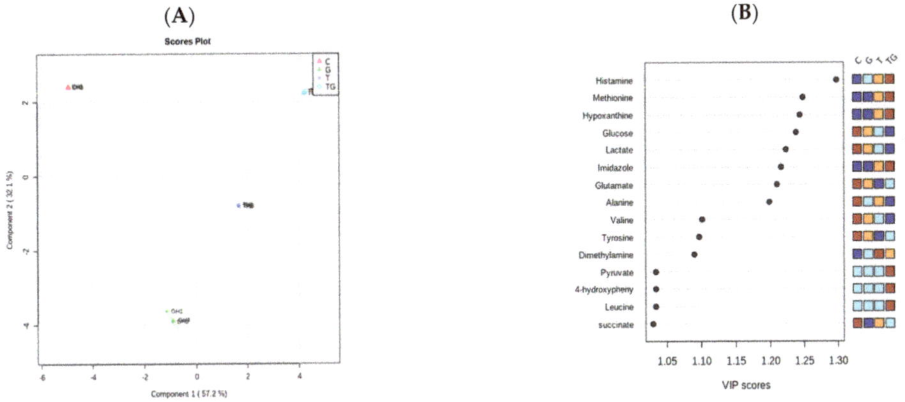

Figure 13. Supervised partial least squares discriminant analysis (PLS-DA) of the metabolite profiling of HCT-166 cells treated with Terr (T), gemcitabine (G), and Terr + GCB (TG) under hypoxic conditions. (**A**) Two-dimensional score plot for HCT-116 cells treatments. (**B**) VIP score plot for treatments of HCT-116 cells.

The same multivariate analysis methods (hierarchical cluster analysis and partial least squares discriminant analysis) were applied to assess the difference in the metabolites released from HCT-116 cells under both normoxic and hypoxic conditions. The heat map showed the difference in metabolites' concentration between hypoxic and normoxic treatments (Figure 14). Moreover, PLSD-A separated the hypoxic metabolites and normoxic metabolites into two clusters (Figure 15A). The most significantly different metabolites between hypoxic and normoxic cells were determined using VIP scores (Figure 15B).

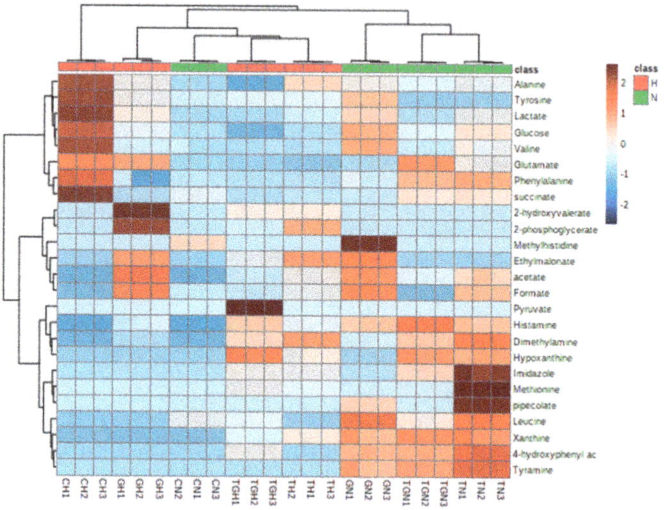

Figure 14. Unsupervised hierarchical clustering and heat map analysis of extracellular metabolites released from HCT-116 comparing normoxic and hypoxic conditions.

(A) (B)

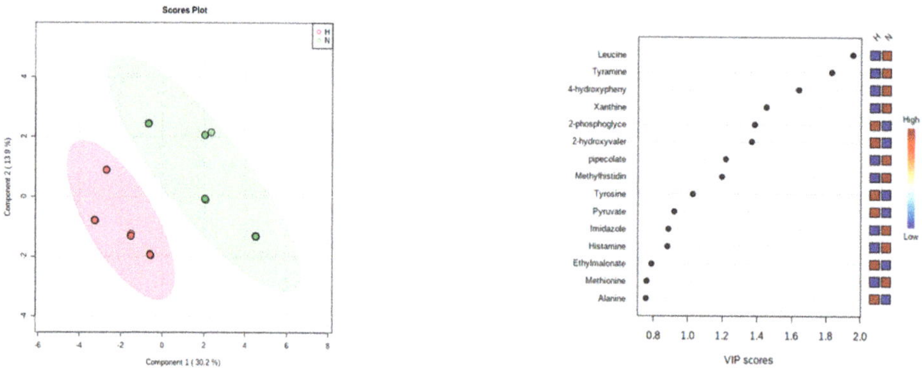

Figure 15. Supervised partial least squares discriminant analysis (PLS-DA) of the metabolite profiling of HCT-166 cells treated with Terr (T), gemcitabine(G), and Terr + GCB (TG) under normoxic and hypoxic conditions. (**A**) Two-dimensional score plot for HCT-116 cells treatments. (**B**) VIP score plot for treatments of HCT-116 cells.

3. Discussion

The need for novel anticancer treatment is an emerging matter as cancer is a major health problem worldwide, and the currently available chemotherapeutic options are becoming increasingly susceptible to resistance [1,2]. Terrein (Terr) is a bioactive marine metabolite isolated from the fungal strain of *Penicillium* species SF-7181 and *Aspergillus terreus* [39]. It exerts its activity via different mechanisms such as angiogenesis inhibition, cell cycle regulation, apoptosis, and autophagy induction [6–9]. However, the exact role of Terr as an antitumor agent remains unclear. Gemcitabine (GCB) is generally considered to be the drug of choice for pancreatic adenocarcinoma and has also been used in other types of cancers such as colorectal cancer; however, its major drawback is its susceptibility to both intrinsic and acquired chemoresistance [40]. Therefore, improving the therapeutic effect of GCB is crucial.

In the current work, Terr showed relatively high IC_{50}s against three different colorectal cancer cell lines under normoxic and hypoxic conditions. Yet, several similar studies showed that Terr induces cancer cell death; however, it is highly cell type-dependent, as well as dose and time dependent [12,16]. On the other hand, GCB showed much higher anticancer potencies against the same set of cell lines compared to Terr under normoxic and hypoxic conditions. The combination indices for GCB with Terr were indicative of a synergistic interaction in HCT-116 and SW620; however, it was antagonistic in HT-29 in both normoxia and hypoxia.

To explain the characteristics of the interaction between GCB and Terr treatments as well as their combination, apoptosis, autophagy, and cell cycle interference were assessed using the flowcytometry technique. This allowed us to determine if the cell death was due to programmed/non-programmed (apoptosis vs necrosis) cell death, autophagy induction/suppression, or simply interference with cell cycle progression (antiproliferative properties). It is worth mentioning that the role of autophagy in cancer is very controversial as it is often referred to as either inducing cell death by suppressing tumorigenesis or facilitating tumorigenesis [41,42]. Apoptotic cell death in GCB singular treatment was significantly higher compared to control untreated cells in all cell lines assessed as expected. Yet, after 24 h of treatment in HCT-116 and SW620 cell lines, the combination treatment significantly increased apoptotic cell death compared to other treatments. The opposite was evident in the HT-29 cell line under both normoxic and hypoxic conditions. Both cases are in alignment with combination indices calculated in these cell lines (synergistic versus antagonistic). It is worth mentioning that apoptosis was induced, and autophagy was suppressed after 24 h in HCT-116 and SW620 cells under combination conditions (GCB + Terr). This might be attributed to suppressed autophagy leading to apoptosis and cell death. Furthermore, the gene expression profile of apoptosis and autophagy genes confirmed this pattern. In the previous literature, cells such as HCT-116 can proceed via apoptosis and suppress autophagy through the regulation of certain pathways, such as PI3K/AKT/mTOR [43]. It is prevalent in colorectal cancer cells and exhibits the antagonistic effect between autophagy and apoptosis as a survival mechanism due to crucial environmental factors [44]. Concerning cell cycle analysis, all cell lines that were tested showed a significant increase in the S phase population after 24 and 48 h of treatment with GCB, while Terr induced a significant increase in G_2/M phase arrest in the same set of cell lines. Previous studies showed similar findings and have shown that GCB induces S-phase cell cycle arrest and regulates cell cycle-related proteins, while Terr induces G_2/M phase cell cycle arrest [45,46].

RT-qPCR analysis for single and combined treatment was conducted for apoptosis-related genes (BCL2, BIRC5, TP53, and FOXO3), autophagy-related genes (ATG5 and Beclin-1), cell cycle-related genes (CCND1, CDK4, and MCM7), cellular growth-related genes (AKT1, TGF-B1, HIF1-a, and PRKDC) and cellular proliferation-related genes (PCNA and RAD18). Most of the apoptotic genes that were studied were under-expressed in all cell lines that were tested for all treatments under both normoxic and hypoxic conditions. The anti-apoptotic gene, BCL2, which plays a role in programmed cell death as an antiapoptotic protein [47], was over-expressed when HCT-116 cells were treated with Terr alone under normoxic conditions; however, it was significantly downregulated when the cells were treated with the combination treatment (GCB + Terr). This observation is supported by similar combination studies that showed the downregulation of BCL2 when HCT-116 was treated with combination treatment [48].

Similar results were observed with the autophagy-regulating genes; they were found to be under-expressed in all cell lines tested for all treatments under both normoxic and hypoxic conditions. ATG5, which regulates autophagy by forming autophagic vesicles and controls mitochondrial quality after oxidative damage, was over-expressed when HCT-116 cells were treated with Terr under normoxic conditions. According to the literature, Beclin-1 is a gene that plays a significant role in regulating autophagy, proliferation, and apoptosis in colorectal cancer cells (HCT-116 and SW620). The inhibition of Beclin-1 leads to the suppression of autophagy and proliferation as well as the promotion of apoptosis, which are observed in the results [49]. The exact mechanism by which Beclin-1 promotes apoptosis and suppresses autophagy, however, remains unclear.

Concerning cell cycle regulator genes, they followed the same pattern, and most of the genes that were tested were under-expressed in all cell lines that were tested for all treatments under both normoxic and hypoxic conditions. The only exception was CDK4, which is an important gene that encodes proteins for the cell cycle G1 phase progression. CDK4 was only over-expressed when the cells were treated with Terr under normoxic conditions in the HCT-116 cell line. However, the combination treatment (GCB + Terr) led to the downregulation of CDK4 in HCT-116. CDK4 is found to be amplified in colorectal cancer cells compared to normal cells [50,51], and evidence has shown that inhibiting certain CDKs such as CDK1, 2, 4/6, and 9 is useful in enhancing colorectal cancer cell (HCT-116) death [52]. According to previous studies, the inhibition of CDKs has also proven to be beneficial in suppressing colorectal cancer cells from proliferation through cell cycle arrests, and in some cases, can also lead to apoptotic cell death [53].

Most of the cellular growth genes that were tested were observed to be under-expressed in all cell lines in response to treatment. The AKT1 gene was of great interest and a known target for terrein, as it works by regulating various processes such as metabolism, proliferation, cell survival, and cell growth. In HCT-116, AKT1 was over-expressed when the cells were treated with Terr in normoxia and under-expressed for other treatments in the same condition. The contrary was evident in hypoxia. Hypoxia plays a major role in tumor cell behavior and the way it responds to treatment [54]. Stegeman et. al. showed that hypoxia stimulates AKT expression and activation in vivo and in vitro. The current study might prove that terrein-induced pAKT inhibition can overcome the influence of hypoxia and diminish cell survival in hypoxic cells, rather than in normoxic conditions [55]. In HT-29, AKT1 was over-expressed after treatment with GCB in normoxia, which might explain the antagonistic interaction with terrein. In SW620, AKT1 was over-expressed after single as well as combination treatment under both normoxic and hypoxic conditions. Still, this can explain the very high resistance fraction (R-value) to treatment in this cell line. AKT is known to regulate cellular proliferation through the degradation of CDK inhibitors, therefore promoting cell cycle progression and inhibiting apoptosis by inactivating pro-apoptotic molecules [56,57]. As a result, AKT plays a vital role as a signaling biomarker, which integrates many potential oncogenic signals [58]. On the other hand, recent studies showed that overactivation of AKT can also increase cell resistance to oxidative stress and allow cells to be more viable in high reactive oxygen species (ROS) conditions [59]. On the other hand, several studies showed that AKT nucleus translocation can induce cell death via apoptin due to the activity of some anticancer drugs [60]. This proves that the overall outcome of AKT activation or inhibition depends on the signaling context as well as the topological characteristics [58]. Most of the cell proliferation genes that were studied were under-expressed in response to single and combined treatment. Yet, this might indicate the antiproliferative effect rather than the cytotoxic properties of treatments under investigation, especially if we noticed the high resistance fraction in all treatment conditions.

Only HCT-116 cells showed a synergistic versus antagonistic interaction between terrein and GCB under normoxic versus hypoxic conditions, respectively. Yet, it was further assessed via a metabolomic study using ^{1}H-NMR comparative analysis and profiling under normoxic versus hypoxic conditions. Lately, many studies have been trying to investigate colorectal cancer metabolic profiling compared to normal tissues [61]. Hirayama et al. found that due to hypoxia, there is a significant variation in the energy metabolism in the colorectal cancer tissues [62]. Denkert et al. reported that intermediates of the lipids and tricarboxylic acid (TCA) cycle were downregulated in tumor tissue, while urea cycle metabolites, purines, pyrimidines, and amino acids were upregulated compared to normal tissue [63]. In the current study, the extracellular metabolites of the different treatment conditions (single versus combination treatment) were found to be significantly different from one another when compared under different oxygen conditions. According to the VIP scores, the metabolites that were believed to exhibit a significant role in the metabolic shift when comparing the different oxygen conditions as well as when testing normoxia and hypoxia per se were leucine, tyramine, 4-hydroxyphenyl acetate, xanthine, and tyrosine. A study carried out by Hirayama et al. showed that tumors can have a tumor-specific metabolism that grants them more prevalent proliferation, at the same time keeping some metabolic characteristics of the tissues from which they originated. To put it in another way, cancer cells are progressed via metabolic adaptation that includes the upregulation of glucose consumption and increase in amino acids, while preserving the tissue-specific dependency of aerobic respiration characterized by TCA intermediate and nucleotide levels [62]. Another study performed by Frezza et al. suggests that hypoxic HCT116 cells could depend on catabolic processes to make up the energetic defect created by the loss of mitochondrial activity and that cannot be made up by the increased glycolytic flux [64]. Herein, these metabolic profiling data are in alignment with autophagy and apoptosis results. Mitochondrial membrane and integrity are among the early flip points in apoptosis/autophagy balance [44,65]. The significantly different profiles of energy metabolites in both normoxic and hypoxic conditions mirrored the different profiles in autophagy/apoptosis balance in the HCT-116 cells [66]. In our study, according to the PLS-DA, there was no overlap in metabolites clustering among the different treatments. Further analysis was conducted via a metabolite–gene–disease interaction network, which illustrated that some of the genes that were tested were directly linked to colorectal cancer in alignment with the metabolites that were found earlier, such as AKT1 and BCL2.

In conclusion, terrein possesses a controversial role in influencing the anticancer properties of gemcitabine in colorectal cancer cells under normoxic versus hypoxic conditions ranging from antagonism to synergism. However, this influence is evident via significant changes in cell proliferation patterns, cell cycle progression, apoptosis, and autophagy with mirrored confirmatory gene expression profiles. AKT1 seems crucial in all these processes in terms of activation and expression. On top of all these, the metabolic profile of energy and mitochondrial function was significantly and differentially affected by single and combined treatments under normoxia versus hypoxic conditions. It is recommended to further study these effects under more complicated tissue culture conditions such as a 3D culture system or even in vivo animal models to add the dimension of tissue penetration to the current research outcome.

4. Materials and Methods

4.1. Chemicals and Drugs

Gemcitabine (GCB), terrein (Terr), and sulforhodamine-B were purchased from Sigma Chemical Co. (St. Louis, MO, USA). RPMI-164 media, DMEM, fetal bovine serum, and other cell culture materials were purchased from ATCC (Houston, TX, USA). Other reagents used were of the highest analytical grade.

4.2. Cell Culture

Colorectal cancer cell lines HCT-116 (Accession number: CRL-3504), HT-29 (Accession number: HTB-38), and SW620 (Accession number: CVCL_0547) were obtained from ATCC (Houston, TX, USA). Cells were maintained in RPMI-1640 and DMEM supplemented with 100 μg/mL streptomycin, 100 units/mL penicillin, and 10% heat-inactivated fetal bovine serum in a humidified, 5% (*v/v*) CO_2 atmosphere at 37 °C [67]. All cell lines and cell line materials were confirmed to be mycoplasma free.

4.3. Cytotoxicity Assay

The cytotoxicity of GCB and Terr was tested against HCT-116, HT-29, and SW620 cells using the SRB assay as previously described. Exponentially growing cells were harvested using 0.25% Trypsin-EDTA and plated in 96-well plates, at concentrations of 1000–2000 cells/well. Cells were exposed to GCB, Terr, and GCB + Terr for 72 h (normoxia and hypoxia) and subsequently fixed with TCA (10%) for 1 h at 4 °C. After the plates were washed several times, the cells were exposed to a 0.4% SRB solution for 10 min in the dark and subsequently washed with 1% glacial acetic acid. After leaving the plates to dry overnight, Tris-HCl was used to dissolve the SRB-stained cells, and their color intensity was measured at 540 nm using a microplate reader [68].

4.4. Data Analysis

The viability dose–response curve of the compounds was analyzed using the Emax model (Equation (1)).

$$\% \; Cell\ viability = (100 - \mathrm{R}) \times \left(1 - \frac{[\mathrm{D}]^m}{K_d^m + [\mathrm{D}]^m}\right) + \mathrm{R} \qquad (1)$$

where R is the residual unaffected fraction (the resistance fraction) which is deduced from fitting concentration versus viability on the E_{max} equation (Equation (1)) described above, [D] is the drug concentration used, K_d is the drug concentration that produces a 50% reduction in the maximum inhibition rate and m is a Hill-type coefficient. IC_{50} was defined as the drug concentration required to reduce optical density to 50% of that of the control (i.e., $K_d = IC_{50}$ when R = 0 and $E_{max} = 100 - R$) [68].

4.5. Apoptosis

Annexin V conjugates allow for the identification of cell surface changes that occur early during the apoptotic process using flow cytometry. Early in the apoptotic process, phosphatidylserine emerges from within the cytoplasmic membrane and becomes exposed on the cell surface, which is thought to be important for macrophage recognition of cells undergoing apoptosis. The binding of Annexin V to phosphatidylserine is calcium-dependent, reversible, and specific with a K_d of approximately $5 \times 10 - 10$ M [69].

4.6. Assessment of Active Caspase-3 Concentration

To assess the effect of GCB, terrein, and their combination on apoptosis, the active caspase-3 concentration was measured using a Quantikine® caspase-3 ELISA Kit (R&D Systems, Inc., Minneapolis, MN, USA). Briefly, the cells were exposed to the predetermined IC_{50}s of test compounds (single or combined treatments) or drug-free media (control group) for 24 h. Cells were harvested and washed twice with PBS, then incubated with the biotin-ZVKD-fmk inhibitor for 1 h. Cells were transferred into the wells of a microplate pre-coated with a monoclonal antibody specific for caspase-3. Following a wash to remove any unbound substances, streptavidin conjugated to horseradish peroxidase was added to the wells and bound to the biotin on the inhibitor. Following a wash to remove any unbound streptavidin–HRP, a substrate solution was added to the wells. The enzyme reaction yields a blue product that turned yellow when a stop solution was added. The optical density of each well was determined within 30 min, using a microplate reader set to 450 nm with a wavelength correction at 540 nm or 570 nm. The concentrations of active caspase-3 were calculated from a standard curve constructed with known concentrations of active caspase-3. Caspase concentration was expressed as ng/mg protein. Proteins were determined by the Bradford method using purified bovine serum albumin as a standard protein.

4.7. Autophagy

Acridine orange (AO) is a cell-permeable green fluorophore that can become hydronated and consequently absorbed by acidic vesicular organelles. Its metachromatic shift from green to red fluorescence is highly dependent on its concentration, which causes AO to fluoresce from green to red in acidic organelles, such as lysosomes. Lysosomes tend to increase in number and volume when autophagy occurs; AO staining is a quick and reliable method for the assessment of autophagy [70].

4.8. Cell Cycle Analysis

Propidium iodide (PI) is a dye that can be used to stain DNA content by intercalating into a double-stranded nucleic acid, producing a highly fluorescent signal when excited at 488 nm with a broad emission centered around 600 nm. The stoichiometric nature of PI ensures accurate quantification of DNA content and reveals the distribution of cells in the G1, S, and G2 cell cycle stages, and even in the sub-G1 cell death stage, which is characterized by DNA fragmentation. Since PI can also bind to double-stranded RNA, it is necessary to treat the cells with RNase for optimal DNA resolution [71].

4.9. Gene Expression Analysis

To assess the gene expression of GCB and Terr and their combination, total RNA extraction from cells was performed using the easy-BLUE Kit® (Qiagen Inc., Valencia, CA, USA). Reverse transcription was undertaken to construct a cDNA library from different treatments using a High-Capacity cDNA Reverse Transcription Kit (Applied Biosystems, Foster City, CA, USA). The archived cDNA libraries were then subjected to quantitative real-time PCR reactions [72] using SYBR-green fluorophore (Fermentas Inc., Glen Burnie, MD, USA). Primer sequences were as shown in Table 3.

Table 3. Primer sequences of target genes were used for the qPCR analysis.

Classification	Primer	Direction	Code
Apoptosis regulators	BCL2	Forward	GAT-TGT-GGC-CTT-CTT-TGA-G
		Reverse	CAA-ACT-GAG-CAG-AGT-CTT-C
	BIRC5	Forward	AGG-ACC-ACC-GCA-TCT-CTA-CAT
		Reverse	AAG-TCT-GGC-TCG-TTC-TCA-GTG
	TP53	Forward	TTC-CTC-CAA-CCA-AGA-ACC-AGA
		Reverse	GCT-CAG-TAG-GTG-ACT-CTT-CAC-T
	FOXO3	Forward	ACG-GCT-GAC-TGA-TAT-GGC-AG
		Reverse	CGT-GAT-GTT-ATC-CAG-CAG-GTC
Autophagy	ATG5	Forward	AGA-AGC-TGT-TTC-GTC-CTG-TGG
		Reverse	AGG-TGT-TTC-CAA-CAT-TGG-CTC
	Beclin1	Forward	AGC-TGC-CGT-TAT-ACT-GTT-CTG
		Reverse	ACT-GCC-TCC-TGT-GTC-TTC-AAT-CTT
Cell cycle regulators	CCND1	Forward	TGT-TCG-TGG-CCT-CTA-AGA-TGA-AG
		Reverse	AGG-TTC-CAC-TTG-AGC-TTG-TTC-AC
	CDK4	Forward	CTG-GTG-TTT-GAG-CAT-GTA-GAC-C
		Reverse	AAA-CTG-GCG-CAT-CAG-ATC-CTT
	MCM7	Forward	GGG-CTC-CAG-ATT-CAT-CAA-AT
		Reverse	ATA-CCA-GTG-ACG-CTG-ACG-TG
Cell growth	AKT1	Forward	GGA-TGT-GGA-CCA-ACG-TGA-G
		Reverse	AGC-GGA-TGA-TGA-AGG-TGT-TG
	TGF-β1	Forward	GGT-ACC-TGA-ACC-CGT-GTT-GCT
		Reverse	TGT-TGC-TGT-ATT-TCT-GGT-ACA-GCT-C
	HIF1-a	Forward	GAA-CGT-CGA-AAA-GAA-AAG-TCT-CG
		Reverse	CCT-TAT-CAA-GAT-GCG-AAC-TCA-CA
	PRKDC	Forward	GAC-ATC-TCC-TGA-GCT-CTG-AC
		Reverse	CTC-TTG-TTC-CCC-AAC-AGT-CT
Cell proliferation	PCNA	Forward	CGG-ATA-CCT-TGG-CGC-TAG-TA
		Reverse	TCT-CGG-CAT-ATA-CGT-GCA-AA
	RAD18	Forward	CTC-AGT-GTC-CAA-CTT-GCT-GTG
		Reverse	GAA-GAG-GAA-GAA-GCA-GGA-GAT

4.10. Metabolomics Analysis

4.10.1. Sample Processing for NMR Spectroscopy

The lyophilized extracellular cell media were mixed with the internal reference 3-(Trimethylsilyl)-1- propane sulfonic acid-d6 sodium salt (DSS-d6, dissolved in methanol-d4,

10 mM) to reach the final concentration of 1 mM. From each sample, 600 μL was placed in 5 mm NMR tubes for NMR analyses. Three biological replicates from each sample were analyzed [73].

4.10.2. NMR Measurement

The ^{1}H-NMR experiments were carried out using a Bruker NMR spectrophotometer (Bruker Biospin GmbH, Karlsruhe, Germany) operating at 600 MHz and a temperature of 25 °C.

4.10.3. NMR Spectral Processing

Metabolite annotation was conducted using ChenomX NMR Suite 8.6 (ChenomX Inc., Edmonton, AB, Canada), and phase and baseline corrections were performed initially. The identification was then verified by Human Metabolome Database (http://www.hmdb.ca/) accessed on 1 April 2021, Madison Metabolomics Consortium Database (http://mmcd.nmrfam.wisc.edu) accessed on 1 April 2021. The metabolites with corresponding concentrations were subjected to multivariate analysis [73].

4.10.4. Multivariate Analysis

Metabolite data from all replicates were then imported to MetaboAnalyst 5.0 platform (http://www.metaboanalyst.ca/) accessed on 5 May 2021, for multivariate analysis [74]. Hierarchical cluster analyses were performed to visualize the grouping resulting from the difference between the metabolites released from the HCT-116 cells treated with terrein, GCB, and GCB + Terr under both normoxic and hypoxic conditions. Partial least squares discriminant analysis (PLS-DA) was performed to visualize the grouping tendencies in the samples with many variables. Significant metabolites were determined from variable importance in projection scores (VIP) values. VIP values above 1.00 were significant [75].

4.11. Statistical Analysis

Data are presented as mean ± SD. Analysis of variance (ANOVA) with Tukey's Honest Significance post hoc test was carried out to test for significance using SPSS® for Windows, Version 17.0.0. $p < 0.05$ was used as the cut-off value for significance.

5. Conclusions

In conclusion, while combining terrein with gemcitabine did not improve gemcitabine's resistance fraction, it did however improve its cytotoxic effect against HCT-116 and SW620 cells. The current work focused on HCT-116 as it was the cell line of interest due to the opposite effect of the combination treatment (GCB + Terr) on the cells when treated under normoxic versus hypoxic conditions. Expression of certain genes was affected due to the variable treatment action, more specifically BCL2, Beclin-1, CDK4, and AKT1. This urged us to further investigate the metabolic profile of the HCT-116 cell line after treatment with terrein, gemcitabine, and their combination, and promising results supported our findings. A difference between the metabolites found under each oxygen condition was found, and this could explain the synergistic effect in normoxia versus the antagonistic effect in hypoxia.

Supplementary Materials: The following supporting information can be downloaded at: https://www.mdpi.com/article/10.3390/md21050271/s1, Figure S1: The effect of terrein on the cytotoxicity of GCB (A) in normoxia and (B) in hypoxia in HCT-116 cell line. The cells were exposed to serial dilution of terrein (○), GCB (•) or Terr/GCB combination (▾) for 72 h. The cell viability was determined using SRB assay; Figure S2: The effect of terrein on the cytotoxicity of GCB (A) in normoxia and (B) in hypoxia in HT-29 cell line. The cells were exposed to serial dilution of terrein (○), GCB (•) or Terr/GCB combination (▾) for 72 h. The cell viability was determined using SRB assay; Figure S3: The effect of terrein on the cytotoxicity of GCB (A) in normoxia and (B) in hypoxia in SW620 cell line. The cells were exposed to serial dilution of terrein (○), GCB (•) or Terr/GCB combination (▾) for 72 h. The cell viability was determined using SRB assay.

Author Contributions: Conceptualization, A.M.A.-A., S.S.E. and N.S.A.; methodology, R.K.A., D.Y.A.S., S.S.E. and A.E.G.; software, R.K.A., D.Y.A.S., N.S.A. and A.E.G.; validation, A.M.A.-A., S.S.E. and N.S.A.; formal analysis, R.K.A. and A.M.A.-A.; investigation, R.K.A., D.Y.A.S., S.S.E. and N.S.A.; resources, A.M.A.-A., S.S.E. and N.S.A.; data curation, R.K.A., D.Y.A.S. and N.S.A.; writing—original draft preparation, R.K.A.; writing—review and editing, A.M.A.-A., S.S.E. and N.S.A.; supervision, A.M.A.-A.; project administration, A.M.A.-A.; funding acquisition, A.M.A.-A. All authors have read and agreed to the published version of the manuscript.

Funding: This research including the APC was funded by Gulf Medical University, Ajman, UAE, grant number GMU/COP/GR/209-10/004.

Institutional Review Board Statement: Ethical review and approval were not applicable for studies not involving humans or animals.

Data Availability Statement: All supporting raw data are available upon request.

Acknowledgments: The authors would like to thank Thumbay Research Institute for Precision Medicine for providing all the technical support required during the study.

Conflicts of Interest: The authors declare no conflict of interest.

References

1. Sung, H.; Ferlay, J.; Siegel, R.L.; Laversanne, M.; Soerjomataram, I.; Jemal, A.; Bray, F. Global cancer statistics 2020: GLOBOCAN estimates of incidence and mortality worldwide for 36 cancers in 185 countries. *CA. Cancer J. Clin.* **2021**, *71*, 209–249. [CrossRef] [PubMed]
2. Nikolaou, M.; Pavlopoulou, A.; Georgakilas, A.G.; Kyrodimos, E. The challenge of drug resistance in cancer treatment: A current overview. *Clin. Exp. Metastasis* **2018**, *35*, 309–318. [CrossRef] [PubMed]
3. Gewirtz, D.A.; Bristol, M.L.; Yalowich, J.C. Toxicity issues in cancer drug development. *Curr. Opin. Investig. Drugs* **2010**, *11*, 612–614.
4. El-Sawy, H.S.; Al-Abd, A.M.; Ahmed, T.A.; El-Say, K.M.; Torchilin, V.P. Stimuli-responsive nano-architecture drug-delivery systems to solid tumor micromilieu: Past, present, and future perspectives. *ACS Nano* **2018**, *12*, 10636–10664. [CrossRef] [PubMed]
5. Nakazawa, M.S.; Keith, B.; Simon, M.C. Oxygen availability and metabolic adaptations. *Nat. Rev. Cancer* **2016**, *16*, 663–673. [CrossRef] [PubMed]
6. Rajabi, M.; Mousa, S.A. The role of angiogenesis in cancer treatment. *Biomedicines* **2017**, *5*, 34. [CrossRef] [PubMed]
7. Bai, J.; Li, Y.; Zhang, G. Cell cycle regulation and anticancer drug discovery. *Cancer Biol. Med.* **2017**, *14*, 348. [CrossRef]
8. Pfeffer, C.M.; Singh, A.T.K. Apoptosis: A target for anticancer therapy. *Int. J. Mol. Sci.* **2018**, *19*, 448. [CrossRef]
9. Li, X.; Xu, H.; Liu, Y.; An, N.; Zhao, S.; Bao, J. Autophagy modulation as a target for anticancer drug discovery. *Acta Pharmacol. Sin.* **2013**, *34*, 612–624. [CrossRef]
10. Liao, W.-Y.; Shen, C.-N.; Lin, L.-H.; Yang, Y.-L.; Han, H.-Y.; Chen, J.-W.; Kuo, S.-C.; Wu, S.-H.; Liaw, C.-C. Asperjinone, a nor-neolignan, and terrein, a suppressor of ABCG2-expressing breast cancer cells, from thermophilic *Aspergillus terreus*. *J. Nat. Prod.* **2012**, *75*, 630–635. [CrossRef]
11. Arakawa, M.; Someno, T.; Kawada, M.; Ikeda, D. A new terrein glucoside, a novel inhibitor of angiogenin secretion in tumor angiogenesis. *J. Antibiot.* **2008**, *61*, 442. [CrossRef] [PubMed]
12. Shibata, A.; Ibaragi, S.; Mandai, H.; Tsumura, T.; Kishimoto, K.; Okui, T.; Hassan, N.M.M.; Shimo, T.; Omori, K.; Hu, G.-F. Synthetic terrein inhibits progression of head and neck cancer by suppressing angiogenin production. *Anticancer Res.* **2016**, *36*, 2161–2168. [PubMed]
13. Chen, Y.-F.; Wang, S.-Y.; Shen, H.; Yao, X.-F.; Zhang, F.-L.; Lai, D. The marine-derived fungal metabolite, terrein, inhibits cell proliferation and induces cell cycle arrest in human ovarian cancer cells. *Int. J. Mol. Med.* **2014**, *34*, 1591–1598. [CrossRef]
14. Zhang, F.; Mijiti, M.; Ding, W.; Song, J.; Yin, Y.; Sun, W.; Li, Z. (+)-Terrein inhibits human hepatoma Bel-7402 proliferation through cell cycle arrest. *Oncol. Rep.* **2015**, *33*, 1191–1200. [CrossRef] [PubMed]
15. Porameesanaporn, Y.; Uthaisang-Tanechpongtamb, W.; Jarintanan, F.; Jongrungruangchok, S.; Thanomsub Wongsatayanon, B. Terrein induces apoptosis in HeLa human cervical carcinoma cells through p53 and ERK regulation. *Oncol. Rep.* **2013**, *29*, 1600–1608. [CrossRef] [PubMed]
16. Wu, Y.; Zhu, Y.; Li, S.; Zeng, M.; Chu, J.; Hu, P.; Li, J.; Guo, Q.; Lv, X.; Huang, G. Terrein performs antitumor functions on esophageal cancer cells by inhibiting cell proliferation and synergistic interaction with cisplatin. *Oncol. Lett.* **2017**, *13*, 2805–2810. [CrossRef]
17. Kim, K.-W.; Kim, H.J.; Sohn, J.H.; Yim, J.H.; Kim, Y.-C.; Oh, H. Terrein suppressed lipopolysaccharide-induced neuroinflammation through inhibition of NF-κB pathway by activating Nrf2/HO-1 signaling in BV2 and primary microglial cells. *J. Pharmacol. Sci.* **2020**, *143*, 209–218. [CrossRef] [PubMed]
18. Bours, V.; Bentires-Alj, M.; Hellin, A.-C.; Viatour, P.; Robe, P.; Delhalle, S.; Benoit, V.; Merville, M.-P. Nuclear factor-κB, cancer, and apoptosis. *Biochem. Pharmacol.* **2000**, *60*, 1085–1089. [CrossRef] [PubMed]

19. Xia, L.; Tan, S.; Zhou, Y.; Lin, J.; Wang, H.; Oyang, L.; Tian, Y.; Liu, L.; Su, M.; Wang, H. Role of the NFκB-signaling pathway in cancer. *Onco. Targets. Ther.* **2018**, *11*, 2063. [CrossRef]
20. Zhao, C.; Guo, L.; Wang, L.; Zhu, G.; Zhu, W. Improving the yield of (+)-terrein from the salt-tolerant *Aspergillus terreus* PT06-2. *World J. Microbiol. Biotechnol.* **2016**, *32*, 77. [CrossRef]
21. Abuhijjleh, R.K.; Shabbir, S.; Al-Abd, A.M.; Jiaan, N.H.; Alshamil, S.; El-Labbad, E.M.; Khalifa, S.I. Bioactive marine metabolites derived from the Persian Gulf compared to the Red Sea: Similar environments and wide gap in drug discovery. *PeerJ* **2021**, *9*, e11778. [CrossRef] [PubMed]
22. Burris, H.A., 3rd; Moore, M.J.; Andersen, J.; Green, M.R.; Rothenberg, M.L.; Modiano, M.R.; Christine Cripps, M.; Portenoy, R.K.; Storniolo, A.M.; Tarassoff, P. Improvements in survival and clinical benefit with gemcitabine as first-line therapy for patients with advanced pancreas cancer: A randomized trial. *J. Clin. Oncol.* **1997**, *15*, 2403–2413. [CrossRef] [PubMed]
23. Neesse, A.; Michl, P.; Frese, K.K.; Feig, C.; Cook, N.; Jacobetz, M.A.; Lolkema, M.P.; Buchholz, M.; Olive, K.P.; Gress, T.M. Stromal biology and therapy in pancreatic cancer. *Gut* **2011**, *60*, 861–868. [CrossRef] [PubMed]
24. Jia, Y.; Xie, J. Promising molecular mechanisms responsible for gemcitabine resistance in cancer. *Genes Dis.* **2015**, *2*, 299–306. [CrossRef] [PubMed]
25. Chen, M.; Xue, X.; Wang, F.; An, Y.; Tang, D.; Xu, Y.; Wang, H.; Yuan, Z.; Gao, W.; Wei, J. Expression and promoter methylation analysis of ATP-binding cassette genes in pancreatic cancer. *Oncol. Rep.* **2012**, *27*, 265–269. [PubMed]
26. Zinzi, L.; Contino, M.; Cantore, M.; Capparelli, E.; Leopoldo, M.; Colabufo, N.A. ABC transporters in CSCs membranes as a novel target for treating tumor relapse. *Front. Pharmacol.* **2014**, *5*, 163. [CrossRef]
27. Quint, K.; Tonigold, M.; Di Fazio, P.; Montalbano, R.; Lingelbach, S.; Rueckert, F.; Alinger, B.; Ocker, M.; Neureiter, D. Pancreatic cancer cells surviving gemcitabine treatment express markers of stem cell differentiation and epithelial-mesenchymal transition. *Int. J. Oncol.* **2012**, *41*, 2093–2102. [CrossRef]
28. Zheng, C.; Jiao, X.; Jiang, Y.; Sun, S. ERK1/2 activity contributes to gemcitabine resistance in pancreatic cancer cells. *J. Int. Med. Res.* **2013**, *41*, 300–306. [CrossRef]
29. Arlt, A.; Gehrz, A.; Müerköster, S.; Vorndamm, J.; Kruse, M.-L.; Fölsch, U.R.; Schäfer, H. Role of NF-κ B and Akt/PI3K in the resistance of pancreatic carcinoma cell lines against gemcitabine-induced cell death. *Oncogene* **2003**, *22*, 3243–3251. [CrossRef]
30. Pan, X.; Arumugam, T.; Yamamoto, T.; Levin, P.A.; Ramachandran, V.; Ji, B.; Lopez-Berestein, G.; Vivas-Mejia, P.E.; Sood, A.K.; McConkey, D.J. Nuclear factor-κB p65/relA silencing induces apoptosis and increases gemcitabine effectiveness in a subset of pancreatic cancer cells. *Clin. Cancer Res.* **2008**, *14*, 8143–8151. [CrossRef]
31. Jung, H.; Kim, J.S.; Kim, W.K.; Oh, K.J.; Kim, J.M.; Lee, H.J.; Han, B.S.; Kim, D.S.; Seo, Y.-S.; Lee, S.C. Intracellular annexin A2 regulates NF-κ B signaling by binding to the p50 subunit: Implications for gemcitabine resistance in pancreatic cancer. *Cell Death Dis.* **2015**, *6*, e1606. [CrossRef]
32. Yu, Y.; Wang, J.; Xia, N.; Li, B.; Jiang, X. Maslinic acid potentiates the antitumor activities of gemcitabine in vitro and in vivo by inhibiting NF-κB-mediated survival signaling pathways in human gallbladder cancer cells. *Oncol. Rep.* **2015**, *33*, 1683–1690. [CrossRef] [PubMed]
33. Simon, P.O., Jr.; McDunn, J.E.; Kashiwagi, H.; Chang, K.; Goedegebuure, P.S.; Hotchkiss, R.S.; Hawkins, W.G. Targeting AKT with the proapoptotic peptide, TAT-CTMP: A novel strategy for the treatment of human pancreatic adenocarcinoma. *Int. J. Cancer* **2009**, *125*, 942–951. [CrossRef] [PubMed]
34. Tréhoux, S.; Duchêne, B.; Jonckheere, N.; Van Seuningen, I. The MUC1 oncomucin regulates pancreatic cancer cell biological properties and chemoresistance. Implication of p42–44 MAPK, Akt, Bcl-2 and MMP13 pathways. *Biochem. Biophys. Res. Commun.* **2015**, *456*, 757–762. [CrossRef] [PubMed]
35. Yang, X.L.; Lin, F.J.; Guo, Y.J.; Shao, Z.M.; Ou, Z.L. Gemcitabine resistance in breast cancer cells regulated by PI3K/AKT-mediated cellular proliferation exerts negative feedback via the MEK/MAPK and mTOR pathways. *Onco. Targets. Ther.* **2014**, *7*, 1033.
36. Wang, R.; Cheng, L.; Xia, J.; Wang, Z.; Wu, Q.; Wang, Z. Gemcitabine resistance is associated with epithelial-mesenchymal transition and induction of HIF-1α in pancreatic cancer cells. *Curr. Cancer Drug Targets* **2014**, *14*, 407–417. [CrossRef]
37. De Castro, F.; Stefano, E.; De Luca, E.; Muscella, A.; Marsigliante, S.; Benedetti, M.; Fanizzi, F.P. A NMR-Based Metabolomic Approach to Investigate the Antitumor Effects of the Novel [Pt (η1-C2H4OMe)(DMSO)(phen)]+(phen= 1, 10-Phenanthroline) Compound on Neuroblastoma Cancer Cells. *Bioinorg. Chem. Appl.* **2022**, *2022*, 8932137. [CrossRef]
38. Shammout, O.D.A.; Ashmawy, N.S.; Shakartalla, S.B.; Altaie, A.M.; Semreen, M.H.; Omar, H.A.; Soliman, S.S.M. Comparative sphingolipidomic analysis reveals significant differences between doxorubicin-sensitive and-resistance MCF-7 cells. *PLoS ONE* **2021**, *16*, e0258363. [CrossRef]
39. Raistrick, H.; Smith, G. Studies in the biochemistry of micro-organisms: The metabolic products of *Aspergillus terreus* Thom. A new mould metabolic product—Terrein. *Biochem. J.* **1935**, *29*, 606. [CrossRef]
40. de Sousa Cavalcante, L.; Monteiro, G. Gemcitabine: Metabolism and molecular mechanisms of action, sensitivity and chemoresistance in pancreatic cancer. *Eur. J. Pharmacol.* **2014**, *741*, 8–16.
41. Yun, C.W.; Lee, S.H. The roles of autophagy in cancer. *Int. J. Mol. Sci.* **2018**, *19*, 3466. [CrossRef] [PubMed]
42. Chen, Q.; Kang, J.; Fu, C. The independence of and associations among apoptosis, autophagy, and necrosis. *Signal Transduct. Target. Ther.* **2018**, *3*, 18. [CrossRef] [PubMed]
43. Zheng, Y.; Tan, K.; Huang, H. Retracted: Long noncoding RNA HAGLROS regulates apoptosis and autophagy in colorectal cancer cells via sponging miR-100 to target ATG5 expression 2019. *J. Cell. Biochem.* **2018**, *120*, 3922–3933. [CrossRef] [PubMed]

44. Su, M.; Mei, Y.; Sinha, S. Role of the crosstalk between autophagy and apoptosis in cancer. *J. Oncol.* **2013**, *2013*, 102735. [CrossRef]
45. Kim, D.; Lee, H.; Park, S.; Lee, S.; Ryoo, I.; Kim, W.; Yoo, I.; Na, J.; Kwon, S.; Park, K. Terrein inhibits keratinocyte proliferation via ERK inactivation and G2/Mcell cycle arrest. *Exp. Dermatol.* **2008**, *17*, 312–317. [CrossRef]
46. Namima, D.; Fujihara, S.; Iwama, H.; Fujita, K.; Matsui, T.; Nakahara, M.; Okamura, M.; Hirata, M.; Kono, T.; Fujita, N. The effect of gemcitabine on cell cycle arrest and microRNA signatures in pancreatic cancer cells. *In Vivo* **2020**, *34*, 3195–3203. [CrossRef]
47. Tsujimoto, Y. Role of Bcl-2 family proteins in apoptosis: Apoptosomes or mitochondria? *Genes Cells* **1998**, *3*, 697–707. [CrossRef]
48. Hakata, S.; Terashima, J.; Shimoyama, Y.; Okada, K.; Fujioka, S.; Ito, E.; Habano, W.; Ozawa, S. Differential sensitization of two human colon cancer cell lines to the antitumor effects of irinotecan combined with 5-aza-2′-deoxycytidine. *Oncol. Lett.* **2018**, *15*, 4641–4648. [CrossRef]
49. Liu, L.; Zhao, W.; Yang, X.; Sun, Z.; Jin, H.; Lei, C.; Jin, B.; Wang, H. Effect of inhibiting Beclin-1 expression on autophagy, proliferation and apoptosis in colorectal cancer. *Oncol. Lett.* **2017**, *14*, 4319–4324. [CrossRef]
50. Mastrogamvraki, N.; Zaravinos, A. Signatures of co-deregulated genes and their transcriptional regulators in colorectal cancer. *NPJ Syst. Biol. Appl.* **2020**, *6*, 23. [CrossRef]
51. Jardim, D.L.; Millis, S.Z.; Ross, J.S.; Woo, M.S.; Ali, S.M.; Kurzrock, R. Cyclin Pathway Genomic Alterations Across 190,247 Solid Tumors: Leveraging Large-Scale Data to Inform Therapeutic Directions. *Oncologist* **2021**, *26*, e78–e89. [CrossRef] [PubMed]
52. Squires, M.S.; Feltell, R.E.; Wallis, N.G.; Lewis, E.J.; Smith, D.-M.; Cross, D.M.; Lyons, J.F.; Thompson, N.T. Biological characterization of AT7519, a small-molecule inhibitor of cyclin-dependent kinases, in human tumor cell lines. *Mol. Cancer Ther.* **2009**, *8*, 324–332. [CrossRef] [PubMed]
53. Thoma, O.-M.; Neurath, M.F.; Waldner, M.J. Cyclin-Dependent Kinase Inhibitors and Their Therapeutic Potential in Colorectal Cancer Treatment. *Front. Pharmacol.* **2021**, *12*, 757120. [CrossRef] [PubMed]
54. Adrian, L.H. Hypoxia-A key regulatory factor in tumor growth. *Nat. Rev. Cancer* **2002**, *2*, 38–47.
55. Stegeman, H.; Kaanders, J.H.; Wheeler, D.L.; van der Kogel, A.J.; Verheijen, M.M.; Waaijer, S.J.; Iida, M.; Grénman, R.; Span, P.N.; Bussink, J. Activation of AKT by hypoxia: A potential target for hypoxic tumors of the head and neck. *BMC Cancer* **2012**, *12*, 463. [CrossRef]
56. Manning, B.D.; Cantley, L.C. AKT/PKB signaling: Navigating downstream. *Cell* **2007**, *129*, 1261–1274. [CrossRef]
57. Plas, D.R.; Thompson, C.B. Akt-dependent transformation: There is more to growth than just surviving. *Oncogene* **2005**, *24*, 7435–7442. [CrossRef]
58. Los, M.; Maddika, S.; Erb, B.; Schulze-Osthoff, K. Switching Akt: From survival signaling to deadly response. *Bioessays* **2009**, *31*, 492–495. [CrossRef]
59. Nogueira, V.; Park, Y.; Chen, C.-C.; Xu, P.-Z.; Chen, M.-L.; Tonic, I.; Unterman, T.; Hay, N. Akt determines replicative senescence and oxidative or oncogenic premature senescence and sensitizes cells to oxidative apoptosis. *Cancer Cell* **2008**, *14*, 458–470. [CrossRef]
60. Maddika, S.; Ande, S.R.; Wiechec, E.; Hansen, L.L.; Wesselborg, S.; Los, M. Akt-mediated phosphorylation of CDK2 regulates its dual role in cell cycle progression and apoptosis. *J. Cell Sci.* **2008**, *121*, 979–988. [CrossRef]
61. Fukui, Y.; Itoh, K. A plasma metabolomic investigation of colorectal cancer patients by liquid chromatography-mass spectrometry. *Open Anal. Chem. J.* **2010**, *4*, 1–9. [CrossRef]
62. Hirayama, A.; Kami, K.; Sugimoto, M.; Sugawara, M.; Toki, N.; Onozuka, H.; Kinoshita, T.; Saito, N.; Ochiai, A.; Tomita, M. Quantitative metabolome profiling of colon and stomach cancer microenvironment by capillary electrophoresis time-of-flight mass spectrometry. *Cancer Res.* **2009**, *69*, 4918–4925. [CrossRef] [PubMed]
63. Denkert, C.; Budczies, J.; Weichert, W.; Wohlgemuth, G.; Scholz, M.; Kind, T.; Niesporek, S.; Noske, A.; Buckendahl, A.; Dietel, M. Metabolite profiling of human colon carcinoma–deregulation of TCA cycle and amino acid turnover. *Mol. Cancer* **2008**, *7*, 72. [CrossRef] [PubMed]
64. Frezza, C.; Zheng, L.; Tennant, D.A.; Papkovsky, D.B.; Hedley, B.A.; Kalna, G.; Watson, D.G.; Gottlieb, E. Metabolic profiling of hypoxic cells revealed a catabolic signature required for cell survival. *PLoS ONE* **2011**, *6*, e24411. [CrossRef]
65. Karantza-Wadsworth, V.; Patel, S.; Kravchuk, O.; Chen, G.; Mathew, R.; Jin, S.; White, E. Autophagy mitigates metabolic stress and genome damage in mammary tumorigenesis. *Genes Dev.* **2007**, *21*, 1621–1635. [CrossRef]
66. Jeon, Y.J.; Khelifa, S.; Ratnikov, B.; Scott, D.A.; Feng, Y.; Parisi, F.; Ruller, C.; Lau, E.; Kim, H.; Brill, L.M.; et al. Regulation of glutamine carrier proteins by RNF5 determines breast cancer response to ER stress-inducing chemotherapies. *Cancer Cell* **2015**, *27*, 354–369. [CrossRef]
67. Moore, G.E.; Mount, D.T.; Wendt, A.C. The growth of human tumor cells in tissue culture. *Surg. Forum* **1958**, *9*, 572–576.
68. Skehan, P.; Storeng, R.; Scudiero, D.; Monks, A.; McMahon, J.; Vistica, D.; Warren, J.T.; Bokesch, H.; Kenney, S.; Boyd, M.R. New colorimetric cytotoxicity assay for anticancer-drug screening. *JNCI J. Natl. Cancer Inst.* **1990**, *82*, 1107–1112. [CrossRef]
69. Lakshmanan, I.; Batra, S.K. Protocol for apoptosis assay by flow cytometry using annexin V staining method. *Bio-Protocol* **2013**, *3*, e374. [CrossRef]
70. Thomé, M.P.; Filippi-Chiela, E.C.; Villodre, E.S.; Migliavaca, C.B.; Onzi, G.R.; Felipe, K.B.; Lenz, G. Ratiometric analysis of Acridine Orange staining in the study of acidic organelles and autophagy. *J. Cell Sci.* **2016**, *129*, 4622–4632. [CrossRef]
71. Shen, Y.; Vignali, P.; Wang, R. Rapid profiling cell cycle by flow cytometry using concurrent staining of DNA and mitotic markers. *Bio-Protocol* **2017**, *7*, e2517. [CrossRef] [PubMed]

72. Longo, M.C.; Berninger, M.S.; Hartley, J.L. Use of uracil DNA glycosylase to control carry-over contamination in polymerase chain reactions. *Gene* **1990**, *93*, 125–128. [CrossRef] [PubMed]
73. Chen, L.; Zhao, X.; Wu, J.; Liu, Q.; Pang, X.; Yang, H. Metabolic characterisation of eight Escherichia coli strains including "Big Six" and acidic responses of selected strains revealed by NMR spectroscopy. *Food Microbiol.* **2020**, *88*, 103399. [CrossRef] [PubMed]
74. Xia, J.; Wishart, D.S. Metabolomic data processing, analysis, and interpretation using MetaboAnalyst. *Curr. Protoc. Bioinf.* **2011**, *34*, 10–14. [CrossRef] [PubMed]
75. Lee, T.-H.; Cheng, M.-L.; Shiao, M.-S.; Lin, C.-N. Metabolomics study in severe extracranial carotid artery stenosis. *BMC Neurol.* **2019**, *19*, 138. [CrossRef]

marine drugs

MDPI

Review

Marine *Aspergillus*: A Treasure Trove of Antimicrobial Compounds

Honghua Li, Yanqi Fu and Fuhang Song *

Key Laboratory of Geriatric Nutrition and Health, Ministry of Education of China, School of Light Industry, Beijing Technology and Business University, Beijing 100048, China; lihonghua@btbu.edu.cn (H.L.); fuyanqi0601@126.com (Y.F.)
* Correspondence: songfuhang@btbu.edu.cn

Abstract: Secondary metabolites from marine organisms are diverse in structure and function. Marine *Aspergillus* is an important source of bioactive natural products. We reviewed the structures and antimicrobial activities of compounds isolated from different marine *Aspergillus* over the past two years (January 2021–March 2023). Ninety-eight compounds derived from *Aspergillus* species were described. The chemical diversity and antimicrobial activities of these metabolites will provide a large number of promising lead compounds for the development of antimicrobial agents.

Keywords: marine *Aspergillus*; secondary metabolites; antimicrobial activity

Citation: Li, H.; Fu, Y.; Song, F. Marine *Aspergillus*: A Treasure Trove of Antimicrobial Compounds. *Mar. Drugs* **2023**, *21*, 277. https://doi.org/10.3390/md21050277

Academic Editor: Dehai Li

Received: 31 March 2023
Revised: 22 April 2023
Accepted: 25 April 2023
Published: 28 April 2023

1. Introduction

Compared with terrestrial fungi, marine fungi are more abundant in species. Due to the complex environment, their metabolites have novel structures and diverse activities [1–4]. As an important member of marine microorganisms, fungi play an important role in the study of active natural products. Marine fungi can be obtained from marine animals, plants, sediments and seawater [5–8]. Therefore, marine fungi have a wide range of sources [6,9–15].

Aspergillus is a genus of fungi widely distributed in marine environments [16–18]. Common species include *A. fumigatus*, *A. niger*, *A. versicolor*, *A. flavus*, *A. ochraceu*, *A. ticus*, *A. terreus*, etc. Marine *Aspergillus* is an important resource in the production of active natural products, such as steroids, flavonoids, azolones, etc. [7,19–22]. These metabolites are structurally diverse and exhibit a wide range of biological activities, including anticancer, antiviral, antibacterial, anti-inflammatory, lipid-lowering and anti-diabetic [22–27].

Due to the wide range of *Aspergillus* sources, the diverse secondary metabolites and the wide biological activities, the research on *Aspergillus* metabolites has attracted much attention. Therefore, a series of excellent reviews on this subject have been published so far [28–39]. In 2016, Fouillaud et al. reviewed the knowledge of anthraquinones and their derivatives derived from filamentous fungi [40]. In 2022, Hafez Ghoran et al. updated this study and summarized and classified the structures and activities of 296 anthraquinones and their derivatives [41]. In 2019, Youssef et al. reviewed the chemical and biological activities of peptides which isolated and identified from marine fungi [22]. 131 peptides were reported from these 17 genera, and about 53% of the isolated peptides showed cytotoxic, antibacterial and antiviral activities. In 2020, Jiang et al. reviewed the chemical structure and bioactive properties of new terpenes from marine derived fungi, as well as the biodiversity of these fungi from 2015 to 2019 [19]. *Penicillium*, *Aspergillus* and *Trichoderma* fungi were the main producers of terpenes. In 2021, Rani et al. reviewed the research status of microbial antibacterial molecules [10]. In 2022, Li et al., reviewed the chemistry and bioactivity of marine-derived bisabolane sesquiterpenoids [1]. In 2013, Lee et al. reviewed the bioactive secondary metabolites of *Aspergillus* derived from marine sources [42]. In 2018, Wang et al. reviewed 232 new bioactive metabolites from *Aspergillus* of marine origin from 2006 to 2016 and classified their bioactivity and chemical structures [43]. In 2020, Xu et al. reviewed the structural diversity and biological activity of 130 heterocyclic alkaloids produced by *Aspergillus* of marine origin from early 2014 to late 2018 [44]. However, there have been no studies on the antimicrobial compounds from marine *Aspergillus* in the last two years despite the fact that over the past two years, reports of antibacterial metabolites from *Aspergillus* have increased [45–51]. It is believed that the study of *Aspergillus* living in marine environments will facilitate the isolation of new fungal species and lead to the discovery of new compounds. Therefore, this review updates current compounds to cover metabolites isolated from marine *Aspergillus* between January 2021 and March 2023. It also provides structural diversity of compounds, as well as detailed information on sources and associated antimicrobial activity. We introduced the structural diversity and antimicrobial activity of 98 compounds isolated from marine-derived *Aspergillus*. This study will contribute to a better understanding of the chemical properties and biological activities of natural products from marine *Aspergillus*, thus facilitating drug discovery and development.

2. *Aspergillus* sp. from Various Marine Sources and Their Antimicrobial Activities

2.1. *Aspergillus sp. from Marine Animals and Their Antimicrobial Activities*

Trypacidin (**1**) was isolated from the *A. fumigatus* HX-1 associated with clams (Figure 1). The anti-*Vibrio harveyi* activity of trypacidin was the same as that of streptomycin sulfate, and the minimum inhibitory concentration (MIC) was 31.25 μg/mL [52].

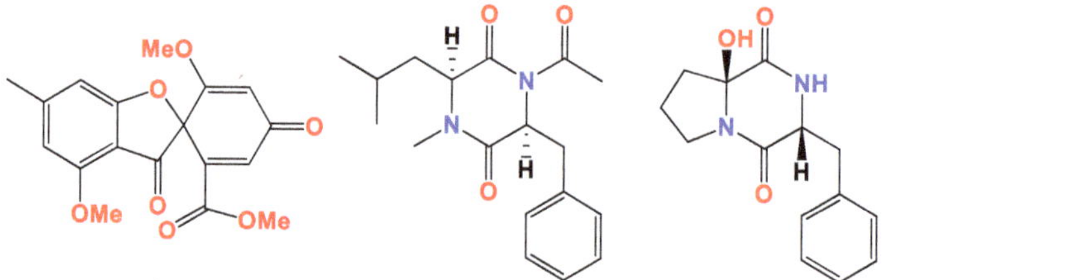

1 **2** **3**

Figure 1. Compounds of *Aspergillus* sp. derived from marine animals.

Two new dipeptides, asperopiperazines A and B (**2** and **3**), were obtained from *Aspergillus* sp. DY001 (Figure 1). The MICs of asperopiperazines A and B against *Escherichia coli* were 8 and 4 μM, and 8 and 8 μM against *S. aureus*, respectively [53].

In conclusion, only two *Aspergillus* species producing antimicrobial compounds are found from marine animals (except sponges and corals). Three compounds from these two *Aspergillus* strains have been reviewed for their antimicrobial activities. Notably, asperopiperazines A and B from *Aspergillus* sp. DY001 showed potent antimicrobial activities against *E. coli* and *S. aureus*.

2.2. *Aspergillus sp. from Marine Plants and Their Antimicrobial Activities*

Six new terpenoids were isolated from a seaward fungus *A. alabamensis* (Figure 2). They are asperalacids A-E and 4-hydroxy-5-(6)-dihydroterrecyclic acid A (**4**). Compound **4** and asperalacids A–D (**5**–**8**) showed antimicrobial activities against plant pathogenic fungi *Penicillium italicum*, *Fusarium graminearum* and *F. oxysporum*, as well as *S. aureus* and the Gram-positive bacteria *Bacillus subtilis*. Both MICs of asperalacids A and D against *F. graminearum* were 200 μg/mL. The MIC of asperalacids B and C against *F. oxysporum* were 100 and 100 μg/mL, and 200 and 25 μg/mL against *F. graminearum*, respectively. The MIC of compound **8** against *P. italicum*, *F. graminearum*, *F. oxysporum* and *S. aureus* were 200, 50, 100 and 25 μg/mL, respectively [54].

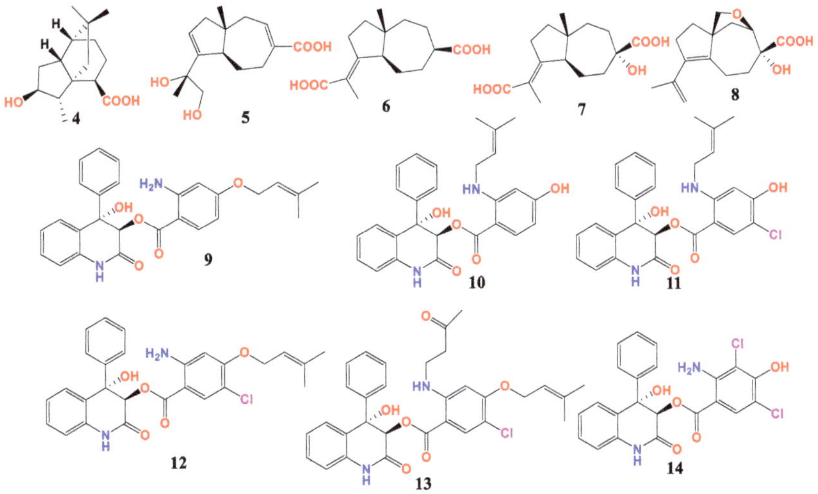

Figure 2. Compounds of *Aspergillus* sp. derived from marine plants.

Eight new benzoic acid-containing alkaloids were isolated and identified from *A. alabamensis*. Among these compounds, asperalins A–F (**9–14**) showed moderate or strong inhibitory activities against some fish pathogens, *Streptococcus parauberis*, *S. iniae* and *Edwardsiella ictalurid* (Figure 2). Asperalins C and D showed strong antibacterial activities against *S. aureus*, *S. parauberis* and *S. iniae*, with MIC values of 10.1, 10.1 and 5.0 μM, respectively. Asperalin E had the strongest inhibitory effect on *S. iniae* with an MIC value of 2.2 μM. Notably, the MICs of asperalin F against four Gram-positive bacteria *S. aureus*, *B. subtilis*, *S. parauberis*, *S. iniae* and one Gram-negative bacterium *E. ictalurid* were 21.8, 87.3, 21.8, 43.6 and 10.9 μM, respectively [55].

In conclusion, *Aspergillus* species and its active metabolites from marine plant sources (except mangrove and seagrasses) were summarized. Eleven antimicrobial compounds were identified in the seagrass-derived fungus *A. alabamensis* during 2022 and 2023. Compounds **4–8** had a weak inhibitory effect on plant pathogens. However, compounds **11–14** showed strong antibacterial effects against *S. aureus*, *S. iniae* and some Gram-positive bacteria.

2.3. *Aspergillus sp. from Mangroves and Their Antimicrobial Activities*

Six antibacterial compounds were isolated from the marine fungus *A. brunneoviolaceus* MF180246 (Figure 3). These compounds included asperbrunneo acid (**15**), secalonic acid H (**16**), chrysoxanthone C (**17**), secalonic acid F1 (**18**), asperdichrome (**19**) and penicillixanthone A (**20**). They showed antibacterial activity against *S. aureus* with MIC values of 200, 50, 50, 25, 25 and 6.25 μg/mL [27].

Six polyhydroxy p-terphenyls (asperterphenyllins A–F) were isolated from the endophytic fungus *A. candidus* LDJ-5 in mangroves. Only asperterphenyllin C (**21**) showed antibacterial activity against *Proteus* sp. with an MIC value of 19 μg/mL [56].

Two new heterodimeric tetrahydroxanthones, aflaxanthones A and B (**22** and **23**), were isolated from *A. flavus* QQYZ. These two compounds showed potential antimicrobial activity and broad spectrum against several pathogenic fungi such as *C. albicans* and *F. oxysporum*, with MIC values in the range of 3.13–50 μM. They also showed moderate antibacterial activity against several bacteria such as *B. subtilis* and methicillin-resistant *S. aureus* (MRSA), with MIC values in the range of 12.5–25 μM [57].

In conclusion, *Aspergillus* and its active metabolites from mangroves were summarized. Due to the special geographical environment, mangroves had a wide variety of organisms, which has been thoroughly examined in previous studies of metabolites. Nine antimicrobial compounds were found in three *Aspergillus* strains from mangrove sources. Most of the compounds showed moderate antimicrobial activities. Among these compounds, compound 20 showed a strong inhibitory effect on *S. aureus*.

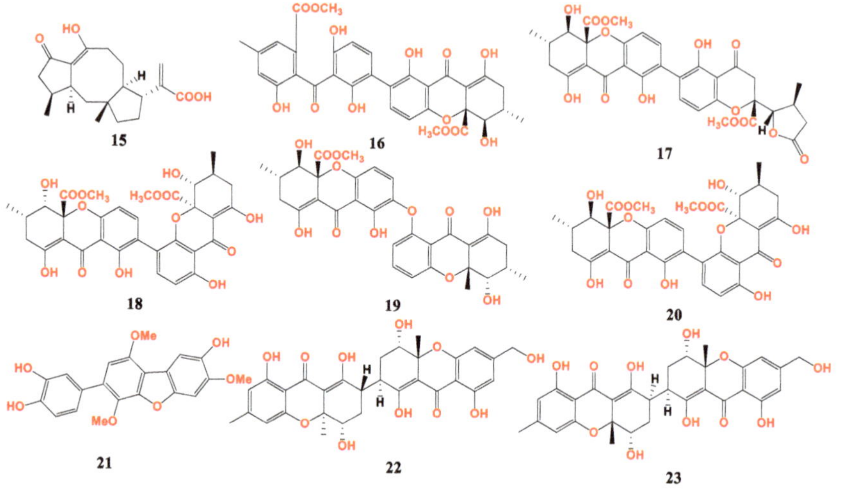

Figure 3. Compounds of *Aspergillus* sp. derived from mangroves.

2.4. Aspergillus sp. Derived from Algae and Their Antimicrobial Activities

Two C_7-alkylated salicylaldehyde derivatives metabolites, namely asperglaucins A and B (**24** and **25**), were isolated from the endophytic fungus *A. chevalieri* SQ-8 (Figure 4). Asperglaucins A and B showed potent antimicrobial activities against plant pathogens *B. cereus* and *Pseudomonas syringae* pv *actinidae* (Psa), with an MIC value of 6.25 μM. Further analysis showed that asperglaucins A and B may change the external structure of *B. cereus* and Psa and cause cell membrane rupture or deformation. The results indicated that asperglaucins A and B may be potential lead compounds of pesticide fungicides [58].

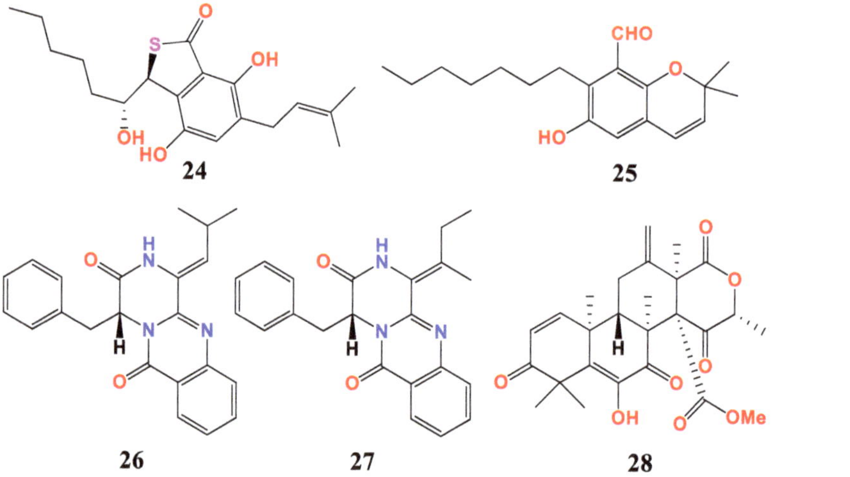

Figure 4. Compounds of *Aspergillus* sp. derived from algae.

Two new diketopiperazines, namely versiamide A (**26**) and 3, 15-dehydroprotuboxepin K (**27**), were isolated from endophytic fungus *A. creber* EN-602 obtained from the marine red algae *Rhodomela confervoides*. Versiamide A and 3, 15-dehydroprotuboxepin K showed inhibitory activities against a variety of aquatic bacteria, with MIC values ranging from 8 to 64 μg/mL. Versiamide A showed antibacterial activity against *Aeromonas hydrophila*, *E. coli*, *Micrococcus luteus* and *P. aeruginosa*, with MIC values of 64, 16, 64 and 64 μg/mL. 3, 15-dehydroprotuboxepin K showed antibacterial activity against *E. tarda*, *E. coli*, *M. luteus*, *P. aeruginosa* and *V. harveyi*, with MIC values of 64, 8, 16, 32 and 64 μg/mL [59].

An antibacterial terpenoid, namely terretonin F (**28**), were isolated from the *Aspergillus* sp. RR-YLW12, which derived from marine red algae *R. confervoide*. Terretonin F showed significant inhibitory activities against *Chattonella marina*, *Heterosigma akashiwo* and *Prorocentrum donghaiense*, with IC_{50} values of 3.1, 5.2 and 10.5 μg/mL, respectively [60].

In conclusion, *Aspergillus* species from marine algae and active metabolites were summarized. Five antimicrobial compounds were found in three fungi strains of algae origin. It should be noted that asperglaucins A and B (**24** and **25**) showed a strong inhibitory effect on *B. cereus*. The possible bacteriostatic mechanism of the compounds was also introduced. At present, the studies on the structure and biological activity of compounds are abundant, but the studies on the mechanism of biological activity are limited.

2.5. Aspergillus sp. from Corals and Their Antimicrobial Activities

Three known metabolites, including demethylincisterol A_2 (**29**), asperophiobolin E (**30**) and butyrolactone I (**31**), were isolated and identified from the soft coral fungus *A. hiratsukae* SCSIO $5B_{n1}003$ (Figure 5). Compounds **29–31** showed potent antibacterial activity against *B. subtilis*, with MIC values of 10.26 ± 0.76, 17.00 ± 1.25 and 5.30 ± 0.29 μM. Meanwhile, asperophiobolin E and butyrolactone I showed weak activity against *S. aureus*, with MIC values of 102.86 ± 4.50 and 59.54 ± 0.50 μM, respectively [61].

Five new antimicrobial α-pyranone methterpenoids H-L (**32–36**) and one known antimicrobial compound, namely neoechinulin A (**37**), were isolated from *A. hiratsukae* SCSIO 7S2001, a fungus derived from ophiophora coral. Methterpenoids H-L and neoechinulin A showed varying degrees of antibacterial activity, with MIC values of 6.25–100 μg/mL. The MIC values of methterpenoid H were 6.25 μg/mL for *Micrococcus lutea* 01, MRSA, and *Streptococcus faecalis*; that of methterpenoid I was 6.25 μg/mL for MRSA; that of methterpenoid G was 12.5 μg/mL for MRSA; that of methterpenoid K was 6.25 μg/mL for *Klebsiella pneumoniae*; that of methterpenoid L was 12.5 μg/mL for *M. lutea*, *S. faecalis* and MRSA; and that of neoechinulin A was 12.5 μg/mL for *S. faecalis*. [62].

Two butenolides, including versicolactone B (**38**) and butyrolactone VI (**39**), were isolated from *Aspergillus terreus* SCSIO41404, a fungus derived from coral. Versicolactone B and butyrolactone VI showed weak antibacterial activity against *Enterococcus faecalis* and *K. pneumoniae* with IC_{50} values of 25 and 50 μg/mL, respectively [63].

Six chlorinated polyketones were isolated from the coral fungus *A. unguis* GXIMD 02505 in the Beibu Gulf. These polyketones included aspergillusethers J and F (**40** and **41**), nornidulin (**42**), aspergillusidones B and C (**43** and **44**) and 1-(2, 6-dihydroxy-4-methoxy-3, 5-dimethylphenyl)- 2-methylbutan-1-one (**45**). Compounds **40–45** exhibited inhibitory activities against marine biofilm-forming bacteria, *Marinobacterium jannaschii*, MRSA, *Microbulbifer variabilis* and *Vibrio pelagius*, with MIC values ranging from 2 to 64 μg/mL [64].

Five antimicrobial cyclic lipopeptides, namely maribasins C-E (**46–48**) and maribasins A and B (**49** and **50**), were isolated from the marine fungus *Aspergillus* sp. SCSIO 41501. These compounds showed strong antifungal activities against five plant pathogenic fungi, with MIC values ranging from 3.12 to 50 μg/disc [34].

In conclusion, coral-derived *Aspergillus* and its active metabolites were summarized. Twenty-two antimicrobial compounds were found in five fungi strains of coral origin. It was a relatively large variety of compounds compared with *Aspergillus* from other origins. Most of the compounds had a wide antimicrobial spectrum against different bacteria and fungi.

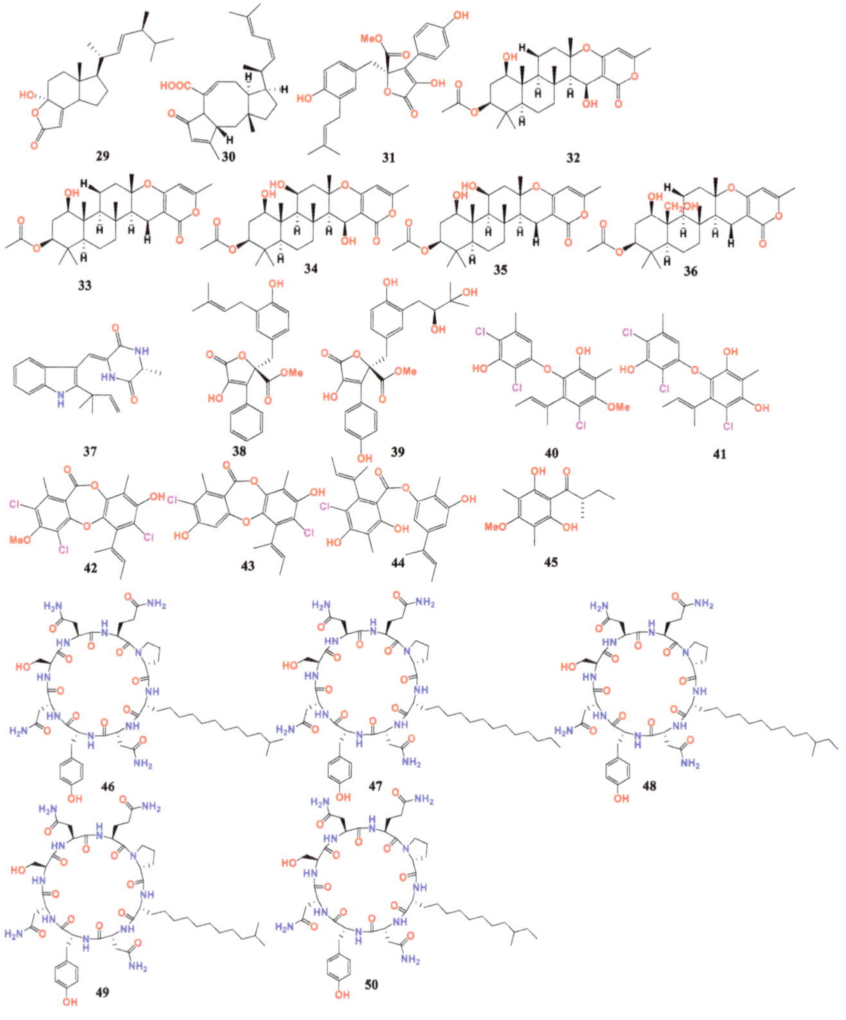

Figure 5. Compounds of *Aspergillus* sp. derived from corals.

2.6. Aspergillus sp. Derived from Sponges and Their Antimicrobial Activities

One hydroxypyrrolidine alkaloid preussin (**51**) was isolated and identified from marine sponge-related fungus *A. candius* KUFA 0062 (Figure 6). Preussin showed inhibition against vancomycin-resistant *Enterococcus* (VRE) and MRSA, as well as *E. faecalis* ATCC29212 and *S. aureus* ATCC 29213 [65].

Four antimicrobial compounds were isolated from the marine sponge-derived fungus *Aspergillu flavus* KUFA1152. These compounds were aspulvinones B′, H, R and S (**52–55**). Aspulvinones B′, H, R and S showed antibacterial activity against some multidrug-resistant strains isolated from the environment, and inhibited the biofilm formation of strains. Aspulvinones B′ and H displayed activity with MIC values of 16 μg/mL for the *S. aureus*, and for *E. faecalis*, MIC values ranged from 16 to 64 μg/mL. Aspulvinones R and S exhibited the potent activity against all Gram-positive strains tested, with MIC values ranging from 4 to 16 μg/mL for *S. aureus* and *E. faecalis*, and from 8 to 16 μg/mL for the VRE and MRSA [66].

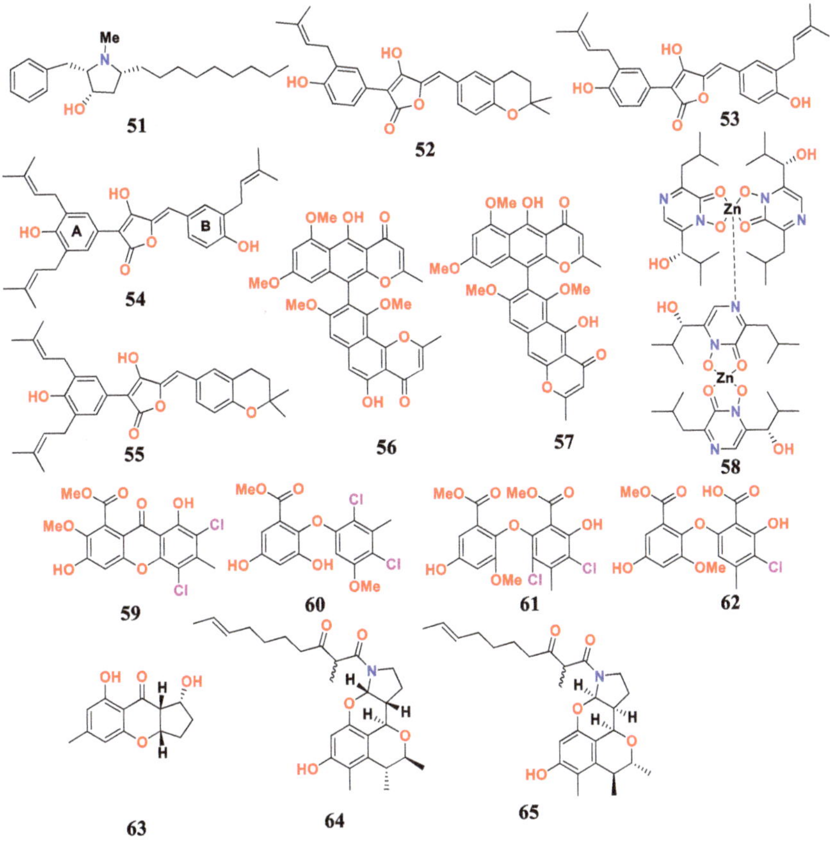

Figure 6. Compounds of *Aspergillus* sp. derived from sponges.

The endophytic fungus *A. niger* L14 has been chemically studied, and two dimers, naphtho-γ-pyrone, fonsecinone A (**56**) and isoaurasperone A (**57**), have been isolated. These compounds had obvious inhibitory effects on human pathogenic bacteria *Helicobacter pylori* 159 and G27 with MIC values ≤ 4 μg/mL, comparable to the antibacterial effect of ampicillin sodium [67].

One antimicrobial compound, namely dizinc hydroxy-neotriamycin (**58**), was isolated from the sponge-related fungus *A. ochraceopetaliformis* SCSIO 41018. Dizinchydroxyneoaspergillin showed potent inhibition against MRSA, *Acinetobacter baumannii, E. faecalis, Staphyloccocus aureus* and *Klebsiella pneumonia*, with MIC values ranging from 0.45 to 7.8 μg/mL [68].

Two new chlorinated biphenyls, including aspergetherins A and C (**59** and **60**), and two known biphenyl derivatives, including methyl 3, 5-dichloroasterric acid (**61**) and methyl chloroasterrate (**62**), were isolated from a marine sponge symbiotic fungus *A. terreus* 164018. The antibacterial activity of these compounds against MRSA was evaluated, with MIC values ranging from 1.0 to 128 μg/mL. Notably, compound **61** had obvious inhibitory effects on two different MRSA strains, with MIC values of 1 and 16 μg/mL [69].

Chemical studies of the natural compounds of the marine fungus *Aspergillus* sp. LS57 had resulted in the isolation of aspergilluone A (**63**). The MIC value of aspergilluone A was 32 μg/mL against *Mycobacterium tuberculosis*, 64 μg/mL against *S. aureus*, and 128 μg/mL against both Gram-positive *B. subtilis* and Gram-negative *E. coli* [70].

Two novel tetracyclic skeleton alkaloids were isolated from *Aspergillus* sp. LS116, which were perinadines B and C (**64** and **65**). Perinadines B and C showed moderate antibacterial activity for *B. subtilis* with MIC values of 32 and 64 μg/mL [71].

In conclusion, *Aspergillus* and its active metabolites of sponge were summarized in this paper. Sponges are the most primitive marine animals with a large number of microorganisms, which are important sources of active natural products. Fifteen antibacterial compounds were found in seven fungi strains derived from sponge. *Aspergillus* derived from sponge was the source of antimicrobial compounds. Most of the compounds had a wide antimicrobial spectrum against a variety of bacteria and fungi. Hydroxy-neotriamycin (**58**) had a strong bacteriostatic effect on a variety of bacterial pathogens.

2.7. Aspergillus sp. from Seawater and Their Antimicrobial Activities

Nine antimicrobial compounds were isolated from marine fungus *A. fumigatus* H22. These compounds included 12,13-dihydroxyfumitremorgin C (**66**), fumitremorgin B (**67**), 13-oxofumitremorgin B (**68**), fumagillin (**69**), helvolic acid (**70**), 6-O-propionyl-16-O-deacetylhelvolic acid (**71**), 16-O-propionyl-6-O-deacetylhelvolic acid (**72**), penibenzophenone E (**73**) and sulochrin (**74**) (Figure 7). Compounds **66** and **68** showed potent antibacterial activity, and **69–74** exhibited strong anti-MRSA activity with MIC values between 1.25 and 2.5 μM. Additionally, compound **66** showed moderate inhibitory activity against *Mycobacterium Bovis*, with an MIC value of 25 μM, and compound **67** showed moderate inhibitory activity against *C. albicans*, with an MIC value of 50 μM [72].

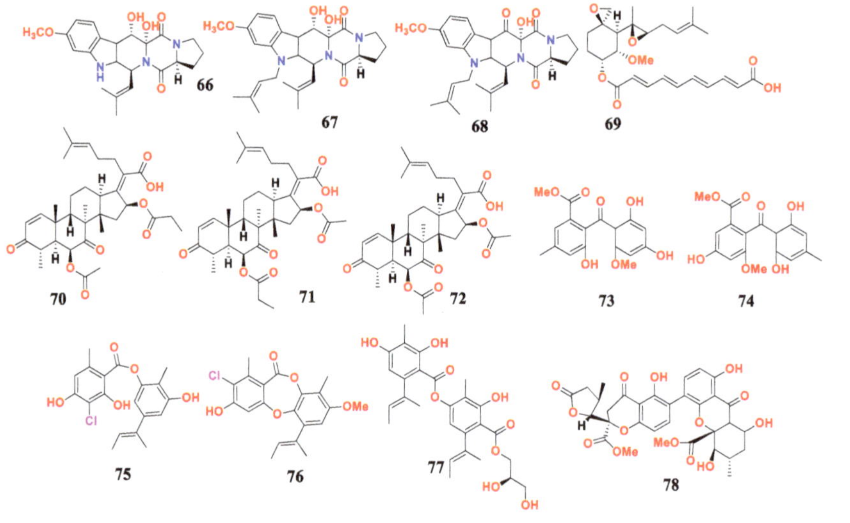

Figure 7. Compounds of *Aspergillus* sp. derived from seawater.

Three novel phenolic polyketones, namely unguidepside C (**75**), aspersidone B (**76**) and agonodepside C (**77**), were isolated from *A. unguis*. These compounds showed a strong activity against Gram-positive bacteria, with MIC ranging from 5.3 to 22.1 μM [73].

Five novel dimeric tetrahydroxanthones, including aculeaxanthones A-E, were extracted from the marine fungus *A. aculeatinus* WHUF0198. Among them, only aculeaxanthone A (**78**) showed activity against *B. subtilis* 168, *S. aureus* USA300, *H. pylori* 159, *H. pylori* 129, *H. pylori* 26695 and *H. pylori* G27, with MIC values of 1.0, 2.0, 2.0, 2.0, 4.0 and 4.0 μg/mL, respectively [74].

In conclusion, *Aspergillus* and its active metabolites from seawater were summarized. Thirteen antimicrobial compounds were found in three fungi strains derived from seawater. Compounds **69–74** exhibited strong anti-MRSA activity and aculeaxanthone A (**78**) showed strong anti-bacterial pathogen activity.

2.8. Aspergillus sp. from Marine Sediments and Their Antimicrobial Activities

Six known compounds, including cyclopiamide (**79**), speradine H (**80**), speradine G (**81**), speradine B (**82**), speradine C (**83**) and cyclopiazonic acid (CPA) (**84**), were isolated from *A. flavus* SCSIO F025 from deep-sea sediments in the South China Sea (Figure 8). Compounds **79–84** showed weak antibacterial activity against *E. coli*, and CPA also exhibited strong antibacterial activity against MRSA, *B. subtilis*, *S. aureus*, *M. luteus* and *Bacillus thuringiensis* [75].

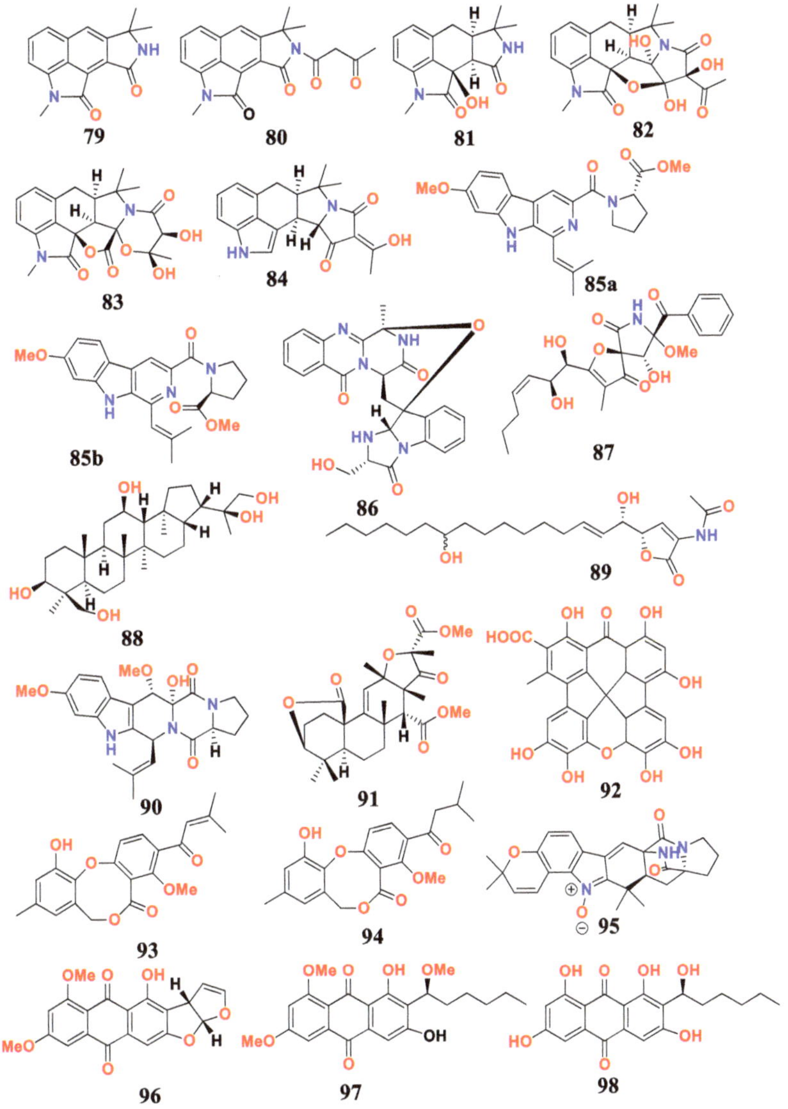

Figure 8. Compounds of *Aspergillus* sp. derived from marine sediments.

Five novel antibacterial metabolites and one known antibacterial compound were all isolated from the deep-sea sediment-derived fungus *A. fumigatus* SD-406. The novel metabolites included secofumitremorgins A and B (**85a** and **85b**), 29-hydroxyfumiquinazoline C (**86**), 10*R*-15-methylpseurotin A (**87**), 1,4,23-trihydroxy-hopan-22,30-diol (**88**) and sphingofungin I (**89**), and one known cyclotryprostatin B (**90**). Compounds **85–90** exhibited inhibitory activities against pathogenic bacteria and plant pathogenic fungi, with MIC values of 4–64 μg/mL [76].

One new metabolite, namely 3, 5-dimethylorsellinic acid-based meroterpenoid (**91**), was isolated from the deep-sea fungus *Aspergillus* sp. CSYZ-1. Compound **91** showed strong antimicrobial activity against *S. aureus* and *H. pylori*, with MIC values of 2–16 and 1–4 μg/mL, respectively [77].

Two novel antibacterial metabolites, including aspergiloxathene A (**92**) and $\Delta^{2'}$-1′-dehydropenicillide (**93**) and one known antibacterial compound, namely dehydropenicillide (**94**), were isolated from *Aspergillus* sp. IMCASMF180035. Aspergiloxathene A exhibited significant inhibition against MRSA and *S. aureus*, with MIC values of 22.40 and 5.60 μM. Dehydropenicillide and $\Delta^{2'}$-1′-dehydropenicillide showed potent antibacterial activities against *H. pylori*, with MIC values of 21.61 and 21.73 μM, respectively [30].

One alkaloid asperthrin A (**95**) had been isolated from the marine endophytic fungus *Aspergillus* sp. YJ191021. The isolated compound had inhibitory effects on *Rhizoctonia solani*, *Xanthomonas oryzae* pv. *Oryzicola* and *Vibrio anguillarum*, with MIC values of 25, 12.5 and 8 μg/mL, respectively [78].

Three antimicrobial compounds were isolated from the fermented extracts of *Aspergillus* sp. WHUF05236. They included 6,8-di-O-methylversicolorin A (**96**), 6,8,1′-tri-O-methylaverantin (**97**) and 6,8-di-O-methylaverantin (**98**). They exhibited antibacterial activity against *H. pylori*, with MIC values ranging from 20.00 to 43.47 μM [79].

In conclusion, *Aspergillus* and its active metabolites from marine sediments were summarized. Twenty antimicrobial compounds were found in six *Aspergillus* strains from marine sediments. According to the literature, more than fifty antimicrobial compounds were produced by *Aspergillus* from marine sediments between 2018 and 2020. Therefore, marine sediments are an important source of secondary metabolites of fungi. Among them, compound **91** showed strong antimicrobial activity against *S. aureus* and *H. pylori*.

Sources and activities of compounds from marine *Aspergillus* were summarized in Table 1. We classified fungi and compounds according to *Aspergillus* origin.

Table 1. Sources and activities of compounds from marine *Aspergillus*.

Sources and *Aspergillus*	Compounds	Activities	References
	Marine animals		
A. fumigatus HX-1	Trypacidin (**1**)	MIC (anti-*V. harveyi*) was 31.25 μg/mL	[52]
Aspergillus sp. DY001	Asperopiperazines A, B (**2**, **3**)	MIC (anti-*E. coli*) were 8 and 4 μM MIC (anti-*S. aureus*) were 8 and 8 μM	[53]
	Marine plants		
A. alabamensis	4-hydroxy-5(6)-dihydroterrecyclic acid A (**4**), asperalacids A–D (**5–8**)	MIC (anti-plant pathogens) was 25–200 μg/mL	[54]
A. alabamensis	asperalins A–F (**9–14**)	MIC (anti-fish pathogens) was 2.2–87.3 μM	[55]

Table 1. *Cont.*

Sources and *Aspergillus*	Compounds	Activities	References
Mangroves			
A. brunneoviolaceus MF180246	asperbrunneo acid (**15**), secalonic acids H, F1 (**16, 18**), chrysoxanthone C (**17**), asperdichrome (**19**), penicillixanthone A (**20**)	MIC (anti-*S. aureus*) were 200, 50, 50, 25, 25, 6.25 μg/mL	[27]
A. candius LDJ-5	asperterphenyllin C (**21**)	MIC (anti-*Proteus* sp.) was 19 μg/mL	[56]
A. flavus QQYZ	aflatoxones A, B (**22, 23**)	MIC (anti-pathogens) was 3.13–50 μM	[57]
Marine algaes			
A. chevalieri SQ-8	asperglaucins A, B (**24, 25**)	MIC (anti-plant pathogens) was 6.25 μM	[58]
A. creber EN-602	versiamide A (**26**), 3, 15-dehydroprotuboxepin K (**27**)	MIC (anti-bacteria) was 8–64 μg/mL	[59]
Aspergillus sp. RR-YLW12	terretonin F (**28**)	IC_{50} (anti-three microalgae) were 3.1, 5.2, 10.5 μg/mL	[60]
Marine corals			
A. hiratsukae SCSIO $5B_{n1}$003	demethylincisterol A_2 (**29**), asperophiobolin E (**30**), butyrolactone I (**31**)	MIC (anti-*B. subtilis*) were 10.26 ± 0.76, 17.00 ± 1.25 and 5.30 ± 0.29 μM	[61]
A. hiratsukae SCSIO 7S2001	methterpenoids H-L (**32–36**) neoechinulin A (**37**)	MIC (anti-bacteria) was 6.25–100 μg/mL	[62]
A. terreus SCSIO41404	versicolactone B (**38**), butyrolactone VI (**39**)	IC_{50} (anti-*E. faecalis*, *K. pneumoniae*) were 25 and 50 μg/mL	[63]
A. unguis GXIMD 02505	**40–45**	MIC (anti-bacteria) was 2–64 μg/mL	[64]
Aspergillus sp. SCSIO 41501	maribasins C–E,A,B (**46–50**)	MIC (anti-plant pathogens) was 3.12–50 μg/disc	[34]
Sponges			
A. candius KUFA 0062	preussin (**51**)	anti-pathogens	[65]
A. flavipes KUFA1152	aspulvinones B′, H, R and S (**52–55**)	MIC (anti-pathogens) was 16–64 μg/mL	[66]
A. niger L14	fonsecinone A (**56**), isoaurasperone A (**57**)	MIC (anti-*H. pylori*) was ≤4 μg/mL	[67]
A. ochraceopetaliformis SCSIO 41018	hydroxy-neotriamycin (**58**)	MIC (anti-pathogens) was 0.45–7.8 μg/mL μM	[68]
A. terreus 164018	aspergetherins A, C (**59, 60**) 3, 5-dichloroasterric acid (**61**), methyl chloroasterrate (**62**)	MIC (anti-MRSA) was 1.0–128 μg/mL	[69]
Aspergillus sp. LS57	aspergilluone A (**63**)	MIC (anti-pathogens) was 32–128 μg/mL	[70]
Aspergillus sp. LS116	perinadines B, C (**64, 65**)	MIC (anti-*B. subtilis*) were 32 and 64 μg/mL	[71]

Table 1. *Cont.*

Sources and *Aspergillus*	Compounds	Activities	References
	Seawater		
A. fumigatus H22	12,13-dihydroxyfumitremorgin C (**66**), fumitremorgin B (**67**)	MIC(anti-*M. Bovis, C. albicans*) were 25 and 50 μM	[72]
A. fumigatus H22	(**66**),13-oxofumitremorgin B (**68**)	antibacterial activity	[72]
A. fumigatus H22	fumagillin (**69**), helvolic acid (**70**), 6-O-propionyl-16-O-deacetylhelvolic acid (**71**), 16-O-propionyl-6-O-deacetylhelvolic acid (**72**), penibenzophenone E (**73**), sulochrin (**74**)	MIC (anti-MRSA) were 1.25 and 2.5	[72]
A. unguis	unguidepside C (**75**), aspersidone B (**76**), agonodepside C (**77**)	MIC (anti-bacteria) was 5.3 to 22.1 μM	[73]
A. aculeatinus WHUF0198	aculeaxanthone A (**78**)	MIC (anti-bacteria) was 1.0 to 4.0 μM	[74]
	Marine sediments		
A. flavus SCSIO F025	cyclopiamide (**79**), speradines G,H,B,C (**80–83**), CPA (**84**)	weak anti-bacteria	[75]
A. fumigatus SD-406	**85–90**	MIC (anti-bacteria and plant pathogens) were 4–64 μg/mL	[76]
Aspergillus sp. CSYZ-1	meroterpenoid (**91**)	MIC (anti-*S. aureus, H. pylori*) were 2–16 and 1–4 μg/mL	[77]
Aspergillus sp. IMCASMF180035	aspergiloxathene A (**92**)	MIC (anti-MRSA, *S. aureus*) were 22.40 and 5.60 μM	[30]
Aspergillus sp. IMCASMF180035	$\Delta^{2'}$-1′-dehydropenicillide (**93**), dehydropenicillide (**94**)	MIC (anti-*H. pylori*) were 21.61 and 21.73 μM	[30]
Aspergillus sp. YJ191021	asperthrins A (**95**)	MIC (anti-plant pathogens) was 8–25μg/mL	[78]
Aspergillus sp. WHUF05236	6, 8-di-O-methylversicolorin A (**96**), 6,8,1′-tri-O-methylaverantin (**97**), 6,8-di-O-methylaverantin (**98**)	MIC (anti-*H. pylori*) was 20.00 to 43.47 μM	[79]

In recent years, marine fungi have attracted the attention of researchers due to their bioactive compounds [10,44,46,80–85]. Combined with a series of previous excellent literature reviews, we conducted a comprehensive literature review of antibacterial compounds produced by Aspergillus fungi of different marine origin during the period of 2021–2023. The reported numbers of *Aspergillus* from marine animals, plants, mangroves, seagrasses, coral, sponge, seawater and marine sediment are shown in Figure 9. The most *Aspergillus* was derived from sponges, accounting for 23.30%. *Aspergillus* derived from marine coral was found in the second place, accounting for 16.7%.

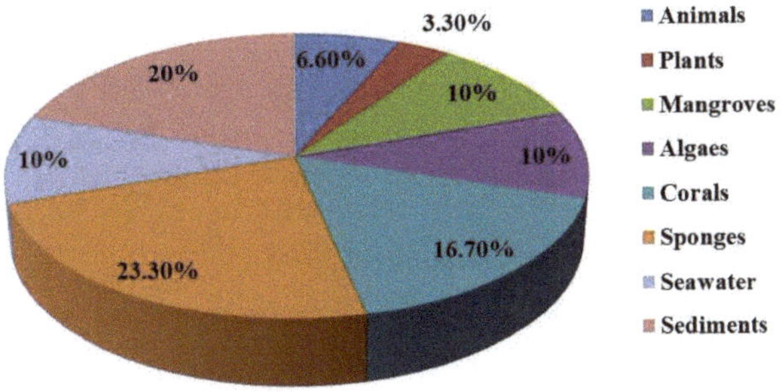

Figure 9. The proportion of Aspergillus from different marine sources.

We summarized ninety-eight antibacterial compounds from *Aspergillus* strains isolated from different marine sources (Figure 10). Among them, twenty-two antimicrobial compounds were found in marine corals from January 2021 to March 2023. Marine sediments had the next highest number of antimicrobial compounds, with twenty compounds. Therefore, in recent years, the antimicrobial compounds of *Aspergillus* from marine sources mainly came from marine corals and marine sediments. Marine natural products are rich in species and play an obvious role in the treatment of pathogen infections [86–92]. More and more novel compounds with different chemical structures and biological activities are being discovered [48,93–99].

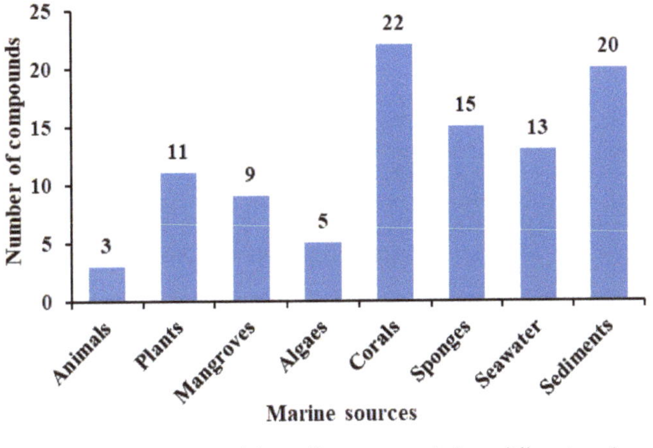

Figure 10. The proportion of *Aspergillus* compounds from different marine sources.

3. Conclusions

This review describes antimicrobial compounds from *Aspergillus* species during January 2021 to March 2023. Ninety-eight compounds derived from *Aspergillus* species were described. Only three compounds with antimicrobial activities are found from marine animals (except sponges and corals). Twenty-two antimicrobial compounds were found in five fungi strains of coral origin. Fifteen antibacterial compounds were found in seven fungi strains derived from sponge. Most of these thirty-seven compounds had a wide antimicrobial spectrum against a variety of bacteria and fungi. Except for the compounds derived from coral and sponge, most of the compounds from other sources showed antibacterial activity, but no fungal inhibitory activity. Most of the compounds had inhibitory effects on *S. aureus*. Some compounds exhibited inhibitory effects on *E. coli* and *B. subtilis*. Among them, compound 91 showed strong antimicrobial activity against *H. pylori*. These active compounds have potential applications in bacterial and fungal infections and will provide reference for the development of novel anti-infective drugs.

Author Contributions: Writing—original draft preparation and editing, H.L.; writing—original draft preparation, Y.F.; writing—review and editing, H.L. and F.S.; supervision, F.S.; funding acquisition, H.L. and F.S. All authors have read and agreed to the published version of the manuscript.

Funding: This research was funded by grants from the General Projects of Science and Technology Program of Beijing Municipal Education Commission (KM202210011008), Research Foundation for Young Teachers of Beijing Technology and Business University (QNJJ2022-21).

Data Availability Statement: Not applicable.

Acknowledgments: The authors gratefully acknowledge the financial supports.

Conflicts of Interest: The authors declare no conflict of interest.

References

1. Li, C.S.; Liu, L.T.; Yang, L.; Li, J.; Dong, X. Chemistry and bioactivity of marine-derived bisabolane sesquiterpenoids: A review. *Front. Chem.* **2022**, *10*, 881767. [CrossRef] [PubMed]
2. Carroll, A.R.; Copp, B.R.; Davis, R.A.; Keyzers, R.A.; Prinsep, M.R. Marine natural products. *Nat. Prod. Rep.* **2020**, *37*, 175–223. [CrossRef]
3. Liu, L.L.; Wu, C.H.; Qian, P.Y. Marine natural products as antifouling molecules—A mini-review (2014–2020). *Biofouling* **2020**, *36*, 1210–1226. [CrossRef] [PubMed]
4. Cardoso, J.; Nakayama, D.G.; Sousa, E.; Pinto, E. Marine-derived compounds and prospects for their antifungal application. *Molecules* **2020**, *25*, 5856. [CrossRef]
5. Bian, C.; Wang, J.; Zhou, X.; Wu, W.; Guo, R. Recent advances on marine alkaloids from sponges. *Chem. Biodivers.* **2020**, *17*, e2000186. [CrossRef]
6. Ge, X.; Wang, Y.; Sun, C.; Zhang, Z.; Song, L.; Tan, L.; Li, D.; Yang, S.; Yu, G. Secondary metabolites produced by coculture of *Pleurotus ostreatus* SY10 and *Pleurotus eryngii* SY302. *Chem. Biodivers.* **2022**, *19*, e202100832. [CrossRef] [PubMed]
7. Li, K.; Chen, S.; Pang, X.; Cai, J.; Zhang, X.; Liu, Y.; Zhu, Y.; Zhou, X. Natural products from mangrove sediments-derived microbes: Structural diversity, bioactivities, biosynthesis, and total synthesis. *Eur. J. Med. Chem.* **2022**, *230*, 114117. [CrossRef]
8. Wiese, J.; Imhoff, J.F. Marine bacteria and fungi as promising source for new antibiotics. *Drug Dev. Res.* **2019**, *80*, 24–27. [CrossRef]
9. Wang, H.N.; Sun, S.S.; Liu, M.Z.; Yan, M.C.; Liu, Y.F.; Zhu, Z.; Zhang, Z. Natural bioactive compounds from marine fungi (2017–2020). *J. Asian Nat. Prod. Res.* **2022**, *24*, 203–230. [CrossRef]
10. Rani, A.; Saini, K.C.; Bast, F.; Varjani, S.; Mehariya, S.; Bhatia, S.K.; Sharma, N.; Funk, C. A review on microbial products and their perspective application as antimicrobial agents. *Biomolecules* **2021**, *11*, 1860. [CrossRef]
11. Wang, W.; Gao, M.; Luo, Z.; Liao, Y.; Zhang, B.; Ke, W.; Shao, Z.; Li, F.; Chen, J. Secondary metabolites isolated from the deep sea-derived fungus *Aspergillus sydowii* C1-S01-A7. *Nat. Prod. Res.* **2019**, *33*, 3077–3082. [CrossRef] [PubMed]
12. Chen, G.; Wang, H.F.; Pei, Y.H. Secondary metabolites from marine-derived microorganisms. *J. Asian Nat. Prod. Res.* **2014**, *16*, 105–122. [CrossRef] [PubMed]
13. Julianti, E.; Abrian, I.A.; Wibowo, M.S.; Azhari, M.; Tsurayya, N.; Izzati, F.; Juanssilfero, A.B.; Bayu, A.; Rahmawati, S.I.; Putra, M.Y. Secondary metabolites from marine-derived fungi and actinobacteria as potential sources of novel colorectal cancer drugs. *Mar. Drugs* **2022**, *20*, 67. [CrossRef] [PubMed]
14. Chen, S.; Cai, R.; Liu, Z.; Cui, H.; She, Z. Secondary metabolites from mangrove-associated fungi: Source, chemistry and bioactivities. *Nat. Prod. Rep.* **2021**, *39*, 560–595. [CrossRef]
15. Chen, Y.; Pang, X.; He, Y.; Lin, X.; Zhou, X.; Liu, Y.; Yang, B. Secondary metabolites from coral-associated fungi: Source, chemistry and bioactivities. *J. Fungi* **2022**, *8*, 1043. [CrossRef]
16. Liu, Z.; Zhao, J.-Y.; Sun, S.-F.; Li, Y.; Liu, Y.-B. Fungi: Outstanding source of novel chemical scaffolds. *J. Asian Nat. Prod. Res.* **2018**, *22*, 99–120. [CrossRef]
17. Schueffler, A.; Anke, T. Fungal natural products in research and development. *Nat. Prod. Rep.* **2014**, *31*, 1425–1448. [CrossRef]
18. Dell'Anno, F.; Rastelli, E.; Buschi, E.; Barone, G.; Beolchini, F.; Dell'Anno, A. Fungi can be more effective than bacteria for the bioremediation of marine sediments highly contaminated with heavy metals. *Microorganisms* **2022**, *10*, 993. [CrossRef]
19. Jiang, M.; Wu, Z.; Guo, H.; Liu, L.; Chen, S. A review of terpenes from marine-derived fungi: 2015–2019. *Mar. Drugs* **2020**, *18*, 321. [CrossRef]
20. Wali, A.F.; Majid, S.; Rasool, S.; Shehada, S.B.; Abdulkareem, S.K.; Firdous, A.; Beigh, S.; Shakeel, S.; Mushtaq, S.; Akbar, I.; et al. Natural products against cancer: Review on phytochemicals from marine sources in preventing cancer. *Saudi Pharm. J.* **2019**, *27*, 767–777. [CrossRef]
21. Qadri, H.; Shah, A.H.; Ahmad, S.M.; Alshehri, B.; Almilaibary, A.; Mir, M.A. Natural products and their semi-synthetic derivatives against antimicrobial-resistant human pathogenic bacteria and fungi. *Saudi J. Biol. Sci.* **2022**, *29*, 103376. [CrossRef] [PubMed]
22. Youssef, F.S.; Ashour, M.L.; Singab, A.N.B.; Wink, M. A comprehensive review of bioactive peptides from marine fungi and their biological significance. *Mar. Drugs* **2019**, *17*, 559. [CrossRef]
23. Qi, J.; Chen, C.; He, Y.; Wang, Y. Genomic analysis and antimicrobial components of M7, an *Aspergillus terreus* strain derived from the south china sea. *J. Fungi* **2022**, *8*, 1051. [CrossRef] [PubMed]

24. Mia, M.M.; Hasan, M.; Miah, M.M.; Hossain, M.A.S.; Islam, S.; Shanta, V. Inhibitory potentiality of secondary metabolites extracted from marine fungus target on avian influenza virus-a subtype H5N8 (Neuraminidase) and H5N1 (Nucleoprotein): A rational virtual screening. *Vet. Anim. Sci.* **2022**, *15*, 100231.
25. Fan, M.; Nath, A.K.; Tang, Y.; Choi, Y.-J.; Debnath, T.; Choi, E.-J.; Kim, E.-K. Investigation of the anti-prostate cancer properties of marine-derived compounds. *Mar. Drugs* **2018**, *16*, 160. [CrossRef]
26. Zhao, H.; Ji, R.; Zha, X.; Xu, Z.; Lin, Y.; Zhou, S. Investigation of the bactericidal mechanism of Penicilazaphilone C on *Escherichia coli* based on 4D label-free quantitative proteomic analysis. *Eur. J. Pharm. Sci.* **2022**, *179*, 106299. [CrossRef] [PubMed]
27. Xu, X.; Han, J.; Zhang, X.; Xu, W.; Yang, J.; Song, F. Investigation on the chemical constituents of the marine-derived fungus strain *Aspergillus brunneoviolaceus* MF180246. *Nat. Prod. Res.* **2022**, 1–6. [CrossRef] [PubMed]
28. Barzkar, N.; Sheng, R.; Sohail, M.; Jahromi, S.T.; Babich, O.; Sukhikh, S.; Nahavandi, R. Alginate lyases from marine bacteria: An enzyme ocean for sustainable future. *Molecules* **2022**, *27*, 3375. [CrossRef] [PubMed]
29. Willems, T.; De Mol, M.L.; De Bruycker, A.; De Maeseneire, S.L.; Soetaert, W.K. Alkaloids from marine fungi: Promising antimicrobials. *Antibiotics* **2020**, *9*, 340. [CrossRef] [PubMed]
30. Song, F.; Lin, R.; Yang, N.; Jia, J.; Wei, S.; Han, J.; Li, J.; Bi, H.; Xu, X. Antibacterial secondary metabolites from marine-derived fungus *Aspergillus* sp. IMCASMF180035. *Antibiotics* **2021**, *10*, 377. [CrossRef]
31. Nweze, J.A.; Mbaoji, F.N.; Huang, G.; Li, Y.; Yang, L.; Zhang, Y.; Huang, S.; Pan, L.; Yang, D. Antibiotics development and the potentials of marine-derived compounds to stem the tide of multidrug-resistant pathogenic bacteria, fungi, and protozoa. *Mar. Drugs* **2020**, *18*, 145. [CrossRef] [PubMed]
32. Chu, Y.C.; Chang, C.H.; Liao, H.R.; Fu, S.L.; Chen, J.J. Anti-cancer and anti-inflammatory activities of three new chromone derivatives from the marine-derived *Penicillium citrinum*. *Mar. Drugs* **2021**, *19*, 408. [CrossRef] [PubMed]
33. Thawabteh, A.M.; Swaileh, Z.; Ammar, M.; Jaghama, W.; Yousef, M.; Karaman, R.; Bufo, S.A.; Scrano, L. Antifungal and antibacterial activities of isolated marine compounds. *Toxins* **2023**, *15*, 93. [CrossRef] [PubMed]
34. Yao, F.-H.; Liang, X.; Cheng, X.; Ling, J.; Dong, J.-D.; Qi, S.-H. Antifungal peptides from the marine gorgonian-associated fungus *Aspergillus* sp. SCSIO41501. *Phytochemistry* **2021**, *192*, 112967. [CrossRef]
35. Sweilam, S.H.; Alqarni, M.H.; Youssef, F.S.; Noureini, S.K. Antimicrobial alkaloids from marine-derived fungi as drug leads versus COVID-19 infection: A computational approach to explore their anti-COVID-19 activity and ADMET properties. *Evid.-Based Complement. Altern. Med.* **2022**, *2022*, 1–19. [CrossRef]
36. Abuhijjleh, R.K.; Shabbir, S.; Al-Abd, A.M.; Jiaan, N.H.; Alshamil, S.; El-labbad, E.M.; Khalifa, S.I. Bioactive marine metabolites derived from the Persian Gulf compared to the Red Sea: Similar environments and wide gap in drug discovery. *PeerJ* **2021**, *9*, e11778. [CrossRef]
37. Giri, A.; Ohshima, T. Bioactive marine peptides: Nutraceutical value and novel approaches. *Adv. Food Nutr. Res.* **2012**, *65*, 73–105. [CrossRef]
38. Elissawy, A.M.; El-Shazly, M.; Ebada, S.S.; Singab, A.B.; Proksch, P. Bioactive terpenes from marine-derived fungi. *Mar. Drugs* **2015**, *13*, 1966–1992. [CrossRef]
39. Jiang, M.; Chen, S.; Li, J.; Liu, L. The biological and chemical diversity of tetramic acid compounds from marine-derived microorganisms. *Mar. Drugs* **2020**, *18*, 114. [CrossRef]
40. Fouillaud, M.; Venkatachalam, M.; Girard-Valenciennes, E.; Caro, Y.; Dufosse, L. Anthraquinones and derivatives from marine-derived fungi: Structural diversity and selected biological activities. *Mar. Drugs* **2016**, *14*, 64. [CrossRef]
41. Hafez Ghoran, S.; Taktaz, F.; Ayatollahi, S.A.; Kijjoa, A. Anthraquinones and their analogues from marine-derived fungi: Chemistry and biological activities. *Mar. Drugs* **2022**, *20*, 474. [CrossRef] [PubMed]
42. Li, Y.X.; Himaya, S.W.; Kim, S.K. Triterpenoids of marine origin as anti-cancer agents. *Molecules* **2013**, *18*, 7886–7909. [CrossRef] [PubMed]
43. Wang, K.W.; Ding, P. New bioactive metabolites from the marine-derived fungi *Aspergillus*. *Mini-Rev. Med. Chem.* **2018**, *18*, 1072–1094. [CrossRef] [PubMed]
44. Xu, K.; Yuan, X.L.; Li, C.; Li, A.X. Recent discovery of heterocyclic alkaloids from marine-derived *Aspergillus* species. *Mar. Drugs* **2020**, *18*, 54. [CrossRef] [PubMed]
45. Al-Rajhi, A.M.H.; Mashraqi, A.; Al Abboud, M.A.; Shater, A.M.; Al Jaouni, S.K.; Selim, S.; Abdelghany, T.M. Screening of bioactive compounds from endophytic marine-derived fungi in saudi arabia: Antimicrobial and anticancer potential. *Life* **2022**, *12*, 1182. [CrossRef]
46. Liu, C.C.; Zhang, Z.Z.; Feng, Y.Y.; Gu, Q.Q.; Li, D.H.; Zhu, T.J. Secondary metabolites from antarctic marine-derived fungus *Penicillium crustosum* HDN153086. *Nat. Prod. Res.* **2019**, *33*, 414–419. [CrossRef]
47. Zheng, Y.Y.; Ma, Z.L.; Wu, J.S.; Shao, C.L.; Yao, G.S.; Wang, C.Y. Induction of secondary metabolite biosynthesis by deleting the histone deacetylase HdaA in the marine-derived fungus *Aspergillus terreus* RA2905. *J. Fungi* **2022**, *8*, 1024. [CrossRef]
48. Zhang, K.; Zhang, X.; Lin, R.; Yang, H.; Song, F.; Xu, X.; Wang, L. New secondary metabolites from the marine-derived fungus *Talaromyces mangshanicus* BTBU20211089. *Mar. Drugs* **2022**, *20*, 79. [CrossRef]
49. Zaman, K.A.U.; Wu, X.; Sarotti, A.M.; Cao, S. New and bioactive polyketides from Hawaiian marine-derived fungus *Trichoderma* sp. FM652. *Nat. Prod. Res.* **2022**, *36*, 5984–5990. [CrossRef]

50. Xie, M.M.; Jiang, J.Y.; Zou, Z.B.; Xu, L.; Zhang, Y.; Wang, C.F.; Liu, C.B.; Yan, Q.X.; Liu, Z.; Yang, X.W. Chemical constituents of the deep-sea-derived fungus *Cladosporium oxysporum* 170103 and their antibacterial effects. *Chem. Biodivers.* **2022**, *19*, e202200963. [CrossRef]
51. Wang, Y.; Chen, W.; Xu, Z.; Bai, Q.; Zhou, X.; Zheng, C.; Bai, M.; Chen, G. Biological secondary metabolites from the lumnitzera littorea-derived fungus *Penicillium oxalicum* HLLG-13. *Mar. Drugs* **2022**, *21*, 22. [CrossRef] [PubMed]
52. Xu, X.; Guo, S.; Chen, H.; Zhang, Z.; Li, X.; Wang, W.; Guo, L. Bioassay-guided isolation and characterization of antibacterial compound from *Aspergillus fumigatus* HX-1 associated with Clam. *3 Biotech* **2021**, *11*, 193. [CrossRef] [PubMed]
53. Youssef, D.T.A.; Shaala, L.A.; Genta-Jouve, G. Asperopiperazines A and B: Antimicrobial and cytotoxic dipeptides from a tunicate-derived fungus *Aspergillus* sp. DY001. *Mar. Drugs* **2022**, *20*, 451. [CrossRef] [PubMed]
54. Hu, Z.; Zhu, Y.; Chen, J.; Chen, J.; Li, C.; Gao, Z.; Li, J.; Liu, L. Sesquiterpenoids with phytotoxic and antifungal activities from a pathogenic fungus *Aspergillus alabamensis*. *J. Agric. Food Chem.* **2022**, *70*, 12065–12073. [CrossRef]
55. Hu, Z.; Zhu, Y.; Chen, J.; Chen, J.; Li, C.; Gao, Z.; Li, J.; Liu, L. Discovery of novel bactericides from *Aspergillus alabamensis* and their antibacterial activity against fish pathogens. *J. Agric. Food Chem.* **2023**, *71*, 4298–4305. [CrossRef]
56. Zhou, G.; Zhang, X.; Shah, M.; Che, Q.; Zhang, G.; Gu, Q.; Zhu, T.; Li, D. Polyhydroxy p-terphenyls from a mangrove endophytic fungus *Aspergillus candidus* LDJ-5. *Mar. Drugs* **2021**, *19*, 82. [CrossRef]
57. Zang, Z.; Yang, W.; Cui, H.; Cai, R.; Li, C.; Zou, G.; Wang, B.; She, Z. Two antimicrobial heterodimeric tetrahydroxanthones with a 7,7′-linkage from mangrove endophytic fungus *Aspergillus flavus* QQYZ. *Molecules* **2022**, *27*, 2691. [CrossRef]
58. Lin, L.-B.; Gao, Y.-Q.; Han, R.; Xiao, J.; Wang, Y.-M.; Zhang, Q.; Zhai, Y.-J.; Han, W.-B.; Li, W.-L.; Gao, J.-M. Alkylated salicylaldehydes and prenylated indole alkaloids from the endolichenic fungus *Aspergillus chevalieri* and their bioactivities. *J. Agric. Food Chem.* **2021**, *69*, 6524–6534. [CrossRef]
59. Li, H.-L.; Yang, S.-Q.; Li, X.-M.; Li, X.; Wang, B.-G. Structurally diverse alkaloids produced by *Aspergillus creber* EN-602, an endophytic fungus obtained from the marine red alga *Rhodomela confervoides*. *Bioorg. Chem.* **2021**, *110*, 104822. [CrossRef]
60. Fang, S.-T.; Liu, X.-H.; Yan, B.-F.; Miao, F.-P.; Yin, X.-L.; Li, W.-Z.; Ji, N.-Y. Terpenoids from the marine-derived fungus *Aspergillus* sp. RR-YLW-12, associated with the red alga *Rhodomela confervoides*. *J. Nat. Prod.* **2021**, *84*, 1763–1771. [CrossRef]
61. Zeng, Q.; Chen, Y.; Wang, J.; Shi, X.; Che, Y.; Chen, X.; Zhong, W.; Zhang, W.; Wei, X.; Wang, F.; et al. Diverse secondary metabolites from the coral-derived fungus *Aspergillus hiratsukae* SCSIO 5Bn$_1$003. *Mar. Drugs* **2022**, *20*, 150. [CrossRef] [PubMed]
62. Chen, X.Y.; Zeng, Q.; Chen, Y.C.; Zhong, W.M.; Xiang, Y.; Wang, J.F.; Shi, X.F.; Zhang, W.M.; Zhang, S.; Wang, F.Z. Chevalones H-M: Six new alpha-pyrone meroterpenoids from the gorgonian coral-derived fungus *Aspergillus hiratsukae* SCSIO 7S2001. *Mar. Drugs* **2022**, *20*, 71. [CrossRef] [PubMed]
63. Peng, Q.; Chen, W.; Lin, X.; Xiao, J.; Liu, Y.; Zhou, X. Butenolides from the coral-derived fungus *Aspergillius terreus* SCSIO41404. *Mar. Drugs* **2022**, *20*, 212. [CrossRef] [PubMed]
64. Zhang, Y.; Li, Z.; Huang, B.; Liu, K.; Peng, S.; Liu, X.; Gao, C.; Liu, Y.; Tan, Y.; Luo, X. Anti-osteoclastogenic and antibacterial effects of chlorinated polyketides from the Beibu gulf coral-derived fungus *Aspergillus unguis* GXIMD 02505. *Mar. Drugs* **2022**, *20*, 178. [CrossRef] [PubMed]
65. Buttachon, S.; Ramos, A.A.; Inacio, A.; Dethoup, T.; Gales, L.; Lee, M.; Costa, P.M.; Silva, A.M.S.; Sekeroglu, N.; Rocha, E.; et al. Bis-indolyl benzenoids, hydroxypyrrolidine derivatives and other constituents from cultures of the marine sponge-associated fungus *Aspergillus candidus* KUFA. *Mar. Drugs* **2018**, *16*, 119. [CrossRef]
66. Machado, F.P.; Kumla, D.; Pereira, J.A.; Sousa, E.; Dethoup, T.; Freitas-Silva, J.; Costa, P.M.; Mistry, S.; Silva, A.M.S.; Kijjoa, A. Prenylated phenylbutyrolactones from cultures of a marine sponge-associated fungus *Aspergillus flavipes* KUFA1152. *Phytochemistry* **2021**, *185*, 112709. [CrossRef]
67. Liu, J.; Yu, R.; Jia, J.; Gu, W.; Zhang, H. Assignment of absolute configurations of two promising anti-helicobacter pylori agents from the marine sponge-derived fungus *Aspergillus niger* L14. *Molecules* **2021**, *26*, 5061. [CrossRef]
68. Guo, C.; Wang, P.; Pang, X.; Lin, X.; Liao, S.; Yang, B.; Zhou, X.; Wang, J.; Liu, Y. Discovery of a dimeric zinc complex and five cyclopentenone derivatives from the sponge-associated fungus *Aspergillus ochraceopetaliformis*. *ACS Omega* **2021**, *6*, 8942–8949. [CrossRef]
69. Li, J.X.; Xu, Q.H.; Shang, R.Y.; Liu, Q.; Luo, X.C.; Lin, H.W.; Jiao, W.H. Aspergetherins A-D, new chlorinated biphenyls with anti-MRSA activity from the marine sponge symbiotic fungus *Aspergillus terreus* 164018. *Chem. Biodivers.* **2023**, e202300010. [CrossRef]
70. Liu, Y.; Ding, L.; He, J.; Zhang, Z.; Deng, Y.; He, S.; Yan, X. A new antibacterial chromone from a marine sponge-associated fungus *Aspergillus* sp. LS57. *Fitoterapia* **2021**, *154*, 105004. [CrossRef]
71. Liu, Y.; Ding, L.; Shi, Y.; Yan, X.; Wu, B.; He, S. Molecular networking-driven discovery of antibacterial perinadines, new tetracyclic alkaloids from the marine sponge-derived fungus *Aspergillus* sp. *ACS Omega* **2022**, *7*, 9909–9916. [CrossRef] [PubMed]
72. Zhang, R.; Wang, H.; Chen, B.; Dai, H.; Sun, J.; Han, J.; Liu, H. Discovery of anti-MRSA secondary metabolites from a marine-derived fungus *Aspergillus fumigatus*. *Mar. Drugs* **2022**, *20*, 302. [CrossRef] [PubMed]
73. Anh, C.V.; Kwon, J.-H.; Kang, J.S.; Lee, H.-S.; Heo, C.-S.; Shin, H.J. Antibacterial and cytotoxic phenolic polyketides from two marine-derived fungal strains of *Aspergillus unguis*. *Pharmaceuticals* **2022**, *15*, 74. [CrossRef]
74. Wu, J.; Shui, H.; Zhang, M.; Zeng, Y.; Zheng, M.; Zhu, K.K.; Wang, S.B.; Bi, H.; Hong, K.; Cai, Y.S. Aculeaxanthones A-E, new xanthones from the marine-derived fungus *Aspergillus aculeatinus* WHUF0198. *Front. Microbiol.* **2023**, *14*, 1138830. [CrossRef]

75. Xiang, Y.; Zeng, Q.; Mai, Z.-M.; Chen, Y.-C.; Shi, X.-F.; Chen, X.-Y.; Zhong, W.-M.; Wei, X.-Y.; Zhang, W.-M.; Zhang, S.; et al. Asperorydines N-P, three new cyclopiazonic acid alkaloids from the marine-derived fungus *Aspergillus flavus* SCSIO F025. *Fitoterapia* **2021**, *150*, 104839. [CrossRef]
76. Yan, L.-H.; Li, X.-M.; Chi, L.-P.; Li, X.; Wang, B.-G. Six new antimicrobial metabolites from the deep-sea sediment-derived fungus *Aspergillus fumigatus* SD-406. *Mar. Drugs* **2021**, *20*, 4. [CrossRef]
77. Cen, S.; Jia, J.; Ge, Y.; Ma, Y.; Li, X.; Wei, J.; Bai, Y.; Wu, X.; Song, J.; Bi, H.; et al. A new antibacterial 3,5-dimethylorsellinic acid-based meroterpene from the marine fungus *Aspergillus* sp. CSYZ-1. *Fitoterapia* **2021**, *152*, 104908. [CrossRef]
78. Yang, J.; Gong, L.; Guo, M.; Jiang, Y.; Ding, Y.; Wang, Z.; Xin, X.; An, F. Bioactive indole diketopiperazine alkaloids from the marine endophytic fungus *Aspergillus* sp. YJ191021. *Mar. Drugs* **2021**, *19*, 157. [CrossRef]
79. Lv, H.; Zhang, J.; Xue, Y.; Li, S.; Sun, X.; Jia, J.; Bi, H.; Wang, S.; Su, H.; Zhu, M.; et al. Two new austocystin analogs from the marine-derived fungus *Aspergillus* sp. WHUF05236. *Chem. Biodivers.* **2022**, *19*, e202200207. [CrossRef]
80. Quang, T.H.; Phong, N.V.; Anh, L.N.; Hanh, T.T.H.; Cuong, N.X.; Ngan, N.T.T.; Trung, N.Q.; Nam, N.H.; Minh, C.V. Secondary metabolites from a peanut-associated fungus *Aspergillus niger* IMBC-NMTP01 with cytotoxic, anti-inflammatory, and antimicrobial activities. *Nat. Prod. Res.* **2022**, *36*, 1215–1223. [CrossRef]
81. Tian, Y.; Li, Y. A review on bioactive compounds from marine-derived chaetomium species. *J. Microbiol. Biotechnol.* **2022**, *32*, 541–550. [CrossRef] [PubMed]
82. Xu, J.; Yi, M.; Ding, L.; He, S. A review of anti-inflammatory compounds from marine fungi, 2000–2018. *Mar. Drugs* **2019**, *17*, 636. [CrossRef] [PubMed]
83. Sun, L.; Wang, H.; Yan, M.; Sai, C.; Zhang, Z. Research advances of bioactive sesquiterpenoids isolated from marine-derived *Aspergillus* sp. *Molecules* **2022**, *27*, 7376. [CrossRef] [PubMed]
84. Arockianathan, P.M.; Mishra, M.; Niranjan, R. Recent status and advancements in the development of antifungal agents: Highlights on plant and marine based antifungals. *Curr. Top. Med. Chem.* **2019**, *19*, 812–830. [CrossRef]
85. Sharma, D.; Bisht, G.S. Recent updates on antifungal peptides. *Mini-Rev. Med. Chem.* **2020**, *20*, 260–268. [CrossRef]
86. Montuori, E.; de Pascale, D.; Lauritano, C. Recent discoveries on marine organism immunomodulatory activities. *Mar. Drugs* **2022**, *20*, 422. [CrossRef]
87. Hou, X.; Zhang, X.; Xue, M.; Zhao, Z.; Zhang, H.; Xu, D.; Lai, D.; Zhou, L. Recent advances in sorbicillinoids from fungi and their bioactivities (Covering 2016–2021). *J. Fungi* **2022**, *8*, 62. [CrossRef]
88. Lima, R.N.; Porto, A.L.M. Recent advances in marine enzymes for biotechnological processes. *Adv. Food Nutr. Res.* **2016**, *78*, 153–192.
89. Wang, Y.-N.; Meng, L.-H.; Wang, B.-G. Progress in research on bioactive secondary metabolites from deep-sea derived microorganisms. *Mar. Drugs* **2020**, *18*, 614. [CrossRef]
90. Hang, S.; Chen, H.; Wu, W.; Wang, S.; Fang, Y.; Sheng, R.; Tu, Q.; Guo, R. Progress in isoindolone alkaloid derivatives from marine microorganism: Pharmacology, preparation, and mechanism. *Mar. Drugs* **2022**, *20*, 405. [CrossRef]
91. Yang, X.; Liu, J.; Mei, J.; Jiang, R.; Tu, S.; Deng, H.; Liu, J.; Yang, S.; Li, J. Origins, structures, and bioactivities of secondary metabolites from marine-derived *Penicillium* fungi. *Mini-Rev. Med. Chem.* **2021**, *21*, 2000–2019. [CrossRef] [PubMed]
92. Li, X.; Xu, J.; Wang, P.; Ding, W. Novel indole diketopiperazine stereoisomers from a marine-derived fungus *Aspergillus* sp. *Mycology* **2023**, *14*, 1–10. [CrossRef] [PubMed]
93. Zhao, L.; Lin, X.; Fu, J.; Zhang, J.; Tang, W.; He, Z. A novel bi-functional fibrinolytic enzyme with anticoagulant and thrombolytic activities from a marine-derived fungus *Aspergillus versicolor* ZLH-1. *Mar. Drugs* **2022**, *20*, 356. [CrossRef] [PubMed]
94. Xu, J.; Liu, P.; Li, X.; Gan, L.; Wang, P. Novel stemphol derivatives from a marine fungus *Pleospora* sp. *Nat. Prod. Res.* **2018**, *33*, 367–373. [CrossRef]
95. Song, Z.; Gao, J.; Hu, J.; He, H.; Huang, P.; Zhang, L.; Song, F. One new xanthenone from the marine-derived fungus *Aspergillus versicolor* MF160003. *Nat. Prod. Res.* **2020**, *34*, 2907–2912. [CrossRef]
96. Wu, J.S.; Shi, X.H.; Yao, G.S.; Shao, C.L.; Fu, X.M.; Zhang, X.L.; Guan, H.S.; Wang, C.Y. New thiodiketopiperazine and 3,4-dihydroisocoumarin derivatives from the marine-derived fungus *Aspergillus terreus*. *Mar. Drugs* **2020**, *18*, 132. [CrossRef]
97. Xu, Y.; Huang, R.; Liu, H.; Yan, T.; Ding, W.; Jiang, Y.; Wang, P.; Zheng, D.; Xu, J. New polyketides from the marine-derived fungus *Letendraea* Sp. 5XNZ4-2. *Mar. Drugs* **2019**, *18*, 18. [CrossRef]
98. Xu, X.; Li, J.; Zhang, K.; Wei, S.; Lin, R.; Polyak, S.W.; Yang, N.; Song, F. New isocoumarin analogues from the marine-derived fungus *Paraphoma* sp. CUGBMF180003. *Mar. Drugs* **2021**, *19*, 313. [CrossRef]
99. Hu, J.; Li, Z.; Gao, J.; He, H.; Dai, H.; Xia, X.; Liu, C.; Zhang, L.; Song, F. New diketopiperazines from a marine-derived fungus strain *Aspergillus versicolor* MF180151. *Mar. Drugs* **2019**, *17*, 262. [CrossRef]

marine drugs

Article

Cytosporin Derivatives from Arctic-Derived Fungus *Eutypella* sp. D-1 via the OSMAC Approach

Hao-Bing Yu [1,†], Zhe Ning [1,†], Bo Hu [1], Yu-Ping Zhu [2], Xiao-Ling Lu [3], Ying He [1], Bing-Hua Jiao [3] and Xiao-Yu Liu [1,*]

1 Naval Medical Center of PLA, Department of Marine Biomedicine and Polar Medicine, Naval Medical University, Shanghai 200433, China; yuhaobing1986@126.com (H.-B.Y.); ningzhe95@163.com (Z.N.); hb8601@163.com (B.H.); yinghe_hys@163.com (Y.H.)
2 Basic Medical Experimental Teaching Center, College of Basic Medical Sciences, Naval Medical University, Shanghai 200433, China; zhuyuping72@hotmail.com
3 Department of Biochemistry and Molecular Biology, College of Basic Medical Sciences, Naval Medical University, Shanghai 200433, China; luxiaoling80@126.com (X.-L.L.); jiaobh@live.cn (B.-H.J.)
* Correspondence: biolxy@163.com; Tel.: +86-21-81883267
† These authors contributed equally to this work.

Abstract: A chemical investigation of the Arctic-derived fungus *Eutypella* sp. D-1 based on the OSMAC (one strain many compounds) approach resulted in the isolation of five cytosporin polyketides (compounds **1–3** and **11–12**) from rice medium and eight cytosporins (compounds **2** and **4–11**) from solid defined medium. The structures of the seven new compounds, eutypelleudesmane A (**1**), cytosporin Y (**2**), cytosporin Z (**3**), cytosporin Y_1 (**4**), cytosporin Y_2 (**5**), cytosporin Y_3 (**6**), and cytosporin E_1 (**7**), were elucidated by analyzing their detailed spectroscopic data. Structurally, cytosporin Y_1 (**4**) may be a key intermediate in the biosynthesis of the isolated cytosporins, rather than an end product. Compound **1** contained a unique skeleton formed by the ester linkage of two moieties, cytosporin F (**12**) and the eudesmane-type sesquiterpene dihydroalanto glycol. Additionally, the occurrence of cyclic carbonate moieties in compounds **6** and **7** was found to be rare in nature. The antibacterial, immunosuppressive, and cytotoxic activities of all compounds derived from *Eutypella* sp. D-1 were evaluated. Unfortunately, only compounds **3**, **6**, **8**, and **10–11** displayed immunosuppressive activity, with inhibitory rates of 62.9%, 59.5%, 67.8%, 55.8%, and 68.7%, respectively, at a concentration of 5 μg/mL.

Keywords: cytosporin; arctic fungus; *Eutypella* sp.; immunosuppressive activity

Citation: Yu, H.-B.; Ning, Z.; Hu, B.; Zhu, Y.-P.; Lu, X.-L.; He, Y.; Jiao, B.-H.; Liu, X.-Y. Cytosporin Derivatives from Arctic-Derived Fungus *Eutypella* sp. D-1 via the OSMAC Approach. *Mar. Drugs* **2023**, *21*, 382. https://doi.org/10.3390/md21070382

Academic Editor: Dehai Li

Received: 14 June 2023
Revised: 26 June 2023
Accepted: 27 June 2023
Published: 28 June 2023

1. Introduction

Fungi have proven to be a valuable source of new secondary metabolites with a wide spectrum of biological activities [1]. Natural products from Polar fungi remain the non-negligible sources of pharmacologically active compounds [1]. Cytosporins are a family of hexahydrobenzopyran metabolites derived from fungi with a distinct heptene side chain residue [2]. Initially isolated from endophytic *Cytospora* sp. in 1996, cytosporins were recognized as inhibitors of angiotensin II binding inhibitors [2]. To date, nearly 30 natural cytosporins of this structural class have been predominantly isolated from four genera of fungi: *Cytospora* sp. [2], *Pestalotiopsis* sp. [3], *Eutypella* sp. [4,5], and *Pseudopestalotiopsis* sp. [6]. The cytosporin family exhibits diverse bioactive effects, including cytotoxic, antibacterial, and antagonistic activity [2,4]. Besides cytosporins, *Eutypella* species have been extensively investigated as a rich source of various bioactive compounds, pimarane diterpenes, γ-lactones, benzopyrans, *ent*-eudesmanes, cytochalasins, and dipeptides, which display a spectrum of bioactivities [7–9].

The OSMAC approach has emerged as a powerful tool in the field of natural product biodiscovery, stimulating the production of a wider range of new metabolites [10]. During our exploration of structurally diverse bioactive natural products from polar fungi, we discovered a series of terpenoids with unique skeleton characteristics from the talented Arctic fungus strain *Eutypella* sp. D-1 [5,7,9]. This strain has proven to be a prolific source of metabolites with diverse biological activities [7,9]. To enhance the chemical diversity of *Eutypella* sp. D-1, we employed the one strain many compounds (OSMAC) strategy, utilizing different culture conditions. Through high-performance liquid chromatography (HPLC) analysis, some structural analogs during fermentation on two distinct media, solid rice medium and defined solid medium, were dramatically discovered. Subsequent chemical investigation led to the isolation of 12 cytosporin derivatives, including seven new cytosporins—eutypelleudesmane A (**1**), cytosporin Y (**2**), cytosporin Z (**3**), cytosporin Y_1 (**4**), cytosporin Y_2 (**5**), cytosporin Y_3 (**6**), and cytosporin E_1 (**7**)—together with five known biogenetic-related analogs—cytosporin X (**8**), cytosporin E (**9**), cytosporin L (**10**), and cytosporins D and F (**11–12**) (Figure 1). Herein, we present the detailed purification, structure elucidation, and bioactive evaluation of these compounds.

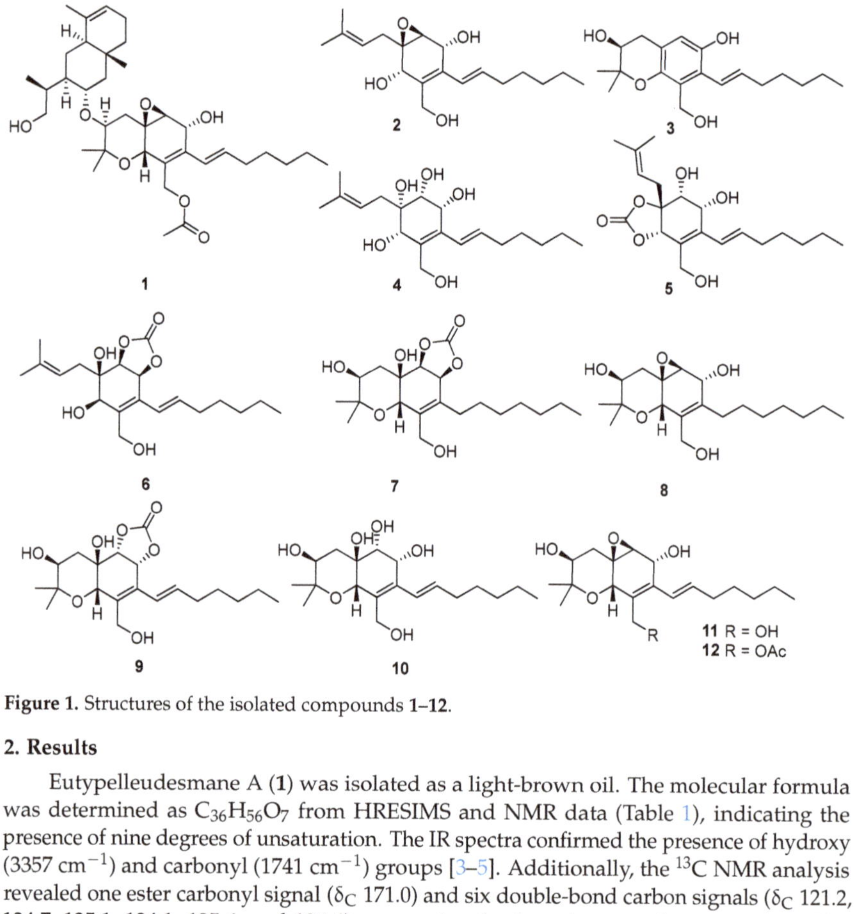

Figure 1. Structures of the isolated compounds **1–12**.

2. Results

Eutypelleudesmane A (**1**) was isolated as a light-brown oil. The molecular formula was determined as $C_{36}H_{56}O_7$ from HRESIMS and NMR data (Table 1), indicating the presence of nine degrees of unsaturation. The IR spectra confirmed the presence of hydroxy (3357 cm^{-1}) and carbonyl (1741 cm^{-1}) groups [3–5]. Additionally, the ^{13}C NMR analysis revealed one ester carbonyl signal (δ_C 171.0) and six double-bond carbon signals (δ_C 121.2, 124.7, 125.1, 134.1, 135.6, and 136.4), accounting for four degrees of unsaturation. The remaining five degrees of unsaturation were attributed to the pentacyclic ring structure present in the molecule.

Table 1. ^{1}H (500 MHz) and ^{13}C NMR (125 MHz) spectroscopic data of **1** in $CDCl_3$.

Position	δ_C	δ_H, mult. (J in Hz)	Position	δ_C	δ_H, mult. (J in Hz)
2	76.6, C		20	14.0, CH_3	0.89, t (7.0)
3	73.8, CH	3.72, m	21	171.0, C	
4α	35.5, CH_2	1.72, dd (12.5, 5.0)	22	20.9, CH_3	2.08, s
4β		2.27, t (12.0)	23	37.5, CH_2	1.35, m
5	55.7, C		24	22.7, CH_2	1.99, m
6	59.7, CH	3.32, s	25	121.2, CH	5.31, brs
7	64.6, CH	4.73, s	26	134.1, C	
8	135.6, C		27	46.3, CH	1.97, m
9	125.1, C		28α	28.5, CH_2	1.66, m
10	67.5, CH	4.38, s	28β		1.23, m
11	16.0, CH_3	1.30, s	29	49.7, CH	1.50, m
12	27.7, CH_3	1.29, s	30	66.7, CH	3.87, td (11.0, 5.0)
13a	61.5, CH_2	4.67, d (12.5)	31α	49.4, CH_2	1.15, m
13b		4.81, d (12.5)	31β		1.79, dd (12.5, 5.0)
14	124.7, CH	6.33, d (16.0)	32	33.7, C	
15	136.4, CH	6.16, m	33	39.0, CH	1.87, m
16	33.5, CH_2	2.17, m	34	67.5, CH_2	3.67, dd (7.5, 4.0)
17	28.8, CH_2	1.42, m	35	11.9, CH_3	1.01, d (7.0)
18	31.4, CH_2	1.27, m	36	16.5, CH_3	0.79, s
19	22.5, CH_2	1.28, m	37	21.2, CH_3	1.61, s

Upon comparing the 1D NMR data of compound **1** and cytosporin F (**12**), it was observed that one set of signals was similar to compound **12**, while the remaining signals resembled a derivative of eudesmane-type sesquiterpene, dihydroalanto glycol. By utilizing 2D NMR correlations (Figure 2), these two structural fragments, labeled as A and B, were deduced. The COSY spectrum revealed the presence of seven continuous spin systems: (a) C-3–C-4, (b) C-6–C-7, (c) C-14–C-15–C-16–C-17–C-18–C-19–C-20, (d) C-23–C-24–C-25, (e) C-27–C-28–C-29–C-30–C-31, (f) C-29–C-33–C-34, and (g) C-33–C-35 (Figure 2). Fragment A, comprising C-2 to C-22, exhibited similarity to compound **12** based on a comparison of their 1D NMR spectra. HMBC correlations from H-4α to C-2, C-5, C-6, and C-10; from H-6 to C-8; from H-7 to C-5, C-8, and C-9; from H-10 to C-5, C-6, C-8, and C-9; from H_3-11 and H_3-12 to C-2 and C-3; and from H_2-13 to C-8, C-9, and C-10 were detected. These correlations, along with the chemical shift of C-2 (δ_C 76.6) and C-10 (δ_C 67.5), indicated the formation of two six-membered rings by connecting C-5 (δ_C 55.7) with C-10 and C-2 with C-10 via an *O*-atom, as well as the location of the two methyl groups CH_3-11 and CH_3-12 both at C-2 and one methylene group CH_2-13 at C-9. The presence of an oxirane resulting from the conjugation of C-5 and C-6 via *O*-atom was supported by the downfield shift of C-5 and C-6 (δ_C 59.7) [3,4]. Furthermore, the direct linkage between C-8 and C-14 was established by HMBC correlations from H-14 to C-7, C-8, and C-9. An additional acetyl group was identified to be connected to C-13 based on the HMBC correlations from H-13 and H-22 to C-21 and the chemical shift of C-13 (δ_C 61.5). Fragment B, spanning from C-23 to C-37, exhibited characteristics of a eudesmane-type sesquiterpene moiety, as deduced from the analysis of the remaining ^{1}H and ^{13}C NMR data. HMBC correlations from H-31α and H-31β to C-27 and C-32; from H_2-23 to C-27 and C-32; and from H_3-36 to C-23, C-27, C-31, and C-32 confirmed the presence of a linkage of C-23, C-27, and C-31 via the quaternary carbon C-32, and placed the methyl group CH_3-36 at C-32 as well. The methyl group CH_3-37 was demonstrated to be connected to C-25 and C-27 via C-26 by the HMBC correlations from H_3-37 to C-25, C-26, and C-27. The linkage of fragments A with B through C-3 (δ_C 73.8) and C-30 (δ_C 66.7) via an *O*-atom was supported by the downfield resonance of C-3 and C-30, along with HMBC correlations from H-30 to C-3. Additionally, the connection of two hydroxyl groups with a downfield carbon shift at C-7 (δ_C 64.6) and C-34 (δ_C 67.5) were determined to satisfy the molecular formula. Consequently, the planar structure of **1** was established as depicted.

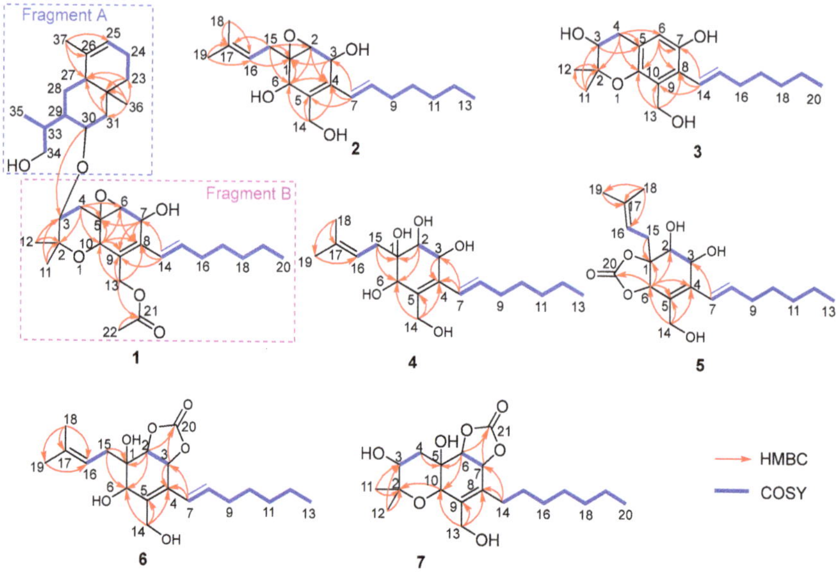

Figure 2. Key COSY and HMBC correlations of 1–7.

The relative configuration of 1 was established by analyzing coupling constants and NOESY experiments [9]. The *trans* configuration of the conjugated C-14/C-15 double bond was inferred based on the large coupling constant (16.0 Hz) and the NOESY correlations of H-14/H_2-16. The observed similarity in the NMR chemical shift values and NOESY correlations of H-7/H-10, H-10/H_3-12, H_3-12/H-4β, H-4α/H-6, and H-3/H_3-11 indicated that the relative configurations of fragment A in 1 were identical to those of 12 [3–5]. Additional NOESY correlations of H-30/H_3-35, H-30/H-36β, and H-31β/H-36β and those of H-27/H-29, H-27/H-31α, and H-27/H-33 indicated the β-orientation and α-orientation of these protons in fragment B, respectively (Figure 3). Furthermore, the absence of a NOESY correlation between H-3 and H-30 supported the *trans* relationship between these two protons [5]. Thus, the relative structure of 1 was determined. Furthermore, the characteristic positive Cotton effect at 242 nm in the CD spectrum of 1 was virtually identical to that of cytosporins D and F (11–12) (Figure 4) [5], which suggested the absolute configuration of 1 was assigned as 3*S*,5*R*,6*S*,7*R*,10*S*,27*S*,29*S*,30*R*,32*S*,33*R*.

Cytosporin X (2) was obtained as a light-brown oil and determined to have a molecular formula of $C_{19}H_{30}O_4$ based on HRESIMS and NMR data, corresponding to an unsaturation index of 5. The presence of hydroxy functionality was indicated by IR absorption bands at 3359 cm^{-1}. The ^{13}C NMR (Table 2) and DEPT spectra revealed the presence of 19 carbons, including six double-bond carbon signals (δ_C 117.3, 124.6, 131.4, 131.6, 135.4, and 135.9) and five oxygenated carbon signals (δ_C 57.4, 59.3, 62.2, 64.3, and 69.5). The COSY spectrum of 2 showed three distinct spin systems: C-2/C-3, C-7/C-8/C-9/C-10/C-11/C-12/C-13, and C-15/C-16 (Figure 2). HMBC correlations from H-2 to C-4 and C-6; from H-3 to C-1 and C-4; from H-6 to C-1, C-2, and C-4; and from H_2-14 to C-4, C-5, and C-6, along with the comparison of the chemical shifts of C-1 (δ_C 59.3) and C-2 (δ_C 57.5) to those of cytosporins D and F [3,4], determined the oxirane-fused cyclohexene moiety with one methylene group (CH_2-14) attached at C-5. The isoamylene group was connected to C-1 based on the HMBC correlations from H_2-15 to C-1, C-2, and C-6, as well as from H_3-18 and H_3-19 to C-16 and C-17. Further HMBC correlations from H-7 to C-3, C-4, and C-5 established the connectivity of C-4 and C-7. With this assignment secured, each of the three oxygenated carbon at C-3 (δ_C 64.3), C-6 (δ_C 69.5), and C-14 (δ_C 62.2) had to be substituted with a hydroxy group to satisfy the molecular formula. The relative stereocenter of 2 was determined from NOESY correlations and coupling constants in comparison with those of **11** and **12** [3,4]. The conjugated C-7/C-8 double bond was assigned as *trans* upon its large coupling constant (16.0 Hz). The NOESY correlations of H-2/H_2-15 and H-6/H_2-15 in $CDCl_3$ and 3-OH/H-6 in DMSO-d_6 (Figure S16), combined with the similarity between the calculated and the experimental ECD spectra, confirmed the absolute configurations of **2** as 1*R*,2*S*,3*R*,6*R* (Figure 4).

Table 2. ^{1}H (500 MHz) and ^{13}C NMR (125 MHz) spectroscopic data of **2** and **3** in $CDCl_3$.

	2			**3**	
Position	**δ_C**	**δ_H, mult. (*J* in Hz)**	**Position**	**δ_C**	**δ_H, mult. (*J* in Hz)**
1	59.3, C		2	77.2, C	
2	57.5, CH	3.29, s	3	69.7, CH	3.79, t (5.0)
3	64.3, CH_2	4.72, s	4α	31.5, CH_2	3.03, dd (17.0, 5.0)
4	131.6, C		4β		2.72, dd (17.0, 5.0)
5	131.4, C		5	118.8, C	
6	69.5, CH	4.45, s	6	115.1, CH	6.61, s
7	124.6, CH	6.28, d (16.0)	7	146.7, C	
8	135.4, CH	6.05, m	8	123.6, C	
9	33.5, CH_2	2.15, m	9	126.8, C	
10	28.9, CH_2	1.41, m	10	144.8, C	
11	31.5, CH_2	1.28, m	11	22.4, CH_3	1.36, s
12	22.5, CH_2	1.29, m	12	24.9, CH_3	1.32, s
13	14.0, CH_3	0.88, t (6.0)	13	58.8, CH_2	4.66, s
14a	62.2, CH_2	4.57, d (12.0)	14	122.7, CH	6.35, d (16.5)
14b		4.06, d (12.0)	15	140.0, CH	5.95, m
15α	29.7, CH_2	2.82, dd (15.0, 8.0)	16	33.4, CH_2	2.27, m
15β		2.30, dd (15.0, 8.0)	17	29.0, CH_2	1.50, m
16	117.3, CH	5.20, t (7.0)	18	31.5, CH_2	1.35, m
17	135.9, C		19	22.6, CH_2	1.35, m
18	18.0, CH_3	1.66, s	20	14.1, CH_3	0.91, t (7.0)
19	25.9, CH_3	1.73, s			

Cytosporin Y (**3**) exhibited a negative HRESIMS with a pseudomolecular ion at *m/z* 319.1912 $[M - H]^-$, consistent with the molecular formula of $C_{19}H_{28}O_4$. The similarity of the ^{1}H and ^{13}C NMR data between **3** and **11** indicated that compound **3** was the derivative of **11**. The presence of a pentasubstituted benzene moiety (δ_H 6.61 (1H, s); δ_C 115.1 (CH), 118.8 (C), 123.6 (CH), 126.8 (C), 144.8 (C), and 146.7 (C)) instead of the oxirane-fused cyclohexene moiety in **11** was suggested by the ^{1}H and ^{13}C NMR spectra. This was further confirmed by the further HMBC correlations from H-4β to C-5, C-6, and C-10; from H-6 to C-7, C-8, and C-10; from H_2-13 to C-8, C-9, and C-10; and from H-14 to C-7, C-8, and C-9 (Figure 2). Additionally, one hydroxy group was attached to C-7, as evidenced by its chemical shift (δ_C 146.7) and the molecular formula. The conjugated C-14/C-15 double bond in **3** was assigned as *trans* based on the similar ^{1}H NMR chemical shift values and coupling constants (16.5 Hz) observed in **3** and **2**. To determine the absolute configuration at C-3 in compound **3**, the specific rotation ($[\alpha]_D^{25}$ +12.3, MeOH, *c* 0.1 and $[\alpha]_D^{25}$ +1.9, $CDCl_3$, *c* 0.1) was measured. The configuration of C-3 could be assigned as *S* by comparison to the literature data for synthetic (*R*)-2,2-dimethylchromane-3,7-diol ($[\alpha]_D^{20}$ −1.2, MeOH, *c* 0.03) [11] and (*S*)-2,2-dimethylchromane-3,7-diol ($[\alpha]_D^{23}$ +11.5, $CHCl_3$, *c* 1.0) [12] (differences in measured versus literature values likely stem from the different concentration and solvent).

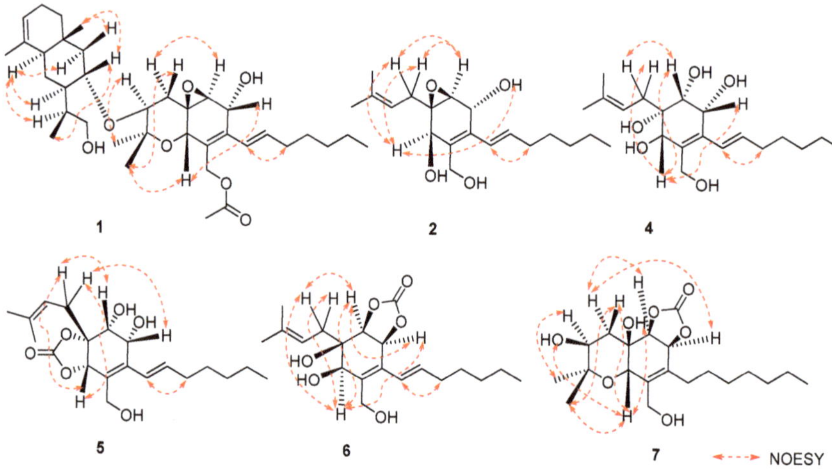

Figure 3. Key NOESY correlations of **1**, **2**, and **4**–**7**.

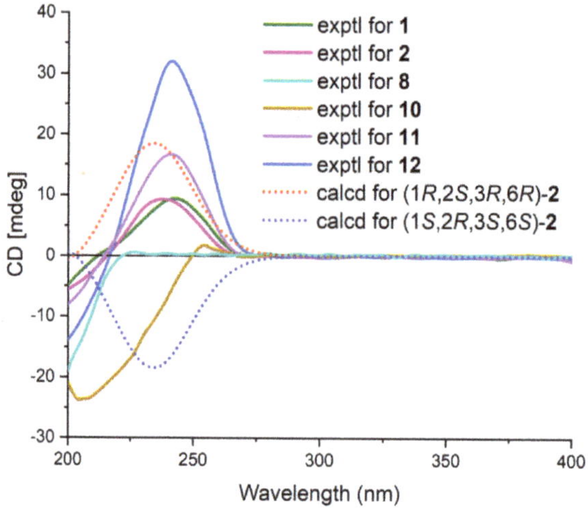

Figure 4. ECD spectra of **1**, **8**, and **10**–**12** and calculated and experimental ECD spectra of **2**.

Cytosporin Y1 (**4**), in the form of a light-yellow oil, had a molecular formula of $C_{19}H_{32}O_5$ based on HRESIMS (*m/z* 363.2139 [M + Na]$^+$), which is larger than that of cytosporin Y (**2**) by 18 amu. The NMR data of **4** (Table 3) were nearly identical to those of **2**, indicating the same carbon skeleton. Considering the degrees of unsaturation of **4**, the observed downfield shift of one quaternary carbon (δ_C 59.3) and one methine (δ_H/δ_C 3.29/57.5) in **2** to δ_C 74.3 and δ_H/δ_C 3.76/75.0 in **4**, respectively, suggested that **4** was the oxirane ring-opening product of **2**. This hypothesis was further supported by further COSY and key HMBC correlations, as shown in Figure 2. The *E*-geometry of the $\Delta^{7,8}$ double bond was deduced from a NOESY correlation between H-7 and H_2-9, as well as the coupling constants (16.5 Hz). Additional NOESY correlations of H-2/H_2-15, H-2/H-6, H-3/H-6, and H-6/H_2-15 indicated the same orientation of these protons. Furthermore, the comparison of the calculated and the experimental ECD spectra confirmed the absolute configurations of **4** as 1*S*,2*R*,3*R*,6*S* (Figure 5).

Table 3. ^{1}H (500 MHz) and ^{13}C NMR (125 MHz) spectroscopic data of **4** and **5** in MeOD-d_4.

	4		**5**	
Position	**δ_C**	**δ_H, mult. (J in Hz)**	**δ_C**	**δ_H, mult. (J in Hz)**
1	74.3, C		84.4, C	
2	75.0, CH	3.76, d (4.5)	71.3, CH	3.94, d (4.5)
3	69.6, CH	4.49, d (4.5)	68.6, CH	4.41, d (4.5)
4	135.8, C		138.3, C	
5	134.5, C		127.3, C	
6	73.6, CH	3.88, s	76.9, CH	5.15, s
7	126.6, CH	6.24, d (16.0)	124.8, CH	6.41, d (16.0)
8	137.2, CH	6.01, m	136.7, CH	6.12, dt (16.0, 7.0)
9	35.0, CH_2	2.17, m	33.1, CH_2	2.21, m
10	30.5, CH_2	1.46, m	28.6, CH_2	1.47, m
11	32.9, CH_2	1.28, m	31.2, CH_2	1.33, m
12	23.9, CH_2	1.34, m	22.2, CH_2	1.33, m
13	14.7, CH_3	0.91, t (7.0)	13.0, CH_3	0.91, t (7.0)
14a	61.1, CH_2	4.25, d (13.0)	58.1, CH_2	4.13, d (13.0)
14b		4.38, d (13.0)		4.51, d (13.0)
15a	35.7, CH_2	2.60, m	31.8, CH_2	2.53, dd (15.0, 8.5)
15b				2.66, dd (15.0, 7.0)
16	119.8, CH	5.40, m	115.3, CH	5.25, m
17	136.1, C		137.7, C	
18	26.7, CH_3	1.76, s	24.9, CH_3	1.77, s
19	18.6, CH_3	1.70, s	16.9, CH_3	1.68, s
20			154.8, C	

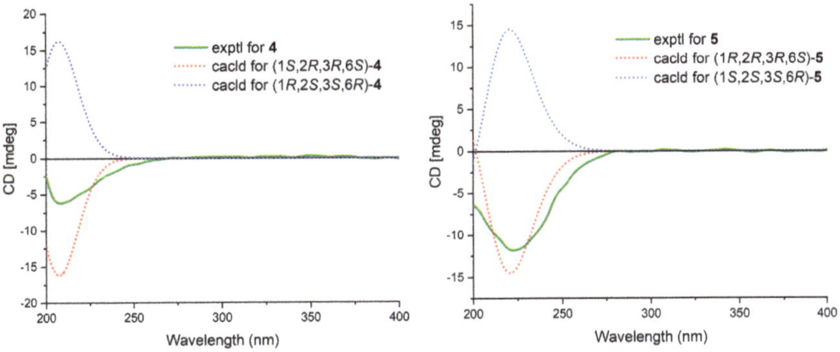

Figure 5. Calculated and experimental ECD spectra of **4** and **5**.

Cytosporin Y2 (**5**) was obtained as a light-yellow oil. Extensive NMR analyses and HRESIMS data (m/z 389.1928 [M + Na]$^+$) led to the determination of its molecular formula as $C_{20}H_{30}O_6$. The overall NMR data of **5** indicated a structure similar to **4**, with the notable difference of an additional quaternary carbon. This carbon was identified as a carbonate moiety based on the strong IR absorption at 1647 cm^{-1} and the diagnostic ^{13}C NMR signal at δ_C 154.8 [4]. Another significant difference was observed for C-1 and C-2, resonating at δ_C 74.3 and 75.0 in compound **4**, whereas in compound **5**, these signals resonated at δ_C 84.4 and 71.3, respectively (Table 3). This observation, along with the key HMBC correlations from H-2 to C-20, led to the linkage of the carbonyl to both oxygen atoms at C-1 and C-2 to form a cyclic carbonate moiety. The relative configurations of **5** were determined via a detailed analysis of the NOESY correlations of H-2/H-15a, H-3/H-15b, H-6/H-15a, H-6/H-15b, and H-7/H_2-9, as well as the coupling constants (16.5 Hz) of H-7/H_2-9. Furthermore, the calculated ECD spectrum of **5** exhibited a close resemblance to the experimental one, confirming the absolute configuration as 1R,2R,3R,6S (Figure 5).

Cytosporin Y3 (**6**) was isolated as a light-yellow oil. Its molecular formula was determined to be $C_{20}H_{30}O_6$, the same as that of **5**, based on HRESIMS data. A comparison of the IR, UV, and NMR data (Table 4) of **6** with those of **5** suggested that **6** was an isomer of **5**. Further analysis of the ^{13}C NMR chemical shift of C-2 (δ_C 79.2) and C-3 (δ_C 75.2), along with the unambiguous HMBC correlations from H-2 and H-3 to C-20 of **2**, revealed that the cyclic carbonate moiety was fused with C-2-C-3 in **6**. The relative configuration and *E*-geometry of the $\Delta^{7,8}$ double bond in **6** were determined from the NOESY correlations of H-2/H-3, H-2/H-6, H-3/H-6, H-2/H-15b, H-6/H-15a, H-6/H-15b, and H-7/H_2-9. The absolute configuration of **6** was subsequently determined to be 1*R*,2*S*,3*S*,6*R* based on the opposite CD spectra (Figure 6) and a comparison of the specific rotation ($[\alpha]_D^{20}$ +37.8, MeOH, *c* 0.1) with that of **4** ($[\alpha]_D^{20}$ −15.8, MeOH, *c* 0.1) and **5** ($[\alpha]_D^{20}$ −60.1, MeOH, *c* 0.1) (Figure 6).

Table 4. ^1H (500 MHz) and ^{13}C NMR (125 MHz) spectroscopic data of **6** and **7** in $CDCl_3$.

	6			**7**	
Position	δ_C	δ_H, **mult. (*J* in Hz)**	**Position**	δ_C	δ_H, **mult. (*J* in Hz)**
1	72.3, C		2	77.7, C	
2	79.2, CH	4.71, dd (8.0, 2.0)	3	71.6, CH	3.96, dd (12.0, 5.0)
3	75.2, CH	5.55, d (8.0)	4α	42.7, CH_2	1.89, dd (12.0, 5.0)
4	129.1, C		4β		2.25, d (12.0)
5	139.4, C		5	68.0, C	
6	71.5, CH	4.14, d (2.0)	6	81.3, CH	4.66, dd (8.0, 2.0)
7	126.1, CH	6.52, d (16.0)	7	78.5, CH	5.23, d (8.0)
8	136.6, CH	6.03, dt (16.0, 7.0)	8	133.3, C	
9	34.9, CH_2	2.21, dd (14.0, 7.0)	9	135.6, C	
10	30.4, CH_2	1.45, m	10	69.7, CH	4.17, d (2.0)
11	32.9, CH_2	1.34, m	11	16.9, CH3	1.26, s
12	23.9, CH_2	1.34, m	12	28.5, CH_3	1.21, s
13	14.7, CH_3	0.91, t (7.0)	13a	60.2, CH_2	4.09, d (12.0)
14a	60.8, CH_2	4.29, d (12.5)	13b		4.29, d, (12.0)
14b		4.43, d (12.5)	14	31.0, CH_2	2.26, m
15a	35.0, CH_2	2.51, dd (15.0, 6.5)	15	30.0, CH_2	1.49, m
15b		2.71, dd (15.0, 8.5)	16	31.1, CH_2	2.25, m
16	118.8, CH	5.38, m	17	30.5, CH_2	1.35, m
17	136.9, C		18	33.3, CH_2	1.31, m
18	26.7, CH_3	1.77, s	19	24.0, CH_2	1.32, m
19	18.6, CH_3	1.71, s	20	14.7, CH_3	0.9, t (7.0)
20	156.8, C		21	156.3, C	

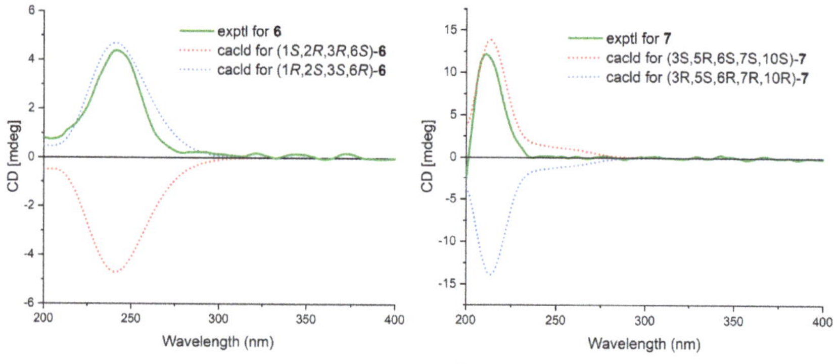

Figure 6. Calculated and experimental ECD spectra of **6** and **7**.

Cytosporin E1 (7) was also purified as a light-yellow oil and exhibited a HRESIMS ion peak at m/z 407.2034 $[M + Na]^+$, consistent with the molecular formula $C_{20}H_{32}O_7$ with five degrees of unsaturation. The ^{1}H and ^{13}C NMR data of **7** (Table 4) closely resembled those of the known compound cytosporin E (**9**), except for two additional methylenes (δ_C/δ_H 31.0/2.26 and 30.0/1.49) in **7** and the absence of two olefinic methines (δ_C/δ_H 123.5/6.39 and 137.1/6.03) in **9**. These observations indicated that C-8 in **7** was substituted by a heptane subunit instead of the 1-heptene part in **9**, which was also confirmed by COSY correlations of H_2-14 (δ_H 2.26)/H_2-15 (δ_H 1.49), H_2-15/H_2-16 (δ_H 2.25), H_2-16/H_2-17 (δ_H 1.35), H_2-17/H_2-18 (δ_H 1.31), H_2-18/H_2-19 (δ_H 1.32), and H_2-19/H_3-20 (δ_H 0.90), as well as HMBC correlations from H_2-14 to C-7 (δ_C 78.5) and C-8 (δ_C 133.3). The relative configuration of **7** was inferred to be different from that of compound **9** through a comparison of the ^{13}C NMR data between **7** (C-6 δ_C 81.3 and C-7 δ_C 78.5) and **9** (C-6 δ_C 81.0 and C-7 δ_C 75.3), as well as the analysis of NOESY correlations of H-3/H_3-11, H-4α/H-6, H-4α/H-7, H-4β/H-10, H-4β/H_3-12, and H-10/H_3-12 in MeOD-d_4 and 3-OH/H-10 and 5-OH/H-10 in DMSO-d_6 (Figure S75). The absolute configurations of **7** were subsequently assigned as 3*S*,5*R*,6*S*,7*S*,10*S* based on the similarity of its calculated and the experimental ECD spectra (Figure 6).

In addition to the seven new compounds **1–7**, the five known cytosporins—cytosporin X (**8**) [13], cytosporin E (**9**) [4], cytosporin L (**10**) [14], cytosporin D (**11**) [4], and cytosporin F (**12**) [3]—were also isolated and identified through a comparison of its NMR spectroscopic data with reported values in the literature.

Structurally, considering the close relationship in biosynthesis among compounds **1–12**, a biosynthetic pathway different from the previous literature for these compounds is proposed (Figure 7) [3]. The possible precursor originated from phenylmethanol [3]. The subsequent addition of an isoprenyl unit, followed by hydroxylation and the addition of an aliphatic chain, would give the intermediate i. The hydroxylation of the C-1/C-6 double bond in i gave rise to the key intermediate **4**. Compound **2** was derived from **4** via a dehydration cyclization reaction. Compound **3** was generated from i via the epoxidation of the C-16/C-17 double bond and a cyclization reaction. Compounds **5** and **6** were derived from the dehydration reaction of compound **4** with carbonic acid by different attack directions and substitution positions, respectively. Compounds **7**, **10**, and **11** were obtained from **6**, **4**, and **2** via the same cyclization reaction as **3**, respectively. Compound **9** was derived from **10** via the carbonic acid substitution, while compound **8** was formed through the hydrogenation of **11**. The cyclization of compound **2**, followed by an acetylation reaction, resulted in the formation of compound **12**. Another possible precursor, the eudesmane-type sesquiterpene dihydroalanto glycol, was generated from farnesyl pyrophosphate with two steps of cyclization, dehydrogenation, and hydroxylation reaction [15]. Then, **1** was formed from the above two precursors, ii and **12**, via a condensation reaction.

All the isolated compounds **1–12** were evaluated for their cytotoxicity against four human cancer cell lines, including DU145, SW1990, Huh7, and PANC-1, and antibacterial activity against *Staphylococcus aureus*, *Escherichia coli*, and *Bacillus subtilis*. Unfortunately, all compounds were not active during the above test, with IC_{50} values higher than 50 μM or MIC values higher than 128 μg/mL. Additional immunosuppressive activity against ConA-induced T cell proliferation for **1–12** was also tested. However, only compounds **3**, **6**, **8**, and **10–11** displayed immunosuppressive activity, demonstrating inhibitory rates of 62.9%, 59.5%, 67.8%, 55.8%, and 68.7%, respectively, at a concentration of 5 μg/mL.

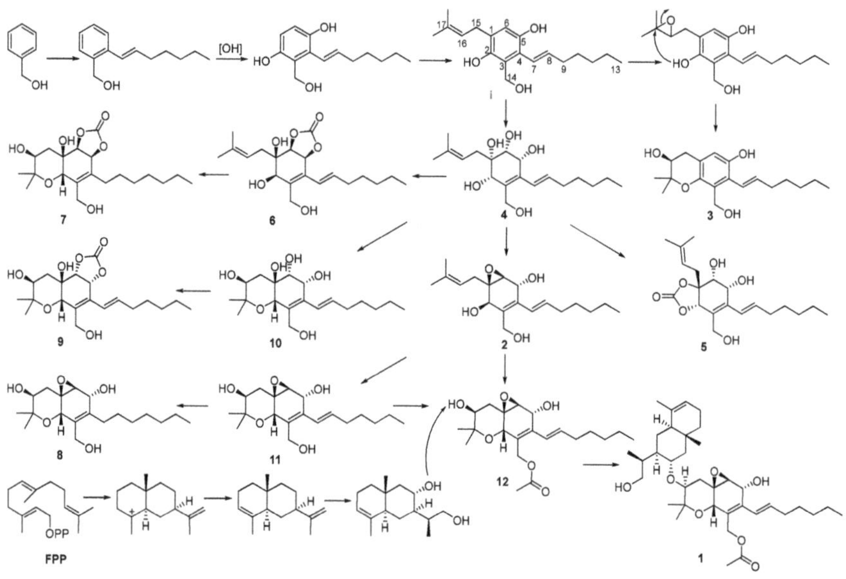

Figure 7. Proposed biogenesis pathway of **1–12**.

3. Materials and Methods

3.1. General Experimental Procedures

Specific rotations and IR (KBr) data were measured on a PerkinElmer model 341 polarimeter (Perkin-Elmer Inc., Waltham, MA, USA) and Jasco FTIR400 spectrometer (Jasco Inc., Tokyo, Japan), respectively. CD and UV spectra were obtained on a Jasco J-715 spectropolarimeter (Jasco Inc., Tokyo, Japan) and UV-8000 spectrophotometer (Shanghai Metash instruments Co., Shanghai, China) in MeOH, respectively. 1D and 2D NMR spectra were acquired using a Bruker AMX-500 instrument (500 MHz for ^{1}H NMR, 125 MHz for ^{13}C NMR) (Bruker Biospin Corp., Billerica, MA, USA) at room temperature. HRESIMS data were measured on an Agilent 6210 LC/MSD TOF mass spectrometer (Agilent Technologies Inc. Lake Forest, CA, USA). HPLC separation was performed using a YMC-Pack Pro C18 (5 µm) column (YMC Co. Ltd., Kyoto, Japan) using a Waters 1525 separation module equipped with a Waters 996 Photodiode Array (PDA) detector (Waters Corp., Milford, MA, USA). Column chromatographic purifications were performed on silica gel 60 (200–300 mesh, Qingdao Ocean Chemical Co., Qingdao, China), ODS (50 µm, YMC Co. Ltd., Kyoto, Japan), and Sephadex LH-20 (Pharmacia Co., Piscataway, NJ, USA).

3.2. Fungal Material

The fungus *Eutypella* sp. D-1 (GenBank accession number FJ430580) was separated from the sample collected near London Island of Kongsfjorden in the Ny-Ålesund District of the Arctic area and recognized based on 18S rDNA gene sequence analysis. The strain (No. D-1) was deposited in the Department of Marine Biomedicine and Polar Medicine, Naval Medical Center of PLA, Naval Medical University.

3.3. Fermentation, Extraction, and Isolation

The fungal strain *Eutypella* sp. D-1 was cultivated in seed medium (PDB 100 mL) in 250 mL Erlenmeyer flasks on a rotatory shaker (180 rpm) at 20 °C for 3 days. Subsequently, seed medium (10 mL) was transferred into 60 × 250 mL Erlenmeyer flasks (40 g of rice and 60 mL of water) and 60 plates of about 20 cm diameter (sucrose 51.4 g, $NaNO_3$ 3.3 g, $K_2HPO_4 \cdot 3H_2O$ 0.07 g, $MgSO_4 \cdot 7H_2O$ 0.4 g, KCl 0.625 g, yeast extract 0.7 g, $CoCl_2 \cdot 6H_2O$ 0.003125 g, $FeSO_4$ 0.01875 g, $CaCl_2$ 0.0065 g, and L-ornithine hydrochloride 15 g, and agar

20.0 g, dissolved in 1 L of water), respectively, and then cultured under static conditions at 20 °C for 45 days.

The rice fermentation was combined and then extracted with CH_2Cl_2−MeOH (1:1, 1 L) three times. The organic solvent was concentrated under reduced pressure and partitioned with EtOAc and H_2O to yield the EtOAc extract (24.5 g). The EtOAc extract was subjected to vacuum liquid chromatography (VLC) on silica gel via gradient elution using CH_2Cl_2/MeOH (80:1, 60:1, 40:1, 20:1, 15:1, 10:1, 0:1, *v*/*v*) as the solvent to give seven fractions (A–G). Fraction B (1.23 g) was chromatographed on a Sephadex LH-20 column using CH_2Cl_2−MeOH (1:1) as mobile phase to afford three subfractions (Fr. B1−B3), and subfraction B1 was further purified by reversed-phase HPLC eluting 43% MeCN/H_2O at a flow rate of 2 mL/min to afford **1** (3.2 mg, t_R = 16.6 min). Compounds **2** (7.3 mg, t_R = 23.4 min) and **3** (1.4 mg, t_R = 46.6 min) were isolated using reversed-phase HPLC (63% MeOH/H_2O) from subfraction B2. Fraction F (3.75 g) was separated using MPLC on an ODS (50 μm) column to give seven fractions (Fr. F1–F7). Fr. F3 was subjected to reversed-phase HPLC (65% MeOH/H_2O, 2 mL/min) to afford **11** (15.3 mg, t_R = 15.3 min). Fr. F4 was separated with reversed-phase HPLC (40% CH_3CN/H_2O, 2 mL/min) to give **12** (40.1 mg, t_R = 11.3 min).

The defined medium fermentation was combined and then extracted with CH_2Cl_2−MeOH (1:1, 1 L) three times. The organic solvent was concentrated under reduced pressure to yield the extract (6.86 g). The extract was subjected to silica gel VLC, eluting with a gradient of petroleum ether/EtOAc (100:1, 80:1, 50:1, 30:1, 20:1, 10:1, 5:1, 3:1, 2:1, 1:1, *v*/*v*) to obtain 20 fractions (Fr.A−Fr.T). Fraction O (0.4 g) was subjected to an ODS (50 μm) column via MPLC (MeOH/H_2O, 50–100%) to give eight fractions, Fr. O1−Fr.O8. Fr. O6 (17.1 mg) was then purified with semipreparative HPLC (MeOH/H_2O, 63:37, *v*/*v*; 2.0 mL/min) at 250 nm to afford **5** (6.2 mg, t_R = 32.1 min). Fr. P (0.42 g) was separated with MPLC (MeOH/H_2O, 60–100%) to afford five fractions, Fr. P1−Fr. P5. Fr. P4 (27.2 mg) and Fr. P5 (19.5 mg) were purified with HPLC on an RP C18 column to give **4** (7.4 mg, MeCN/H_2O 40:60, 2.0 mL/min, t_R = 24.9 min) and **6** (6.3 mg, MeCN/H_2O 50:50, 2.0 mL/min, t_R = 30.1 min), respectively. Fr. Q (0.15 g) was separated with reversed-phase ODS (50 μm) MPLC eluting with a MeOH/H_2O gradient (from 60% to 100%) to afford six subfractions, Fr.Q1−Fr.Q6. Fr. Q5 (17.2 mg) was purified on an RP C18 column with HPLC (80% MeOH/H_2O, 2.0 mL/min), yielding **2** (2.4 mg, t_R = 28.8 min). Fr. R (1.04 g) was chromatographed over ODS via MPLC using a gradient elution of MeOH−H_2O (from 50% to 100%) to get five fractions (Fr. R1−R5). Fr. R3 (475.1 mg) was then subjected to a silica gel CC (petroleum ether/EtOAc, 3:1, *v*:*v*) to give five fractions, Fr. R3a−Fr. R3e. Fr. R3c (168.0 mg) was then purified with semipreparative HPLC on an RP C18 ODS (CH_3CN/H_2O, 30:70, *v*/*v*; 2.0 mL/min) to afford **8** (3.5 mg, t_R = 52.2 min) and **11** (116.8 mg, t_R = 60.5 min). Fr. R3d (275.7 mg) was further purified with 37% CH_3CN via HPLC (2.0 mL/min) to afford **7** (6.8 mg, t_R = 33.0 min) and **9** (174.0 mg, t_R = 39.8 min). Fr. S (0.55 g) was chromatographed over ODS using a gradient elution of MeOH/H_2O (from 50% to 100%) to obtain three fractions (Fr. S1−S3). Fr. S3 (322.0 mg) was further purified with 35% CH_3CN via HPLC to afford compound **10** (243.8 mg, t_R = 18.6 min).

Eutypelleudesmane A (**1**): light-brown oil; $[\alpha]_D^{25}$ −23.0 (*c* 0.10, MeOH); UV (MeOH) (log ε) λ_{max} 241 (4.07) nm; CD (MeOH) (Δε) 242 (+17.1); IR (KBr) ν_{max} 3357, 2956, 2929, 2873, 1741, 1650, 1455, 1438, 1376, 1232, 1153, 1116, 1068, 1024, 958, 883,850,719 cm^{-1}; ^1H and ^{13}C NMR data, see Table 1; HRESIMS *m/z* 601.4074 $[M + H]^+$ (calcd for $C_{36}H_{57}O_7$, 601.4104).

Cytosporin Y (**2**): light brown oil; $[\alpha]_D^{25}$ +14.0 (*c* 0.10, MeOH); UV (MeOH) (log ε) λ_{max} 241 (3.89) nm; CD (MeOH) (Δε) 238 (+9.1); IR (KBr) ν_{max} 3359, 2956, 2927, 2857, 1454, 1376, 1261, 1014, 842, 725 cm^{-1}; ^1H and ^{13}C NMR data, see Table 2; HRESIMS *m/z* 367.2125 $[M + COOH]^-$ (calcd for $C_{20}H_{31}O_6$, 367.2121).

Cytosporin Z (**3**): light-brown oil; $[\alpha]_D^{25}$ +12.3 (*c* 0.10, MeOH), $[\alpha]_D^{25}$ +1.9 (*c* 0.1, $CDCl_3$); UV (MeOH) (log ε) λ_{max} 210 (5.37), 312 (3.13) nm; IR (KBr) ν_{max} 3378, 2954, 2927, 2856, 1708, 1614, 1513, 1434, 1380, 1369, 1255, 1218, 1184, 1143, 1064, 1029, 977, 852 cm^{-1}; ^1H and ^{13}C NMR data, see Table 2; HRESIMS *m/z* 319.1912 $[M - H]^-$ (calcd for $C_{19}H_{27}O_4$, 319.1909).

Cytosporin Y_1 (**4**): light-yellow oil; $[\alpha]_D^{25}$ −15.8 (*c* 0.1, MeOH); UV (MeOH) λ_{max} (log ε) 221 (3.71), 240 (3.77) nm; IR ν_{max} 3367, 2955, 2926, 2857, 1743, 1601, 1378, 1072, 1023 cm^{-1}; CD (MeOH) (Δε) 208 (−6.4); ^1H NMR and ^{13}C NMR, see Table 3; HRESIMS *m/z* 363.2139 $[M + Na]^+$ (calcd for $C_{19}H_{32}O_5Na$, 363.2142).

Cytosporin Y_2 (**5**): light-yellow oil; $[\alpha]_D^{25}$ −60.1 (*c* 0.1, MeOH); UV (MeOH) λ_{max} (log ε) 214 (3.96), 244 (4.28) nm; IR ν_{max} 3381, 2956, 2926, 2857, 1786, 1647, 1344, 1219, 1049, 1023, 1001, 822, 760 cm^{-1}; CD (MeOH) (Δε) 222 (−13.2); ^1H NMR and ^{13}C NMR, see Table 3; HRESIMS *m/z* 389.1928 $[M + Na]^+$ (calcd for $C_{20}H_{30}O_6Na$, 389.1935).

Cytosporin Y_3 (**6**): light-yellow oil, $[\alpha]_D^{25}$ +37.8 (*c* 0.1, MeOH); UV (MeOH) λ_{max} (log ε) 217 (3.64), 241 (3.81) nm; IR ν_{max} 3383, 2956, 2928, 2858, 1783, 1595, 1361, 1270, 1182, 1068, 907, 770, 737 cm^{-1}; CD (MeOH) (Δε) 241 (+4.87); ^1H NMR and ^{13}C NMR, see Table 4; HRESIMS *m/z* 389.1928 $[M + Na]^+$ (calcd for $C_{20}H_{30}O_6Na$, 389.1935).

Cytosporin E_1 (**7**): light-yellow oil, $[\alpha]_D^{25}$ +12.9 (*c* 0.1, MeOH), UV (MeOH) λ_{max} (log ε) 200 (4.10) nm; IR ν_{max} 3393, 2926, 2856, 1783, 1464, 1361, 1184, 1158, 1086, 1058, 1023, 917, 772, 629 cm^{-1}; CD (MeOH) (Δε) 210 (+14.2) nm; ^1H NMR and ^{13}C NMR, see Table 4; HRESIMS *m/z* 363.2145 407.2034 $[M + Na]^+$ (calcd for $C_{20}H_{32}O_7Na$, 407.2040).

3.4. Biological Assay

The antimicrobial activities of compounds **1–12** against *Staphylococcus aureus*, *Escherichia coli*, and *Bacillus subtilis* were evaluated using a previous method [16], and levofloxacin was used as a positive control. The cytotoxicities of compounds **1–12** against DU145, SW1990, Huh7, and PANC-1 human cancer cell lines were determined using the CCK-8 method [17], with cisplatin used as a positive control. The immunosuppressive activities of compounds **1–12** against ConA-induced T cell proliferation were performed as previously described [18], with cyclosporin A as a positive control.

4. Conclusions

In summary, the utilization of the OSMAC (one strain many compounds) culture strategy effectively modified the chemical profile of the Arctic-derived fungus *Eutypella* sp. D-1 when cultivated in different media. This approach resulted in the production of five cytosporin polyketides (compounds **1–3** and **11–12**) from a rice medium and eight cytosporins (compounds **2** and **4–11**) from a solid defined medium. Remarkably, compound **1** contained a unique skeleton formed by the ester linkage of two moieties: cytosporin F (**12**) and the eudesmane-type sesquiterpene dihydroalanto glycol. Compounds **6** and **7**, characterized by a cyclic carbonate-fused cytosporin skeleton, were found to be rare in nature. However, these metabolites only exhibited weak immunosuppressive inhibitory activity against ConA-induced T cell proliferation in the antimicrobial, cytotoxic, and immunosuppressive evaluation. Collectively, this work showcased that changing the fermentation medium could be an effective strategy to trigger the production of secondary metabolites from fungi derived from the polar extreme environment.

Supplementary Materials: The following are available online at https://www.mdpi.com/article/10.3390/md21070382/s1, 1D and 2D NMR, UV, IR, and HRESMS data of **1–7**.

Author Contributions: H.-B.Y. designed this study and drafted the work. H.-B.Y. and Z.N. performed the collection, extraction, isolation, and structure elucidation. Y.-P.Z. and B.H. performed the bioactive evaluation. Y.H. contributed to checking the isolation process. X.-L.L. checked the structure elucidation process. B.-H.J. and X.-Y.L. supervised the laboratory work and contributed to the critical proofreading and revision of the manuscript. All authors have read and agreed to the published version of the manuscript.

Funding: This research was financially supported by the National Key Research and Development Project (Nos. 2022YFC2804500 and 2022YFC2804105), the Natural Science Foundation of Shanghai (No.20ZR1470600), and Shanghai Pujiang Program (No.2020PJD082).

Institutional Review Board Statement: Not applicable.

Data Availability Statement: The data presented in this study are available on request from the corresponding author.

Conflicts of Interest: The authors declare no conflict of interest.

References

1. Arrieche, D.; Cabrera-Pardo, J.R.; San-Martin, A.; Carrasco, H.; Taborga, L. Natural Products from Chilean and Antarctic Marine Fungi and Their Biomedical Relevance. *Mar. Drugs* **2023**, *21*, 98. [CrossRef] [PubMed]
2. Stevens-Miles, S.; Goetz, M.A.; Bills, G.F.; Giacobbe, R.A.; Tkacz, J.S.; Chang, R.S.; Mojena, M.; Martin, I.; Diez, M.T.; Pelaez, F. Discovery of an angiotensin II binding inhibitor from a *Cytospora* sp. using semi-automated screening procedures. *J. Antibiot.* **1996**, *49*, 119–123. [CrossRef] [PubMed]
3. Akone, S.H.; El Amrani, M.; Lin, W.; Lai, D.; Proksch, P. Cytosporins F–K, new epoxyquinols from the endophytic fungus *Pestalotiopsis theae*. *Tetrahedron Lett.* **2013**, *54*, 6751–6754. [CrossRef]
4. Ciavatta, M.L.; Lopez-Gresa, M.P.; Gavagnin, M.; Nicoletti, R.; Manzo, E.; Mollo, E.; Guo, Y.-W.; Cimino, G. Cytosporin-related compounds from the marine-derived fungus *Eutypella scoparia*. *Tetrahedron* **2008**, *64*, 5365–5369. [CrossRef]
5. Zhang, Y.-X.; Yu, H.-B.; Xu, W.-H.; Hu, B.; Guild, A.; Zhang, J.-P.; Lu, X.-L.; Liu, X.-Y.; Jiao, B.-H. Eutypellacytosporins A–D, Meroterpenoids from the Arctic Fungus *Eutypella* sp. D-1. *J. Nat. Prod.* **2019**, *82*, 3089–3095. [CrossRef] [PubMed]
6. Yu, X.; Müller, W.E.G.; Meier, D.; Kalscheuer, R.; Guo, Z.; Zou, K.; Umeokoli, B.O.; Liu, Z.; Proksch, P.J.M.D. Polyketide Derivatives from Mangrove Derived Endophytic Fungus *Pseudopestalotiopsis theae*. *Mar. Drugs* **2020**, *18*, 129–143. [CrossRef] [PubMed]
7. Yu, H.-B.; Wang, X.-L.; Zhang, Y.-X.; Xu, W.-H.; Zhang, J.-P.; Zhou, X.-Y.; Lu, X.-L.; Liu, X.-Y.; Jiao, B.-H. Libertellenones O–S and Eutypellenones A and B, Pimarane Diterpene Derivatives from the Arctic Fungus *Eutypella* sp. D-1. *J. Nat. Prod.* **2018**, *81*, 1553–1560. [CrossRef] [PubMed]
8. Tan, J.J.; Zhang, J.; Li, Y.M. Secondary metabolites from *Eutypella* species and their bioactivities. *Mycosystema* **2017**, *36*, 1181–1191.
9. Yu, H.-B.; Wang, X.-L.; Xu, W.-H.; Zhang, Y.-X.; Qian, Y.-S.; Zhang, J.-P.; Lu, X.-L.; Liu, X.-Y. Eutypellenoids A–C, New Pimarane Diterpenes from the Arctic Fungus *Eutypella* sp. D-1. *Mar. Drugs* **2018**, *16*, 284. [CrossRef] [PubMed]
10. Romano, S.; Jackson, S.A.; Patry, S.; Dobson, A.D.W. Extending the "One Strain Many Compounds" (OSMAC) Principle to Marine Microorganisms. *Mar. Drugs* **2018**, *16*, 244. [CrossRef] [PubMed]
11. Ren, X.-D.; Zhao, N.; Xu, S.; Lü, H.-N.; Ma, S.-G.; Liu, Y.-B.; Li, Y.; Qu, J.; Yu, S.-S. Total synthesis of illicidione A and illihendione A. *Tetrahedron* **2015**, *71*, 4821–4829. [CrossRef]
12. Lim, J.; Kim, I.-H.; Kim, H.H.; Ahn, K.-S.; Han, H. Enantioselective syntheses of decursinol angelate and decursin. *Tetrahedron Lett.* **2001**, *42*, 4001–4003. [CrossRef]
13. Zhang, Y.-H.; Du, H.-F.; Gao, W.-B.; Li, W.; Cao, F.; Wang, C.-Y. Anti-inflammatory Polyketides from the Marine-Derived Fungus *Eutypella* scoparia. *Mar. Drugs* **2022**, *20*, 486. [CrossRef] [PubMed]
14. Liao, H.-X.; Sun, D.-W.; Zheng, C.-J.; Wang, C.-Y. A new hexahydrobenzopyran derivative from the gorgonian-derived Fungus *Eutypella* sp. *Nat. Prod. Res.* **2017**, *31*, 1640–1646. [CrossRef] [PubMed]
15. Allemann, R.K. Chemical wizardry? The generation of diversity in terpenoid biosynthesis. *Pure Appl. Chem.* **2008**, *80*, 1791–1798. [CrossRef]
16. Yu, H.-B.; Jiao, H.; Zhu, Y.-P.; Zhang, J.-P.; Lu, X.-L.; Liu, X.-Y. Bioactive metabolites from the Arctic fungus *Nectria* sp. B-13. *J. Asian Nat. Prod. Res.* **2019**, *21*, 961–969. [CrossRef] [PubMed]
17. Yu, H.-B.; Gu, B.-B.; Iwasaki, A.; Jiang, W.-L.; Ecker, A.; Wang, S.-P.; Yang, F.; Lin, H.-W. Dactylospenes A–E, Sesterterpenes from the Marine Sponge *Dactylospongia elegans*. *Mar. Drugs* **2020**, *18*, 491. [CrossRef] [PubMed]
18. Xu, J.; Zhang, X.; Huang, F.; Li, G.; Leadlay, P.F. Efophylins A and B, Two C2-Asymmetric Macrodiolide Immunosuppressants from *Streptomyces malaysiensis*. *J. Nat. Prod.* **2021**, *84*, 1579–1586. [CrossRef] [PubMed]

marine drugs

Article

Talarolides Revisited: Cyclic Heptapeptides from an Australian Marine Tunicate-Associated Fungus, *Talaromyces* sp. CMB-TU011

Angela A. Salim [1,†], Waleed M. Hussein [1,†], Pradeep Dewapriya [1,‡], Huy N. Hoang [1,2], Yahao Zhou [1], Kaumadi Samarasekera [1,§], Zeinab G. Khalil [1], David P. Fairlie [1,2] and Robert J. Capon [1,*]

1 Institute for Molecular Bioscience, The University of Queensland, St. Lucia, QLD 4072, Australia; a.salim@uq.edu.au (A.A.S.); w.hussein@uq.edu.au (W.M.H.); p.dewapriya@uq.edu.au (P.D.); h.hoang@imb.uq.edu.au (H.N.H.); yahao.zhou@uq.net.au (Y.Z.); kaumadis@sci.pdn.ac.lk (K.S.); z.khalil@uq.edu.au (Z.G.K.); d.fairlie@imb.uq.edu.au (D.P.F.)

2 ARC Centre of Excellence for Innovations in Peptide and Protein Science, The University of Queensland, St. Lucia, QLD 4072, Australia

* Correspondence: r.capon@uq.edu.au; Tel.: +61-7-3346-2979

† These authors contributed equally to this work.

‡ Current address: Queensland Alliance for Environmental Health Science, The University of Queensland, Wooloongabba, QLD 4102, Australia.

§ Current address: Department of Botany, Faculty of Science, University of Peradeniya, Peradeniya, Kandy 20400, Sri Lanka.

Abstract: Application of a miniaturized 24-well plate system for cultivation profiling (MATRIX) permitted optimization of the cultivation conditions for the marine-derived fungus *Talaromyces* sp. CMB-TU011, facilitating access to the rare cycloheptapeptide talarolide A (**1**) along with three new analogues, B–D (**2**–**4**). Detailed spectroscopic analysis supported by Marfey's analysis methodology was refined to resolve *N*-Me-L-Ala from *N*-Me-D-Ala, L-*allo*-Ile from L-Ile and L-Leu, and partial and total syntheses of **2**, and permitted unambiguous assignment of structures for **1** (revised) and **2**–**4**. Consideration of diagnostic ROESY correlations for the hydroxamates **1** and **3**–**4**, and a calculated solution structure for **1**, revealed how cross-ring H-bonding to the hydroxamate moiety influences (defines/stabilizes) the cyclic peptide conformation. Such knowledge draws attention to the prospect that hydroxamates may be used as molecular bridges to access new cyclic peptide conformations, offering the prospect of new biological properties, including enhanced oral bioavailability.

Keywords: talarolides; *Talaromyces*; cycloheptapeptide; *N*-OH glycine; MATRIX; GNPS molecular networking

Citation: Salim, A.A.; Hussein, W.M.; Dewapriya, P.; Hoang, H.N.; Zhou, Y.; Samarasekera, K.; Khalil, Z.G.; Fairlie, D.P.; Capon, R.J. Talarolides Revisited: Cyclic Heptapeptides from an Australian Marine Tunicate-Associated Fungus, *Talaromyces* sp. CMB-TU011. *Mar. Drugs* **2023**, *21*, 487. https://doi.org/10.3390/md21090487

Academic Editor: Dehai Li

Received: 11 August 2023
Revised: 6 September 2023
Accepted: 6 September 2023
Published: 11 September 2023

1. Introduction

During our ongoing investigations into the natural products of Australian marine and terrestrial microbes, we have encountered many new and unusual cyclic and acyclic peptides and depsipeptides, including the antimalarial glyco-cyclohexadepsipeptide-polyketide mollemycin A from a north Queensland marine sediment-derived *Streptomyces* sp. CMB-M0244 [1]; the antitubercular cyclohexapeptide wollamides A–B from a north Queensland desert soil-derived *Streptomyces* sp. MST-115088 [2]; the acyclic peptaibol nonapeptide trichodermamides A–E from a Queensland termite nest-derived fungus *Trichoderma virens* CMB-TN16 [3]; the nitro-depsitetrapeptide-diketopiperazine waspergillamide A from a Queensland mud dauber wasp-derived *Aspergillus* sp. CMB-W031 [4]; the lipocyclopentapeptide scopularides A–H from Queensland mullet gastrointestinal tract-derived *Scopulariopsis* spp. CMB-F458 and CMB-F115, and *Beauvaria* sp. CMB-F585 [5]; and *N*-methylated acyclic undeca- and dodecapeptide talaropeptides A–D [6], and the cycloheptapeptide hydroxamate talarolide A [7], from a Queensland marine tunicate-derived fungus *Talaromyces* sp. CMB-TU011. In the latter case, we took advantage of altering cultivation conditions, with a YES broth cultivation of *Talaromyces* sp. CMB-TU011 yielding the talaropeptides [6] and an M1-saline agar cultivation yielding talarolide A [7].

Notwithstanding that traditional spectroscopic and chemical approaches are generally very effective at assigning structures inclusive of absolute configurations to cyclic peptides, our 2017 account of talarolide A proved challenging, with the proposed structure **1a** inconsistent with a subsequent total synthesis by Brimble et al. [8]. In an effort to address this anomaly, this report describes the application of an innovative miniaturized cultivation profiling methodology (MATRIX) [9] to optimize the production and enable the isolation and characterization of talarolides A–D (**1–4**). With access to larger quantities of talarolide A, we were able to secure superior NMR data, which, together with refinements to the Marfey analysis methodology, as well as partial and total syntheses, allowed us to propose a revised structure **1** for talarolide A and to assign structures to the new analogues talarolides B–D (**2–4**) as shown (Figure 1).

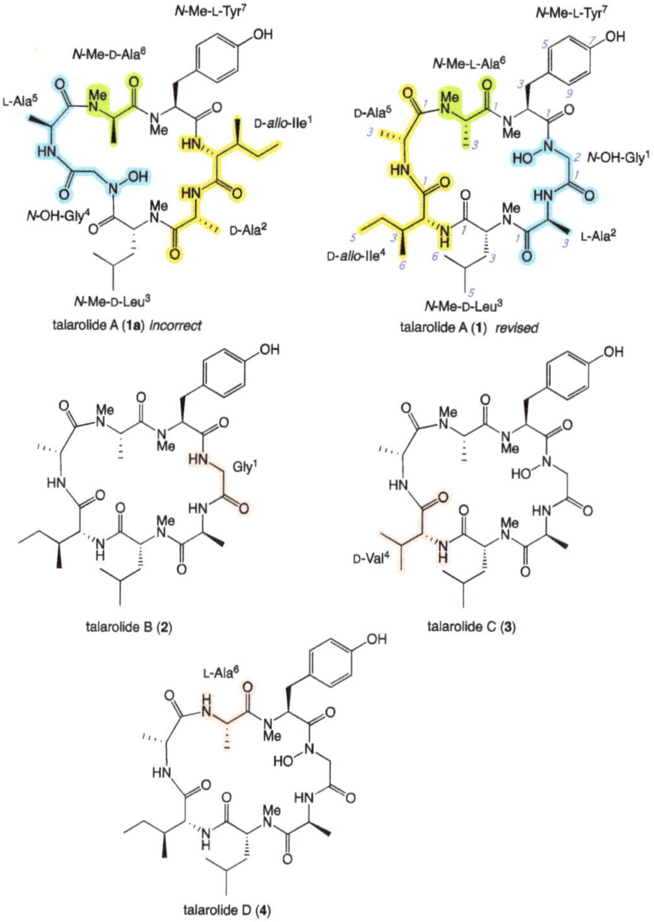

Figure 1. Structures for talarolide A, incorrect (**1a**) [7] and revised (**1**), and new analogues for talarolides B–D (**2–4**) from *Talaromyces* sp. CMB-TU011. Highlights (light blue, green, and yellow) show the difference between the incorrect structure (**1a**) and revised structure (**1**) of talarolide A. Pink highlight in structures **2–4** shows the amino acid variation compared to **1**.

2. Results and Discussion

Since our initial 2017 report on talarolide A [7], we have augmented our microbial biodiscovery efforts by implementing a miniaturized 24-well plate microbioreactor approach to support more comprehensive cultivation profiling (MATRIX) [9], to better optimize production and provide higher yields. Furthermore, we have integrated our MATRIX approach with a chemical profiling strategy employing in situ extraction followed by HPLC-DAD-ESI(+)MS and a UPLC-DAD-QTOF-MS/MS analysis, with the latter visualized as a Global Natural Products Social (GNPS) [10] molecular network, to better detect and prioritize target chemistry (i.e., new from known and rare from common). Applying MATRIX cultivation profiling to *Talaromyces* sp. CMB-TU011 involved 24-well plate cultivations using eleven different media (Table S1) under three conditions (solid agar, and static and shaken broth) (Figure 2B) at 26.5 °C, over 10 days. Following incubation, the resulting 36 individual wells, together with uninoculated media controls, were extracted in situ with EtOAc, and the resulting extracts subjected to chemical profiling. While visualization of the HPLC-DAD-ESIMS data using single ion extraction (SIE, *m/z* 718) detected **1** in most extracts, production levels were highly variable with maximum yields observed under M1-salt, ISP-4, PDA and PYG solid agar, and static and shaken broth conditions returning far lower yields (Figure S1). Significantly, a GNPS analysis of the MATRIX extracts revealed a talarolide molecular family (sodiated adducts) incorporating **1** (*m/z* 740), and nodes for the deoxy analogue **2** (*m/z* 724), lower homologue isomers **3** and **4** (*m/z* 726), and an unidentified minor analogue (*m/z* 752) (Figure 2A). Based on these analyses, a scaled up (×200 plate) 20-day solid phase ISP-4 agar cultivation of CMB-TU011 was extracted and fractionated by solvent partitioning and gel and reversed phase chromatography, to yield talarolides A–D (**1–4**) (Figures 2C and S2). An account of the structure elucidation of **1–4** (including structure revision of **1**) is summarized below.

HRESI(+)MS analysis of **1** revealed a molecular formula ($C_{35}H_{55}N_7O_9$, Δmmu +2.5) requiring 12 double bond equivalents (DBEs), consistent with our earlier 2017 account of talarolide A [7]. Marfey's analysis of **1** returned *N*-Me-L-Tyr, D-*allo*-Ile, *N*-Me-D-Leu, L-Ala, D-Ala, and *N*-Me-D-Ala (Figure S34). While this analysis differed from our earlier assessment of talarolide A (i.e., *N*-Me-L-Ala rather than *N*-Me-D-Ala), on revisiting and repeating our earlier analytical HPLC protocols it became apparent that the relative retention times of Marfey's D-FDAA (or L-FDAA) derivatives of *N*-Me-L-Ala and *N*-Me-D-Ala were very similar, so much so that replicate analyses could experience a reversal in elution times, likely due to subtle variations in eluant composition (i.e., pH) over time. To address this lack of reliability, in this current report, we rely on new analytical HPLC conditions optimized for the unambiguous resolution of Marfey's derivatives of *N*-Me-L-Ala and *N*-Me-D-Ala (Figures 3 and S38). Likewise, we also developed and applied new, superior analytical HPLC conditions optimized for the differentiation of Marfey's derivatives of Leu, Ile, and *allo*-Ile (Figures 4 and S39). With the identity and absolute configuration of the amino acid residues in **1** assigned, we next turned our attention to the amino acid sequence. In our earlier structure elucidation of talarolide A, assignment of the planar sequence of amino acid residues relied on an incomplete set of HMBC correlations and interpretation of the MS/MS fragmentation patterns (the latter challenging for cyclic peptides). Fortunately, the re-isolation of **1** enabled the acquisition of superior NMR (DMSO-d_6) data (Tables 1, 2 and S2, and Figures 5 and S3–S8), which allowed for a comprehensive set of HMBC correlations and unambiguous assembly of the amino acid sequence, as shown. To assign the regiochemistry of the L-Ala and D-Ala residues in **1**, we relied on our earlier 2D C_3 Marfey's analysis [11] where talarolide A was subjected to partial hydrolysis, derivatization, and chromatographic fractionation to yield the dipeptide D-FDAA-D-*allo*-Ile-D-Ala, with the D-Ala configuration confirmed by a subsequent round of hydrolysis and Marfey's analysis [7]. Thus, the revised structure for talarolide A (**1**) is as shown. Of particular interest is the unprecedented *N*-OH-Gly residue and its ability to engage in an extensive network of ROESY interactions (and H-bonding) across the cyclic peptide ring (Figure 5, dashed pink), which presumably also facilitates the observed long-range ROESY linkages (Figure 5, dashed green).

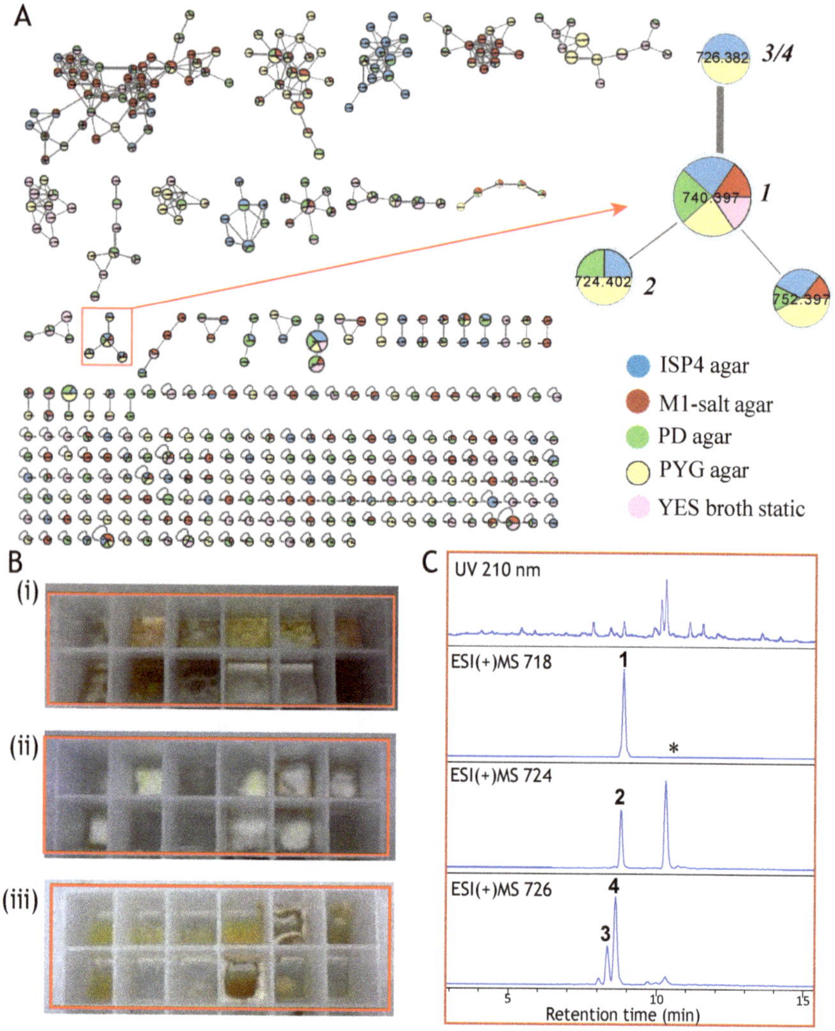

Figure 2. (**A**) GNPS molecular network of *Talaromyces* sp. CMB-TU011 in a selection of five media, with an expansion of the talarolide molecular family. Node segment size correlates with relative yield/metabolite/media; (**B**) images of 24-well plate MATRIX cultivation in 11 different media under three conditions: (i) agar, (ii) static broth, (iii) shaken broth; (**C**) HPLC-DAD-MS chromatograms of CMB-TU011 EtOAc extract obtained from ISP4 agar cultivation, with single ion extractions showing **1**–**4**. (* this peak is not a talarolide analogue).

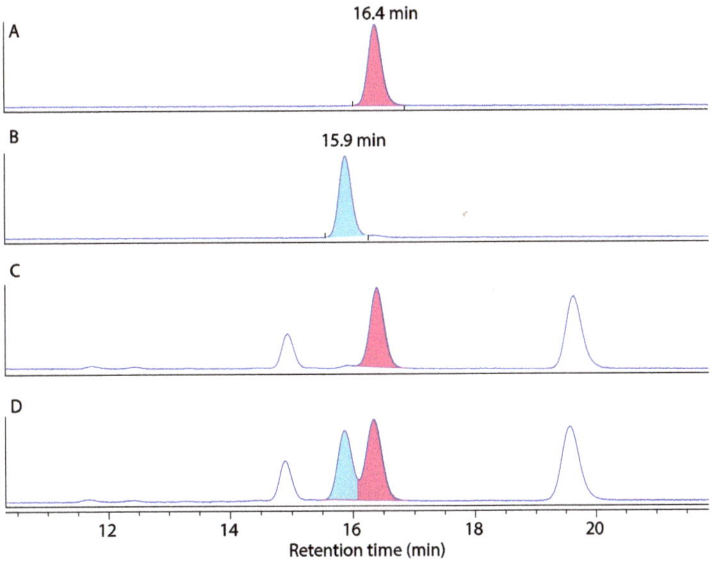

Figure 3. Optimized HPLC conditions for the resolution of Marfey's derivatives L-FDAA-*N*-Me-L-Ala (pink) and L-FDAA-*N*-Me-D-Ala (light blue). (**A**) synthetic L-FDAA-*N*-Me-L-Ala; (**B**) synthetic L-FDAA-*N*-Me-D-Ala; (**C**) L-FDAA-*N*-Me-L-Ala derived from talarolide A (**1**); (**D**) synthetic L-FDAA-*N*-Me-D-Ala co-injected with L-FDAA-*N*-Me-L-Ala derived from talarolide A (**1**).

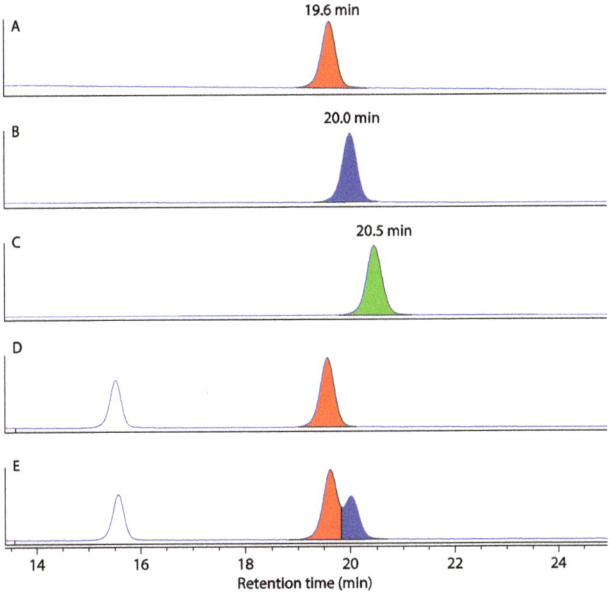

Figure 4. Optimized HPLC conditions for the resolution of Marfey's derivatives L-FDAA-D-*allo*-Ile (red), L-FDAA-D-Ile (blue) and L-FDAA-D-Leu (green). (**A**) synthetic L-FDAA-D-*allo*-Ile; (**B**) synthetic L-FDAA-D-Ile; (**C**) synthetic L-FDAA-D-Leu, (**D**) L-FDAA-D-*allo*-Ile derived from talarolide A (**1**); (**E**) synthetic L-FDAA-D-Ile co-injected with L-FDAA-D-*allo*-Ile derived from talarolide A (**1**).

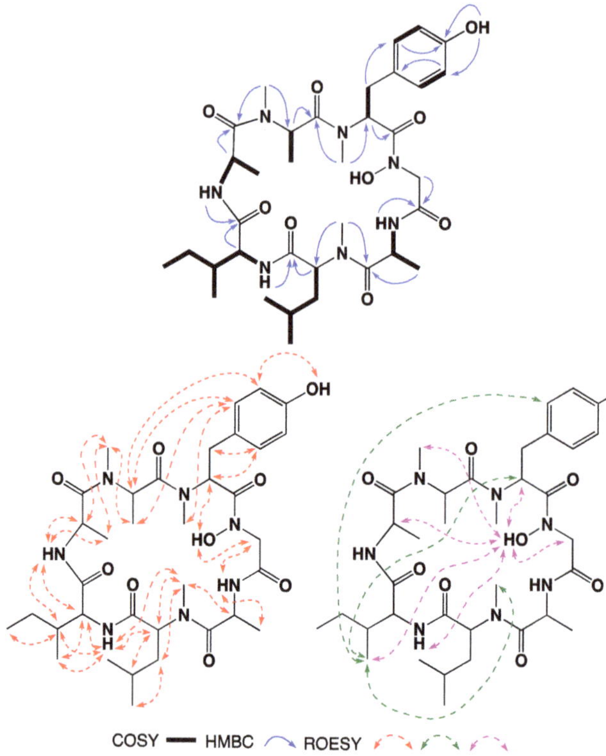

Figure 5. Selected 2D NMR (DMSO-d_6) correlations for talarolide A (**1**).

HRESI(+)MS analysis of **2** revealed a molecular formula ($C_{35}H_{55}N_7O_8$, Δmmu +0.4) consistent with a deoxy analogue of **1**. Indeed, Marfey's analysis of **2** returned *N*-Me-L-Tyr, D-*allo*-Ile, *N*-Me-D-Leu, L-Ala, D-Ala, and *N*-Me-D-Ala (Figure S35), while the NMR (DMSO-d_6) data for **2** (Tables 1, 2 and S3, and Figures 6 and S11–S16) revealed chemical shifts and diagnostic correlations that permitted assignment of the same planar amino acid sequence as **1**, where the *N*-OH-Gly in **1** had been replaced by a Gly residue in **2**. Partial hydrolysis of **2** followed by derivatization with L-FDAA followed by UPLC-DAD-MS analysis detected a dipeptide that co-eluted with an authentic synthetic sample of L-FDAA-D-*allo*-Ile-D-Ala, but not synthetic L-FDAA-D-*allo*-Ile-L-Ala (Figure S41), confirming a D-Ala and L-Ala regiochemistry in **2**, common with that independently established for **1.** To further confirm this assignment, we carried out a successful solid phase peptide synthesis of **2** (Scheme 1), with the synthetic sample proving to be identical to natural talarolide B (Figures S50–S52), including co-elution on HPLC (Figure S48).

HRESI(+)MS analysis of 3 revealed a molecular formula ($C_{34}H_{53}N_7O_9$, Δmmu +3.0) suggestive of a lower homologue (-CH_2) of 1, with Marfey's analysis returning *N*-Me-L-Tyr, *N*-Me-D-Ala, D-Ala, L-Ala, D-Val, and *N*-Me-D-Leu (Figure S36). As with 1, the *N*-OH-Gly residue in 3 was not detectable via Marfey's analysis, although its presence was evident in the NMR (DMSO-d_6) data (Tables 1, 2 and S4, and Figures 4 and S19–S24). Diagnostic 2D NMR correlations (Figure 6) permitted assignment of a planar amino acid sequence comparable to 1, but where the D-*allo*-Ile in 1 was replaced by D-Val in 3. The regiochemistry of the D-Ala and L-Ala residues in 3 was assigned on the basis of biogenetic comparison to 1 and 2, with the structure for talarolide C (3) assigned as shown.

Table 1. ^{1}H NMR (DMSO-d_6) data for talarolides A–D (**1–4**).

	1 δ_H, **Mult, (*J* in Hz)**	**2** δ_H, **Mult, (*J* in Hz)** [e]	**3** δ_H, **Mult, (*J* in Hz)**	**4** δ_H, **Mult, (*J* in Hz)**
		N-OH-Gly¹/Gly¹		
2a	4.75, d (17.1)	4.14 [a]	4.75 [a]	4.80 d (17.0)
2b	3.76, d (17.1)	3.54, dd (17.3, 2.9)	3.76, d (17.2)	3.71, d (17.0)
N-OH	9.31, s		9.41, s	9.13, s
N-H		7.53, dd (7.6, 2.9)		
		L-Ala²		
2	4.49, qd (6.8, 4.1)	4.67, qd (6.7, 5.9)	4.50, qd (6.7, 4.0)	4.49, qd (6.8, 4.1)
3	1.19, d (6.8)	1.20, d (6.7)	1.20, d (6.7)	1.19, d (6.8)
N-H	8.65, d (4.1)	8.59, d (5.9)	8.65, d (4.0)	8.62, d (4.1)
		N-Me-D-Leu³		
2	5.05, dd (11.8, 3.9)	5.05, dd (11.8, 3.8)	5.07, dd (11.6, 3.8)	5.06, dd (11.7, 3.8)
3a	1.79, ddd (14.4, 10.3, 3.9)	1.82, ddd (14.4, 10.5, 3.8)	1.79, ddd (14.4, 10.3, 3.8)	1.79, ddd (13.4, 10.6, 3.9)
3b	1.58, ddd (14.4, 11.8, 3.9)	1.61, ddd (14.4, 11.8, 3.7)	1.57, ddd (14.4, 11.6, 3.9)	1.57, ddd (13.4, 11.7, 3.9)
4	1.37, m	1.39, m	1.38, m	1.38 [a]
5	0.77, d (6.5)	0.79, d (6.5)	0.78, d (6.5)	0.77, d (6.5)
6	0.88, d (6.5)	0.90, d (6.5)	0.89, d (6.5)	0.88, d (6.6)
N-Me	3.00, s	3.11, s	3.01, s	2.98, s
		D-*allo*-Ile⁴/D-Val⁴		
2	4.72 [a]	4.60, dd (9.3, 5.2)	4.58, dd (9.5, 4.9)	4.69, dd (9.5, 3.9)
3	1.95, m	1.95, m	2.17, m	1.91, m
4a	1.42, m	1.56, m	0.92, d (6.8)	1.39 [a]
4b	1.07, m	1.09 [b]		1.05, m
5	0.94, dd (7.3, 7.3)	0.95, dd (7.3, 7.3)	0.86, d (6.8)	0.92, dd (7.3, 7.3)
6	0.81, d (6.9)	0.92, d (6.9)		0.80, d (6.9)
N-H	7.24, d (9.6)	6.96, d (9.3)	7.23, d (9.5)	7.14, d (9.6)
		D-Ala⁵		
2	4.34, qd (7.1, 5.4)	4.42 [c]	4.33, qd (7.1, 5.0)	3.90, m
3	1.12, d (7.1)	1.10 [b], d (7.1)	1.12, d (7.1)	1.13, d (7.1)
N-H	8.87, d (5.4)	8.63, d (5.4)	8.87, d (5.0)	8.58, d (5.2)
		N-Me-L-Ala/L-Ala⁶		
2	4.71 [a]	4.43 [c]	4.75 [a]	4.37, m
3	0.49, d (6.5)	0.58, d (6.5)	0.52, d (6.5)	0.63, d (6.4)
N-Me	2.70, s	2.70, s	2.71, s	
N-H				8.27, d (9.5)
		N-Me-L-Tyr⁷		
2	4.80, dd (10.5, 4.9)	4.15 [a]	4.80, dd (9.8, 5.3)	5.01, dd (8.6, 6.6)
3a	2.84, dd (14.3, 10.5)	2.93, dd (14.1, 11.6)	2.82, dd (14.2, 9.8)	2.81, dd (14.3, 6.6)
3b	2.60, dd (14.3, 4.9)	2.55 [d]	2.62, dd (14.2, 5.3)	2.77, dd (14.3, 8.6)
5/9	6.93, d (8.4)	6.90, d (8.4)	6.94, d (8.4)	6.99, d (8.4)
6/8	6.64, d (8.4)	6.63, d (8.4)	6.63, d (8.4)	6.63, d (8.4)
7-OH	9.20, s	9.25, s	9.18, s	9.15, s
N-Me	2.66, s	2.86, s	2.67, s	2.64, s

[a–c] resonances with the same superscript within a column are overlapping, [d] signal is obscured by DMSO, detected by HSQC. [e] occurs as an equilibrating mixture of major and minor conformers, with the major conformer tabulated.

Table 2. ^{13}C NMR (DMSO-d_6) data for talarolides A–D (**1**–**4**).

	1 δ_C	**2** δ_C	**3** δ_C	**4** δ_C
		N-OH-Gly^1/Gly^1		
1	167.3	169.3	167.3	167.5
2	50.2	41.3	50.1	49.9
		L-Ala^2		
1	174.1	174.4	174.1	174.0
2	45.2	44.6	45.2	45.2
3	15.7	15.8	15.7	15.7
		N-Me-D-Leu^3		
1	169.5	169.2 [a]	169.5	169.5
2	54.5	54.6 [b]	54.4	54.4
3	36.0	36.1	35.9	36.0
4	24.4	24.4	24.4	24.4
5	21.0	20.9	20.9	20.9
6	23.3	23.3	23.3	23.3
N-Me	31.0	30.8	31.0	30.9
		D-*allo*-Ile^4/D-Val^4		
1	172.0	171.2	171.6	171.8
2	53.7	54.6 [b]	55.3	53.8
3	38.5	38.5	31.9	38.8
4	26.2	25.7	19.4	26.4
5	12.0	12.0	17.0	11.9
6	13.7	14.2 [c]		13.7
		D-Ala^5		
1	171.1	171.6	171.0	170.8
2	45.8	45.7	45.8	48.6
3	14.9	14.6	14.9	16.7
		N-Me-L-Ala/L-Ala^6		
1	169.8	169.8	169.8	171.4
2	46.7	49.3	46.7	42.7
3	15.1	14.2 [c]	15.1	18.5
N-Me	28.6 [a]	29.1	28.6 [a]	
		N-Me-L-Tyr^7		
1	168.2	169.2 [a]	168.2	168.3
2	56.6	59.8	56.6	57.1
3	34.1	34.8	34.3	34.5
4	126.6	126.5	126.7	127.2
5/9	130.8	130.3	130.7	130.6
6/8	114.8	114.9	114.8	114.9
7	155.9	156.0	155.9	155.8
N-Me	28.6 [a]	29.5	28.6 [a]	28.6

[a–c] resonances with the same superscript within a column are interchangeable.

HRESI(+)MS analysis of **4** revealed a molecular formula ($C_{34}H_{53}N_7O_9$, Δmmu +3.0) suggestive of an alternate lower homologue (-CH_2) of **1**, with Marfey's analysis returning *N*-Me-L-Tyr, D-Ala, L-Ala, D-*allo*-Ile, and *N*-Me-D-Leu (Figure S37). As with **1**, the *N*-OH-Gly residue in **4** was not detectable via Marfey's analysis, although its presence was evident in the NMR (DMSO-d_6) data (Tables 1, 2 and S5, and Figures 6 and S27–S32). Diagnostic 2D NMR correlations revealed a planar amino acid sequence comparable to **1**, but where the *N*-Me-L-Ala in **1** was replaced by an L-Ala in **4**. The regiochemistry of the D-Ala and L-Ala residues in **4** were assigned on the basis of biogenetic comparison to **1** and **2**, with the structure for talarolide D (**4**) assigned as shown.

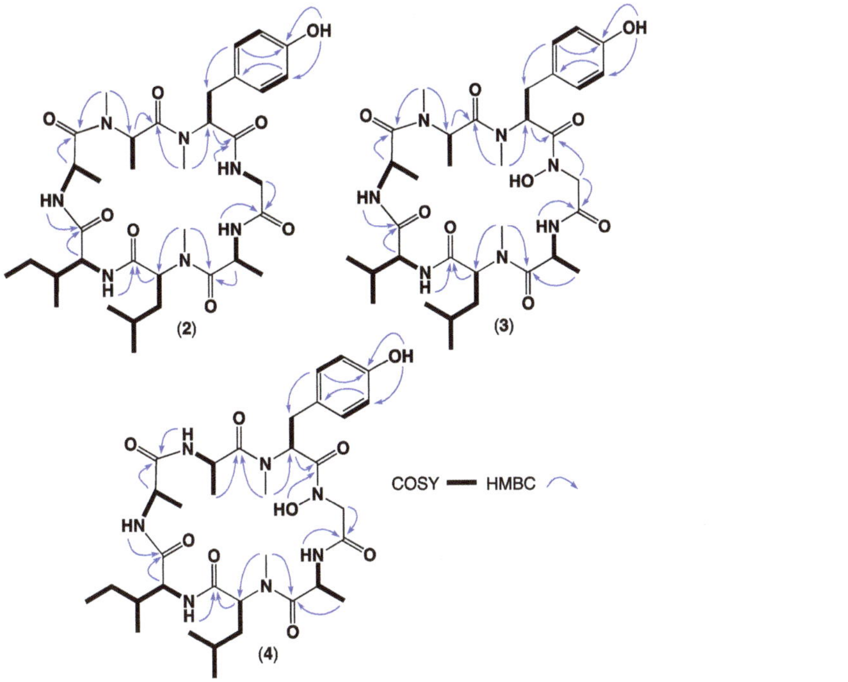

Figure 6. Selected 2D NMR (DMSO-d_6) correlations for talarolides B–D (**2**–**4**).

Of note, both the *N*-OH cyclic peptides **3** and **4** exhibit the same extensive pattern of ROESY correlations (Figures S10, S18 and S26) associated with the *N*-O<u>H</u> moiety evident in **1**, suggesting that all three adopt a common stable conformation dominated by hydrogen bonding to the *N*-O<u>H</u>. Not only is such conformation stabilization not accessible to the cyclic peptide **2**, but the NMR data for **2** reveals two equilibrating conformations (Figure S53), supporting the hypothesis that *N*-hydroxylation can have a pronounced effect on cyclic peptide conformation and stabilization.

In an effort to understand this latter phenomenon, we calculated a solution structure for **1** DMSO-d_6 at 298 K using 2D ROESY NMR spectra, calculated from 41 ROE distance restraints, three backbone φ-dihedral angle restraints derived from $^3J_{NH\text{-}CH\alpha}$, one *cis*-amide between *N*-Me-L-Ala^6-*N*-Me-L-Tyr^7, and one hydrogen bond restraint between *N*-O<u>H</u>-Gly^1 and the D-Ala^5 carbonyl oxygen (Figure 7). This hydrogen bond restraint was supported by the low temperature coefficient for the *N*-O<u>H</u>-Gly^1 in variable temperature ^{1}H NMR experiments (Figure 8). Structures were calculated in XPLOR-NIH using a dynamic simulated annealing protocol in a geometric force field, and energy minimized using the CHARMM force field [12,13]. The 10 lowest energy structures for talarolide A (**1**) had no distance (≥0.2 Å) or dihedral angle (≥2°) violations and were rigid, convergent structures (average pairwise C_a RMSD 0.18 Å) (Figure 7). The structure for **1** supported observations made in the VT (variable temperature) NMR experiments, with the N-O<u>H</u>-Gly^1 to D-Ala^5 carbonyl oxygen hydrogen bond and *cis*-amide bond between *N*-Me-L-Ala^6-*N*-Me-L-Tyr^7 forming a non-classical alpha turn centered at D-Ala^5-*N*-Me-L-Ala^6-*N*-Me-L-Tyr^7, and with L-Ala^2 and *N*-Me-D-Leu^3 forming a distorted beta turn. The D-*allo*-Ile^4 amide proton projects toward the interior of the structure and is shielded from solvent, while the D-Ala^5 amide proton is in close proximity to the *N*-OH-Gly^1 carbonyl oxygen, suggestive of a hydrogen bond and also less accessible to solvent. The opposite side of the molecule features an exposed L-Ala^2 amide proton, making it more accessible to solvent. From these observations, it can be concluded that the presence of the *N*-OH-Gly provides access to a hydrogen bond that defines the overall conformation of the cyclic peptide. It is intriguing to speculate whether this effect is unique to the talarolide scaffold, with its mix of L and D amino acid residues, or whether it is a more general phenomenon. If the latter, it is possible that *N*-hydroxylation could prove to be a valuable molecular tool for accessing new peptide chemical space.

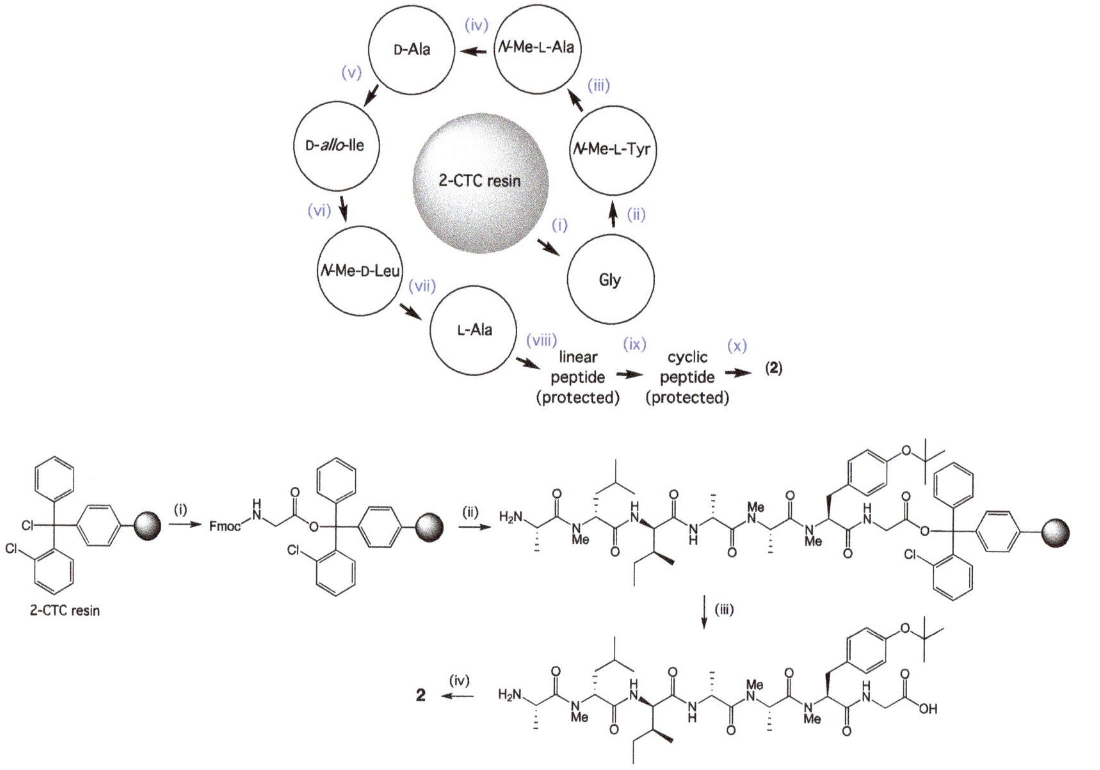

Scheme 1. Top: General outline of the solid phase peptide synthesis (SPPS) of talarolide B (**2**). (**i**) Fmoc-Gly-OH coupling to 2-CTC resin, (**ii–vii**) sequential peptide chain elongation of Fmoc amino acids, (**viii**) cleavage of linear protected peptide from resin, (**ix**) cyclization of linear protected peptide and (**x**) deprotection to yield **2**. Bottom: Experimental details for SPPS of **2**: (**i**) Fmoc-Gly-OH coupling to 2-CTC resin in the presence of DIPEA (2 h), (**ii**) elongation of peptide sequence through a coupling cycle: Fmoc deprotection with 20% of piperidine in DMF (twice, 5 and 10 min), and a 5 min DMF flow-wash followed by coupling with preactivated Fmoc-amino acid (3.2 eq.) over 2 × 30 min, or 2 × 3 h for coupling of Fmoc-amino acids to sterically hindered *N*-Me-amino acids, (**iii**) cleavage of linear protected peptide from resin using 20% HFIP/DCM (3 × 20 min), (**iv**) cyclization of linear protected peptide using HATU, HOBT, and collidine (14 h), followed by deprotection of *N*Me-L-Tyr using 90% formic acid 40 min to give **2** (16 mg, 23% overall yield).

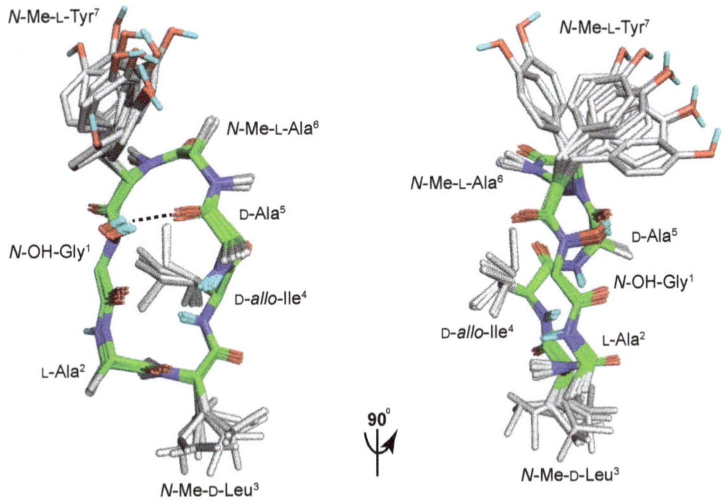

Figure 7. Backbone superimposition of the 10 lowest energy NMR calculated structures for **1** in DMSO-d_6 at 298 K showing hydrogen bonding between *N*-OH-Gly[1] and the carbonyl in D-Ala[5] (dashed line) and a *cis*-amide bond between *N*-Me-L-Ala[6] and *N*-Me-L-Tyr[7] forming a non-classical alpha turn. The D-*allo*-Ile[4] amide is projected inward and shielded from solvent, while L-Ala[2] is solvent exposed. Non-polar hydrogens are omitted for clarity, with backbone carbon atoms (green), sidechain carbon atoms (grey), oxygen atoms (red), nitrogen atoms (blue), and hydrogen atoms (cyan).

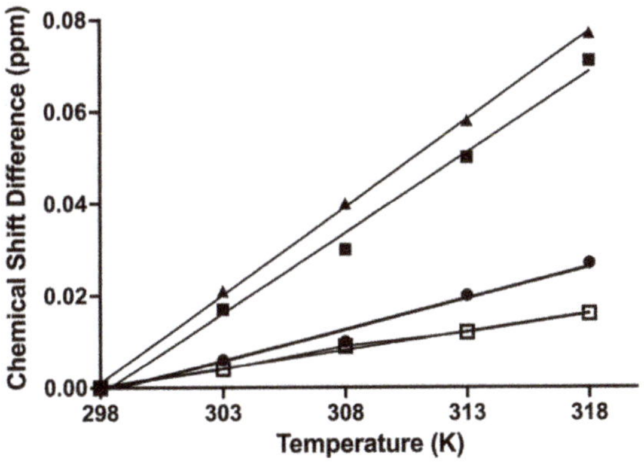

Figure 8. Temperature dependence of the amide NH and OH NMR (DMSO-d_6) chemical shifts for **1**. Line slopes indicating temperature coefficients ($\Delta\delta/T$) for each residue. Circle: *N*-OH-Gly[1] ($\Delta\delta/T$ = 1.4 ppb/K); triangle: L-Ala[2]-NH ($\Delta\delta/T$ = 3.9 ppb/K); black square: D-Ala[5]-NH ($\Delta\delta/T$ = 3.5 ppb/K); opened square: D-*allo*-Ile[4]-NH ($\Delta\delta/T$ = 0.3 ppb/K). Small temperature coefficients ($\Delta\delta/T$) for *N*-OH-Gly[1] and D-*allo*-Ile[4]-NH indicates hydrogen-bonds or solvent shielded [14].

3. Materials and Methods

3.1. General Experimental Procedures

Chiroptical measurements ($[\alpha]_D$) were obtained on a JASCO P-1010 polarimeter in a 100 × 2 mm cell at 25 °C. Nuclear magnetic resonance (NMR) spectra were acquired on a Bruker Avance 600 MHz spectrometer with either a 5 mm PASEL $^1H/D\text{-}^{13}C$ Z-Gradient probe or 5 mm CPTCI $^1H/^{19}F\text{-}^{13}C/^{15}N$/DZ-Gradient cryoprobe. The spectra were acquired at 25 °C in DMSO-d_6 and referenced to residual signals (δ_H 2.50 and δ_C 39.5 ppm) in deuterated solvents. High-resolution ESIMS measurements were obtained on a Bruker micrOTOF mass spectrometer by direct infusion in MeCN at 3 μL/min using sodium formate clusters as an internal calibrant. UPLC-QTOF analysis was performed on a UPLC-QTOF instrument comprising an Agilent 1290 Infinity II UPLC (Agilent Zorbax C_8 RRHD 1.8 μm, 2.1 × 250 mm column, eluting at 0.417 mL/min with a 2.50 min gradient elution from 90% H_2O/MeCN to 100% MeCN with a constant 0.1% formic acid modifier) coupled to an Agilent 6545 QTOF mass detector (Agilent, Mulgrave, Australia). Liquid chromatography-diode array-mass spectrometry (HPLC-DAD-MS) data were acquired on an Agilent 1260 series separation module equipped with an Agilent G6125B series single quad mass detector and diode array detector (Agilent Poroshell 120 SB-C_8 2.7 μm, 3.0 × 150 mm column, eluting at 0.8 mL/min with a 6.25 min gradient elution from 90% H_2O/MeCN to 100% MeCN with a constant 0.05% formic acid modifier). Ultra-high performance liquid chromatograms (UPLCs) were obtained on an Agilent 1290 infinity UPLC system composed of a 1290 infinity binary pump, thermostat, autosampler, and diode array detector (Agilent, Mulgrave, AustraliaPreparative and semi-preparative HPLC were performed using an Agilent 1100 Series diode array and/or multiple wavelength detectors and an Agilent 1100 Series fraction collector (Agilent, Mulgrave, Australia). *N*α-(2,4-dinitro-5-fluorophenyl)-L-alaninamide (L-FDAA, synonym 1-fluoro-2-4-dinitrophenyl-5-L-alanine amide) and *N*α-(2,4-dinitro-5-fluorophenyl)-D-alaninamide (D-FDAA, synonym 1-fluoro-2-4-dinitrophenyl-5-D-alanine amide) were purchased from Merck (Darmstadt, Germany). Amino acids and standards were purchased from BAChem (Torrance, CA, USA) or Merck (Darmstadt, Germany). Analytical-grade solvents were used for solvent extractions. Chromatography solvents were of HPLC grade supplied by Labscan (Bangkok, Thailand) or Merck (Darmstadt, Germany) and filtered/degassed through 0.45 μm polytetrafluoroethylene (PTFE) membrane prior to use. Deuterated solvents were purchased from Cambridge Isotopes (Tewksbury, MA, USA). Microorganisms were manipulated under sterile conditions using a Laftech class II biological safety cabinet and incubated in either MMM Friocell incubators (Lomb Scientific, Taren Point, NSW, Australia) or an Innova 42R incubator shaker (John Morris, Chatswood, NSW, Australia).

3.2. Collection and Taxonomy of *Talaromyces* sp. CMB-TU011

The isolation and taxonomy of *Talaromyces* sp. CMB-TU011 from an unidentified tunicate collected from Tweed Heads, NSW, Australia, has been previously reported [7].

3.3. Cultivation and Fractionation of *Talaromyces* sp. CMB-TU011

A loop of spores from a 7-day old M1-salt culture of CMB-TU011 was streaked on ISP-4 agar plates (×200) and incubated for 20 days at 26.5 °C, after which the combined agar/mycelia was extracted with EtOAc (3 × 500 mL) and concentrated in vacuo to yield an extract (264 mg), which was partitioned between n-hexane and aqueous MeOH to give hexane (70 mg) and MeOH (194 mg) soluble fractions. The MeOH fraction was subjected to gel chromatography (Sephadex® LH-20 (Merck, Darmstadt, Germany) in MeOH) to obtain 15 fractions, which were combined based on HPLC-DAD-MS analysis to yield a talarolides-enriched fraction (30.5 mg). Further semi-preparative HPLC (Agilent Zorbax Eclipse C_8 column, 5 μm, 9.4 × 250 mm, 32% MeCN/H_2O isocratic elution at 3.0 mL/min inclusive of an isocratic 0.01% TFA/MeCN modifier) was used to yield talarolide A (**1**) (t_R 25.6 min, 3.4 mg, 1.3%), talarolide B (**2**) (t_R 21.3 min, 1.1 mg, 0.42%), talarolide C (**3**) (t_R 16.1 min, 1.1 mg, 0.42%), and talarolide D (**4**) (t_R 19.3 min, 0.9 mg, 0.34%). (Note: % yields are calculated as a weight to weight of the EtOAc extract) (Figure S2).

Talarolide A (**1**): white powder; $[\alpha]_D^{25}$ −17 (c 0.12, MeOH); 1D and 2D NMR (DMSO-d_6) see Tables 1, 2 and S2, and Figures S3–S8; UPLC-QTOF (MS/MS) fragmentation see Figure S42; HRESI(+)MS *m*/*z* 740.3978 [M + Na]$^+$ (calcd for $C_{35}H_{55}N_7O_9Na$ 740.3953) (Figure S9).

Talarolide B (**2**): colorless amorphous solid; $[\alpha]_D^{25}$ −15 (c 0.059, MeOH); NMR (DMSO-d_6) see Tables 1, 2 and S3, and Figures S10–S16; UPLC-QTOF (MS/MS) fragmentation see Figure S43; HRESI(+)MS *m/z* 724.4008 $[M + Na]^+$ (calcd for $C_{35}H_{55}N_7O_8Na$ 724.4004) (Figure S17).

Talarolide C (**3**): colorless amorphous solid; $[\alpha]_D^{25}$ −15 (c 0.072, MeOH); NMR (DMSO-d_6) see Tables 1, 2 and S4, and Figures S18–S25; UPLC-QTOF (MS/MS) fragmentation see Figure S44; HRESI(+)MS *m/z* 726.3827 $[M + Na]^+$ (calcd for $C_{34}H_{53}N_7O_9Na$ 726.3797) (Figure S25).

Talarolide D (**4**): colorless amorphous solid; $[\alpha]_D^{25}$ −14 (c 0.054, MeOH); NMR (DMSO-d_6) see Tables 1, 2 and S5, and Figures S26–S32; UPLC-QTOF (MS/MS) fragmentation see Figure S45; HRESI(+)MS *m/z* 726.3827 $[M + Na]^+$ (calcd for $C_{34}H_{53}N_7O_9Na$ 726.3797) (Figure S33).

3.4. *Marfey's Analysis of Talarolides A–D*

3.4.1. Standard Marfey's Hydrolysis and Derivatization Method #1

A sample analyte (50 µg) in 6 M HCl (100 µL) was heated to 100 °C in a sealed vial for 12 h, after which the hydrolysate was concentrated to dryness at 40 °C under a stream of dry N_2. The hydrolysate was then treated with 1 M $NaHCO_3$ (20 µL) and L-FDAA (1-fluoro-2,4-dinitrophenyl-5-L-alanine amide) or D-FDAA (1-fluoro-2,4-dinitrophenyl-5-D-alanine amide) as a 1% (*w/v*) solution in acetone (40 µL) at 40 °C for 1 h, after which the reaction was neutralized with 1 M HCl (20 µL), diluted with MeCN (200 µL) and filtered (0.45 µm PTFE) prior to analysis.

3.4.2. Standard Marfey's HPLC Method #2

An aliquot of Marfey's derivatized analyte (3 µL) (see method #1) was subjected to HPLC-DAD-MS analysis using a binary solvent system (Phase A: 95% H_2O: 5% MeCN: 0.1% formic acid; Phase B: 95% MeOH: 5% MeCN: 0.1% formic acid) on an Agilent Poroshell 120 SB-C_8 2.7 µm, 3.0 × 150 mm column, at 50 °C with a 0.8 mL/min linear gradient over 29 min from 16% to 63% Phase B in A, and with DAD (340 nm) and ESI(±)MS monitoring, supported by single ion extraction (SIE) methodology, and with comparison to authentic standards of Marfey's derivatized amino acids.

3.4.3. Marfey's HPLC Method #3 Optimized for Resolving N-Me-Ala Derivatives

An aliquot of Marfey's derivatized analyte (3 µL) (see method #1) was subjected to UPLC-DAD-MS analysis using the same binary solvent system and detection as described above (method #2), but with an isocratic 0.8 mL/min elution at 23% Phase B in A using an Agilent Poroshell 120 EC-C_{18} 2.7 µm, 3.0 × 150 mm column at 50 °, with comparison to authentic standards of Marfey's derivatized amino acids (Figures 3 and S38).

3.4.4. Marfey's HPLC Method #4 Optimized for Resolving Leu, Ile and *allo*-Ile Derivatives

An aliquot of Marfey's derivatized analyte (3 µL) (see method #1) was subjected to UPLC-DAD-MS analysis using the same binary solvent system and detection as described above (method #2), but with an isocratic 0.8 mL/min elution at 37% Phase B in A, with comparison to authentic standards of Marfey's derivatized amino acids (Figures 4 and S39).

3.4.5. Marfey's Analysis of Talarolides A–D (**1–4**)

Samples of talarolides A–D (**1–4**) (50 µg) were subjected to standard Marfey's hydrolysis and derivatization (method #1), after which individual aliquots of Marfey's derivatized analytes (3 µL) were subjected to each of methods #2, #3, and #4 to unambiguously identify the following amino acid constituents:

Talarolide A (**1**): L-Ala, *N*-Me-D-Leu, D-*allo*-Ile, D-Ala, *N*-Me-L-Ala, *N*-Me-L-Tyr (see Figures 3, 4 and S34)

Talarolide B (**2**): Gly, L-Ala, *N*-Me-D-Leu, D-*allo*-Ile, D-Ala, *N*-Me-L-Ala, *N*-Me-L-Tyr (see Figure S35)

Talarolide C (**3**): L-Ala, *N*-Me-D-Leu, D-Val, D-Ala, *N*-Me-L-Ala, *N*-Me-L-Tyr (see Figure S36)

Talarolide D (**4**): L-Ala, *N*-Me-D-Leu, D-*allo*-Ile, D-Ala, L-Ala, *N*-Me-L-Tyr (see Figure S37)

3.5. *Two-Dimensional Marfey's Analysis of Talarolide B*

3.5.1. Two-Dimensional Marfey's Method #5 Partial Hydrolysis of Talarolide B

A sample of talarolide B (**2**) (50 μg) was subjected to the standard Marfey's hydrolysis conditions (method #1) but with a reduced reaction time from 12 to 3 h, and after derivatization yielded a dipeptide attributed to L-FDAA-D-*allo*-Ile-Ala (unspecified Ala configuration) based on an HPLC-DAD-MS (Marfey's method #2) [t_R 22.3 min, m/z 455 (M + H)] (Figure S40).

3.5.2. Synthesis of Dipeptides L-FDAA-D-*allo*-Ile-L-Ala and L-FDAA-D-*allo*-Ile-D-Ala

Syntheses were performed using standard peptide synthesis on a 2-chlorotrityl chloride (2-CTC) resin (substitution ratio: 1.55 mmol/g, 0.1 mmol scale, 64.5 mg) using hexafluorophosphate azabenzotriazole tetramethyl uranium (HATU) and *N,N*-diisopropylethylamine (DIPEA) coupling, and fluorenylmethoxycarbonyl (Fmoc) protection chemistry.

3.5.3. Coupling of the First Amino Acid

After swelling the 2-CTC resin for 20 min in dry CH_2Cl_2 (2 mL), a solution of either Fmoc-L-Ala-OH or Fmoc-D-Ala-OH (1.2 eq.) and DIPEA (44 μL, 0.25 mmol, 2.5 eq.) in dry CH_2Cl_2 (2 mL) was added to the resin and mixed for 2 h. The resin was filtered and MeOH (200 μL) was added and mixed for 15 min to cap the resin. The resin was washed with dry CH_2Cl_2 (5 × 1 min), 1:1 CH_2Cl_2/MeOH (5 × 1 min) and MeOH (2 × 1 min).

3.5.4. Coupling of Fmoc-D-*allo*-Ile

Coupling of Fmoc-D-*allo*-Ile was achieved by dissolving Fmoc-D-*allo*-Ile (0.32 mmol, 3.2 eq.), in 0.4 M HATU/DMF (0.75 mL, 0.3 mmol, 3.0 eq.), followed by the addition of DIPEA (105 μL, 0.6 mmol, 6.0 eq.). The coupling cycle consisted of Fmoc deprotection with 20% of piperidine in DMF (twice, 5 and 10 min), a 5 min DMF flow-wash, followed by coupling with preactivated Fmoc-D-*allo*-Ile (3.2 eq.) over 2 × 30 min.

3.5.5. Derivatization with L-FDAA

Fmoc deprotection was achieved by the addition of 20% of piperidine in DMF (twice, 5 and 10 min), a 5 min DMF flow-wash, followed by coupling with L-FDAA reagent, 1% solution in acetone (3.2 eq.), in the presence of DIPEA (105 μL, 0.6 mmol, 6.0 eq.) for 1 h. The resin was washed with acetone (5 × 1 min), dry CH_2Cl_2 (5 × 1 min), 1:1 CH_2Cl_2/MeOH (5 × 1 min) and MeOH (2 × 1 min), then dried (vacuum desiccator).

3.5.6. Cleavage of L-FDAA Derivatized Dipeptide from Resin

After swelling the 2-CTC resin for 20 min in dry CH_2Cl_2 (2 mL) the resin was mixed with 20% hexafluoro-2-propanol (HFIP)/CH_2Cl_2 (2 mL × 3 × 20 min) and the combined filtrate evaporated in vacuo to yield analytical samples of L-FDAA-D-*allo*-Ile-L-Ala [HRESI(+)MS m/z 477.1722 [M + Na]$^+$ (calcd for $C_{18}H_{26}N_6O_8Na$ 477.1704) and L-FDAA-D-*allo*-Ile-D-Ala [HRESI(+)MS m/z 477.1692 [M + Na]$^+$ (calcd for $C_{18}H_{26}N_6O_8Na$ 477.1704), both of which were shown to be pure by HPLC-DAD-MS (method as described in general experimental section) (Figures S46 and S47).

3.5.7. Marfey's Method #6 Optimized for L-FDAA-D-*allo*-Ile-Ala Diastereomers

An aliquot of Marfey's derivatized analyte (3 μL) (see method #5) was subjected to UPLC-DAD-MS analysis using the same binary solvent system and detection as described above (method #2), but with an isocratic 0.6 mL/min elution at 37% Phase B in A, and with comparison between natural and synthetic Marfey's derivatized dipeptides (Figure S41).

3.6. Synthesis of Talarolide B

3.6.1. Coupling of the First Amino Acid

After swelling the 2-CTC resin for 20 min in dry CH_2Cl_2 (2 mL), a solution of Fmoc-Gly-OH (1.2 eq.) and DIPEA (44 µL, 0.25 mmol, 2.5 eq.) in dry CH_2Cl_2 (2 mL) was added to the resin and mixed for 2 h. The resin was filtered, then MeOH (200 µL) was added and mixed for 15 min to cap the resin. The resin was washed with dry CH_2Cl_2 (5 × 1 min), 1:1 CH_2Cl_2/MeOH (5 × 1 min) and MeOH (2 × 1 min).

3.6.2. Elongation of Peptide Sequence

Amino acid activation was achieved by dissolving an Fmoc-amino acid (0.32 mmol, 3.2 eq.) in a 0.4 M HATU/DMF solution (0.75 mL, 0.3 mmol, 3.0 eq.), followed by the addition of DIPEA (105 µL, 0.6 mmol, 6.0 eq.). The coupling cycle consisted of Fmoc deprotection with 20% of piperidine in DMF (twice, 5 and 10 min), and a 5 min DMF flow-wash followed by coupling with preactivated Fmoc-amino acid (3.2 eq.) over 2 × 30 min, or 2 × 3 h for coupling of Fmoc-amino acids to sterically hindered *N*-Me-amino acids. Upon completion of the synthesis the resin was washed with DMF, CH_2Cl_2, and MeOH, then dried (vacuum desiccator) as described above in the synthesis of the dipeptides.

3.6.3. Cleavage of Linear Protected Peptide

After swelling the 2-CTC resin for 20 min in dry CH_2Cl_2 (2 mL), the resin was mixed with 20% hexafluoro-2-propanol (HFIP)/CH_2Cl_2 (2 mL × 3 × 20 min and the combined filtrate concentrated in vacuo to give the protected linear peptide (64 mg). The product was confirmed by HPLC-DAD-MS (method as described in general experimental section): t_R = 4.4 min, ESI(+)MS m/z 776 $[M + H]^+$, and was used in the next step without further purification.

3.6.4. Cyclization of Linear Protected Peptide

A solution of the linear protected peptide (0.5 mg/mL in DMF (128 mL, 64 mg, 0.082 mmol) was stirred vigorously and treated by dropwise addition over 30 min with a mixture of 0.4 M HATU (618 µL, 0.247 mmol, 3 eq.), hydroxybenzotriazole (HOBT) (34 mg, 0.247 mmol, 3 eq.) and collidine (33 µL, 30 mg, 0.247mmol, 3 eq.) in DMF (2 mL). After 14 h, HPLC-DAD-MS analysis of the mixture showed the cyclization was completed. The DMF was evaporated, and the residue dissolved in MeCN (10 mL), filtered (0.45 µm filter), and purified by preparative HPLC (Agilent Zorbax Rx-C_8 7 µm, 21.2 × 250 mm column, with a 20 min gradient elution at 20 mL/min from 90% H_2O/MeCN to 100% MeCN with an isocratic 0.01% trifluoroacetic acid/MeCN modifier). After lyophilization, the protected cyclic peptide was obtained as an amorphous powder. The product was confirmed by HPLC-DAD-MS (method as described in general experimental section): t_R = 5.8 min, ESI(+)MS m/z 758.5 $[M + H]^+$

3.6.5. Cyclic Peptide Deprotection

The protected cyclic peptide was mixed with an aqueous solution of 90% formic acid (3 mL) for 40 min, after which it was concentrated under a stream of nitrogen gas and the residue dissolved in MeCN (2 mL) and purified by preparative HPLC (Agilent Zorbax Rx-C_8 7 µm, 21.2 × 250 mm column, with a 20 min gradient elution at 20 mL/min from 90% H_2O/MeCN to 100% MeCN with an isocratic 0.01% trifluoroacetic acid/MeCN modifier) to yield synthetic talarolide B (16 mg, 23% overall yield); NMR (DMSO-d_6), see Figures S50–S52; HRESI(+)MS m/z 724.4008 $[M + Na]^+$ (calcd for $C_{35}H_{55}N_7O_8Na$ 724.4004); identical with natural talarolide B (**2**), including by co-injection HPLC-DAD-MS (Figure S48).

3.7. Three-Dimensional Solution Structure Calculations

The distance restraints used in calculating the structure for talarolide A (**1**) in DMSO-d_6 were derived from ROESY spectra (recorded at 298 K) using mixing time (spin-lock) of 300 ms with 41 NOEs (see Supplementary Materials). NOE cross-peak volumes were classified manually as strong (upper distance constraint ≤ 2.7Å), medium (≤3.5Å), weak (≤5.0Å), or very weak (≤6.0Å). Standard pseudoatom distance corrections were applied for non-stereospecifically assigned protons. To address the possibility of conformational averaging, intensities were classified conservatively and only upper distance limits were included in the calculations to allow the largest possible number of conformers to fit the experimental data. Backbone dihedral angle restraints were inferred from $^3J_{NHCH\alpha}$ coupling constants in 1D spectra, using the Karplus equation [15] with angle ± 30°. There was one *cis*-amides bond between *N*-Me-L-Ala^6-*N*-Me-L-Tyr^7 (i.e., strong CHα-CHα (*i*, *i* + *1*) NOEs); the rest were in *trans* configuration (ψ-angles were set to *trans* (ψ = 180°)). Starting structures with randomized ϕ and ψ angles and extended side chains were generated using an ab initio simulated annealing protocol. The calculations were performed using the standard force-field parameter set (PARALLHDG5.2.PRO) and topology file (TOPALLHDG5.2.PRO) in XPLOR-NIH with in-house modifications to generated N-methylated and *N*-hydroxylated residues. Refinement of structures was achieved using the conjugate gradient Powell algorithm with 4000 cycles of energy minimization and a refined forcefield based on the program CHARMM [12]. Structures were visualized with Pymol and analyzed for distance (>0.2Å) and dihedral angle (>2°) violations using noe.inp files. ^{1}H NMR (DMSO-d_6) variable temperatures for **1** was obtained from 298 K to 318 K in five degrees stepwise on a Bruker Avance III 600 MHz NMR spectrometer. The chemical shift differences of amide N<u>H</u> and *N*-OH were plotted against temperature to generate the temperature coefficient (Δδ/T) using Prism version 10.0.2 (Figure 8).

4. Conclusions

Application of an integrated program of cultivation (MATRIX) and chemical (HPLC-DAD and GNPS) profiling to the marine-derived fungus *Talaromyces* sp. CMB-TU011 enabled access to talarolide A (**1**), along with three new analogues, talarolides B–D (**2**–**4**). Detailed spectroscopic analysis, supported by chemical degradation and derivatization, and partial and total syntheses, permitted assignment of structures to **1** (revised) and **2**–**4**. The talarolides include rare examples of natural cyclic peptides incorporating a hydroxamate moiety, with a solution structure on **1** revealing H-bonding from the *N*-O<u>H</u>-Gly^1 across the macrocyclic ring to the D-Ala^5 carbonyl oxygen, which both defined and stabilized a unique conformation. This contrasts with the deoxy analogue **2** (i.e., Gly^1) where the NMR data indicate two interconverting conformers. Knowledge of the talarolides draws attention to the possible inclusion of hydroxamate moieties in other cyclic peptides (natural and synthetic), as a means to access new and unusual conformations with potentially new biological properties including improved oral bioavailability [16].

Supplementary Materials: The following supporting information can be downloaded at: https://www.mdpi.com/article/10.3390/md21090487/s1. MATRIX study of CMB TU011; NMR spectroscopic data (tabulated data and spectra), Marfey's analysis, and MS/MS spectra of **1**–**4**; NMR spectra comparison of natural and synthetic talarolide B (**2**); 3D calculations of talarolide A (**1**).

Author Contributions: R.J.C. conceptualized the research; P.D. carried out the MATRIX study; P.D. and Z.G.K. analyzed NMR data and assigned the structures; W.M.H. and Y.Z. synthesized the dipeptides and talarolide B; K.S. cultivated the microbe and isolated talarolides, H.N.H. and D.P.F. performed 3D structure calculation; A.A.S. analyzed MS/MS data, revised NMR data, developed revised Marfey's method and prepared Supplementary Materials, R.J.C. writing—original draft, R.J.C. and A.A.S.; writing—review and editing. All authors have read and agreed to the published version of the manuscript.

Funding: D.P.F. and H.N.H. were supported by Australian Research Council (CE200100012) and NHMRC Investigator (2009551) grants.

Institutional Review Board Statement: Not applicable.

Data Availability Statement: Raw NMR data have been deposited in the Natural Product Magnetic Resonance Database Project (NP-MRD).

Acknowledgments: We thank R. Damodar for collection of CMB-TU011 and the University of Queensland (UQ), Institute for Molecular Bioscience for supporting this research. P.D. and K.S. acknowledge UQ for provision of postgraduate research scholarships.

Conflicts of Interest: The authors declare no conflict of interest.

References

1. Raju, R.; Khalil, Z.G.; Piggott, A.M.; Blumenthal, A.; Gardiner, D.L.; Skinner-Adams, T.S.; Capon, R.J. Mollemycin A: An Antimalarial and Antibacterial Glyco-Hexadepsipeptide-Polyketide from an Australian Marine-Derived *Streptomyces* sp. (CMB-M0244). *Org. Lett.* **2014**, *16*, 1716–1719. [CrossRef] [PubMed]
2. Khalil, Z.G.; Salim, A.A.; Lacey, E.; Blumenthal, A.; Capon, R.J. Wollamides: Antimycobacterial Cyclic Hexapeptides from an Australian Soil *Streptomyces*. *Org. Lett.* **2014**, *16*, 5120–5123. [CrossRef] [PubMed]
3. Jiao, W.-H.; Khalil, Z.; Dewapriya, P.; Salim, A.A.; Lin, H.-W.; Capon, R.J. Trichodermides A-E: New Peptaibols Isolated from the Australian Termite Nest-Derived Fungus *Trichoderma virens* CMB-TN16. *J. Nat. Prod.* **2018**, *81*, 976–984. [CrossRef] [PubMed]
4. Quezada, M.; Shang, Z.; Kalansuriya, P.; Salim, A.A.; Lacey, E.; Capon, R.J. Waspergillamide A, a Nitro Depsi-Tetrapeptide Diketopiperazine from an Australian Mud Dauber Wasp-Associated *Aspergillus* sp. (CMB-W031). *J. Nat. Prod.* **2017**, *80*, 1192–1195. [CrossRef] [PubMed]
5. Elbanna, A.H.; Khalil, Z.G.; Bernhardt, P.V.; Capon, R.J. Scopularides Revisited: Molecular Networking Guided Exploration of Lipodepsipeptides in Australian Marine Fish Gastrointestinal Tract-Derived Fungi. *Mar. Drugs* **2019**, *17*, 475. [CrossRef] [PubMed]
6. Dewapriya, P.; Khalil, Z.G.; Prasad, P.; Salim, A.A.; Cruz-Morales, P.; Marcellin, E.; Capon, R.J. Talaropeptides A-D: Structure and Biosynthesis of Extensively N-Methylated Linear Peptides from an Australian Marine Tunicate-Derived *Talaromyces* sp. *Front. Chem.* **2018**, *6*, 394. [CrossRef] [PubMed]
7. Dewapriya, P.; Prasad, P.; Damodar, R.; Salim, A.A.; Capon, R.J. Talarolide A, a Cyclic Heptapeptide Hydroxamate from an Australian Marine Tunicate-Associated Fungus, *Talaromyces* sp. (CMB-TU011). *Org. Lett.* **2017**, *19*, 2046–2049. [CrossRef] [PubMed]
8. Zhang, S.; De Leon Rodriguez, L.M.; Huang, R.; Leung, I.K.H.; Harris, P.W.R.; Brimble, M.A. Total Synthesis of the Proposed Structure of Talarolide, A. *Org. Biomol. Chem.* **2018**, *16*, 5286–5293. [CrossRef] [PubMed]
9. Salim, A.A.; Khalil, Z.G.; Elbanna, A.H.; Wu, T.; Capon, R.J. Methods in Microbial Biodiscovery. *Mar. Drugs* **2021**, *19*, 503. [CrossRef] [PubMed]
10. Aron, A.T.; Gentry, E.C.; McPhail, K.L.; Nothias, L.-F.; Nothias-Esposito, M.; Bouslimani, A.; Petras, D.; Gauglitz, J.M.; Sikora, N.; Vargas, F.; et al. Reproducible Molecular Networking of Untargeted Mass Spectrometry Data Using GNPS. *Nat. Protoc.* **2020**, *15*, 1954–1991. [CrossRef] [PubMed]
11. Vijayasarathy, S.; Prasad, P.; Fremlin, L.J.; Ratnayake, R.; Salim, A.A.; Khalil, Z.; Capon, R.J. C_3 and 2D C_3 Marfey's Methods for Amino Acid Analysis in Natural Products. *J. Nat. Prod.* **2016**, *79*, 421–427. [CrossRef] [PubMed]
12. Brooks, B.R.; Bruccoleri, R.E.; Olafson, B.D.; States, D.J.; Swaminathan, S.; Karplus, M. Charmm: A Program for Macromolecular Energy, Minimization, and Dynamics Calculations. *J. Comput. Chem.* **1983**, *4*, 187–217. [CrossRef]
13. Brunger, A.T. *X-Plor Manual Version 3.1*; Yale University Press: New Haven, CT, USA, 1992.
14. Wishart, D.S.; Sykes, B.D.; Richards, F.M. The Chemical Shift Index: A Fast and Simple Method for the Assignment of Protein Secondary Structure through NMR Spectroscopy. *Biochemistry* **1992**, *31*, 1647–1651. [CrossRef] [PubMed]
15. Pardi, A.; Billeter, M.; Wüthrich, K. Calibration of the Angular Dependence of the Amide Proton-Cα Proton Coupling Constants, $^{3}J_{HN\alpha}$, in a Globular Protein: Use of $^{3}J_{HN\alpha}$ for Identification of Helical Secondary Structure. *J. Mol. Biol.* **1984**, *180*, 741–751. [CrossRef] [PubMed]
16. Nielsen, D.S.; Hoang, H.N.; Lohman, R.-J.; Hill, T.A.; Lucke, A.J.; Craik, D.J.; Edmonds, D.J.; Griffith, D.A.; Rotter, C.J.; Ruggeri, R.B.; et al. Improving on Nature: Making a Cyclic Heptapeptide Orally Bioavailable. *Angew. Chem.* **2014**, *53*, 12059–12063. [CrossRef] [PubMed]

marine drugs

MDPI

Article

Activation of a Silent Polyketide Synthase SlPKS4 Encoding the C_7-Methylated Isocoumarin in a Marine-Derived Fungus *Simplicillium lamellicola* HDN13-430

Jing Yu [1,†], Xiaolin Liu [1,†], Chuanteng Ma [1], Chen Li [1], Yuhan Zhang [2], Qian Che [1], Guojian Zhang [1,3], Tianjiao Zhu [1,*] and Dehai Li [1,3,*]

1 Key Laboratory of Marine Drugs, Chinese Ministry of Education, School of Medicine and Pharmacy, Ocean University of China, Qingdao 266003, China; jingjingyu95@163.com (J.Y.); 17865327317@163.com (X.L.); ma_chuanteng@163.com (C.M.); 17629508296@163.com (C.L.); cheqian064@ouc.edu.cn (Q.C.); zhangguojian@ouc.edu.cn (G.Z.)

2 School of Pharmaceutical Science, Shandong University, Jinan 250100, China; 202100260059@mail.sdu.edu.cn

3 Laboratory for Marine Drugs and Bioproducts, Pilot National Laboratory for Marine Science and Technology (Qingdao), Qingdao 266237, China

* Correspondence: zhutj@ouc.edu.cn (T.Z.); dehaili@ouc.edu.cn (D.L.)

† These authors contributed equally to this work.

Abstract: Coumarins, isocoumarins and their derivatives are polyketides abundant in fungal metabolites. Although they were first discovered over 50 years ago, the biosynthetic process is still not entirely understood. Herein, we report the activation of a silent nonreducing polyketide synthase that encodes a C_7-methylated isocoumarin, similanpyrone B (**1**), in a marine-derived fungus *Simplicillium lamellicola* HDN13-430 by heterologous expression. Feeding studies revealed the host enzymes can change **1** into its hydroxylated derivatives pestapyrone A (**2**). Compounds **1** and **2** showed moderate radical scavenging activities with ED_{50} values of 67.4 μM and 104.2 μM. Our discovery fills the gap in the enzymatic elucidation of naturally occurring C_7-methylated isocoumarin derivatives.

Keywords: isocoumarins; nonreducing polyketide synthase (nrPKS); *Simplicillium lamellicola*; genome mining; silent gene clusters

Citation: Yu, J.; Liu, X.; Ma, C.; Li, C.; Zhang, Y.; Che, Q.; Zhang, G.; Zhu, T.; Li, D. Activation of a Silent Polyketide Synthase SlPKS4 Encoding the C_7-Methylated Isocoumarin in a Marine-Derived Fungus *Simplicillium lamellicola* HDN13-430. *Mar. Drugs* **2023**, *21*, 490. https://doi.org/10.3390/md21090490

Academic Editor: Georg Pohnert

Received: 24 August 2023
Revised: 6 September 2023
Accepted: 12 September 2023
Published: 13 September 2023

1. Introduction

Fungi have proven to be a tremendous source of new bioactive lead compounds with thousands of bioactive compounds isolated [1,2]. Meanwhile, whole-genome sequencing data revealed that the number of biosynthetic gene clusters encoded in fungi is much larger than the types of natural products isolated, which indicates that a major portion of biosynthetic gene clusters are still silent or poorly expressed [3,4]. To activate these silent gene clusters and increase the silent metabolic potential, a variety of techniques have been developed, including epigenetics regulation, co-culture, precursor feeding, heterologous expression, changing fermentation parameters and ribosome engineering, etc. [5,6].

Among these, heterologous expression has unique advantages, especially to achieve the de novo biosynthesis of compounds in a heterologous host, which benefits from genetic tractability, short life-cycle, and high bio-safety [7]. As for the activation of silent gene clusters in fungi, heterologous expression shows special superiority, such as: (1) it is more controllable compared to other activation methods, especially, the activation is orientated instead of randomly; (2) the ideal chassis cells with simple metabolite backgrounds make it easy to perform the isolation of targeted compounds; (3) it is still effective without regulators, selective markers, or strain genetic operating system, which is frequently the major obstacle in non-model fungi [8–10].

The fungal specie, *Simplicillium lamellicola*, has great ecological and commercial importance due to the exceptional bioactivities, particularly in microbial biopesticide [11–14].

However, to our best knowledge, four new compounds, reported in the specie (Figure 1), indicate a great potential for bioactive secondary compounds from *S. lamellicola*. Isolated from a marine sediment collected in Pritz Bay, *S. lamellicola* HDN13-430 was the first strain of this specie isolated from Antactica, which made the genome mining worthwhile. During our ongoing genome mining work on the fungal strain *S. lamellicola* HDN13-430, a nonreducing polyketide synthase (nrPKS), termed SlPKS4, attracts our attention. SlPKS4 showed a low sequence identity to other PKSs. With the expression test of cDNA, the nrPKS, together with the gene cluster, were proven to be completely silent under regular laboratory conditions (Figure S1). Due to the lack of regulators in the native strain, the following heterologous expression of SlPKS4 in *Aspergillus nidulans* lead to the yield of two isocoumarin derivatives: similanpyrone B (**1**) and pestapyrone A (**2**). Compounds **1** and **2** showed radical scavenging activities, while no activity in the antibacterial bioassays was observed. Although compounds **1** and **2** were first described more than 50 years ago, the biosynthetic process is still not entirely understood. We proposed the biosynthetic pathway of compounds **1** and **2**, and conducted phylogenetic analysis with other PKSs responsible for the synthesis of isocoumarin derivatives.

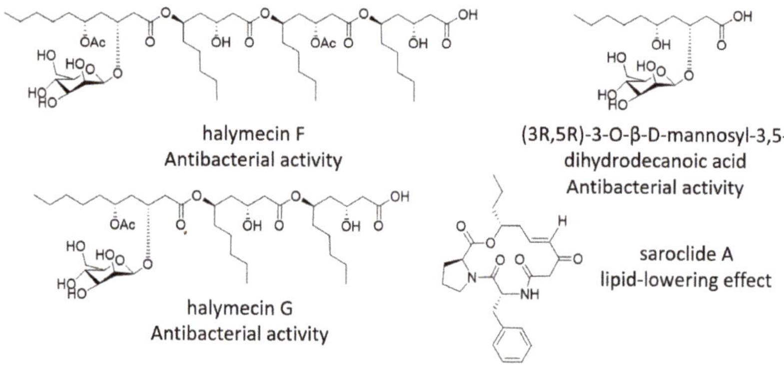

Figure 1. Bioactive compounds reported from *S. lamellicola*.

2. Results

2.1. *Bioinformatic Analysis of the Target nrPKS and the Gene Cluster in S. lamellicola HDN13-430*

To probe the metabolic potential of *S. lamellicola* HDN13-430, the whole-genome sequencing and analysis were performed. The prediction of secondary metabolites using antiSMASH indicated 12 PKSs, 16 NRPSs, 3 terpenes, 6 hybrids and 2 other types biosynthetic gene clusters (Figure S2). During analyzing the PKSs of strain *S. lamellicola* HDN13-430, an nrPKS which we termed as *SlPKS4* (Figure 2), exhibits low identities with known nrPKSs, while the highest similarities of 41.46% and 41.20% at the amino acid level were donated by pkbA [15] and andM [16], which are responsible for the biosynthesis of compounds 3-methylorsellinic and 3,5-dimethylorsellinic acid, respectively. The low-similarity PKS *SlPKS4*, located in a cluster with seven tailoring enzymes termed Sl4001-Sl4007, with proposed functionsas quinone oxidoreductase, dienelactone hydrolase, γ-glutamyl phosphate reductase, hypothetical protein, sulfide quinone reductase, threonine dehydratase, ketol-acid reductoisomerase, respectively (details in Table S2). The proposed functions of enzymes located in the cluster are all uncommon with a rare report in secondary metabolites biosynthesis. However, further analysis of gene transcription status by RT-PCR of six media based on OSMAC (one strain many compounds) and epigenetic regulation strategies, shows that the cluster including *SlPKS4* is totally silent under regular laboratory culture conditions (Figure S1). Also, there are no reports about the construction of a genetic operating system on the specie *S. lamellicola*, which prompts us to investigate the function by heterologous biosynthesis in *A. nidulans*.

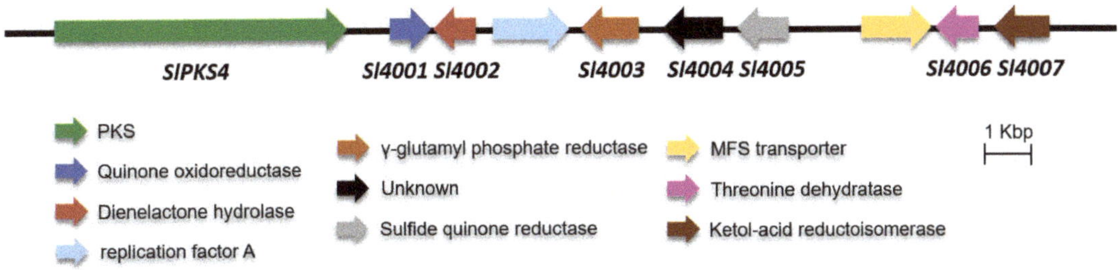

Figure 2. Organization and proposed function of *SlPKS4* and tailoring genes.

2.2. *Heterologous Expression of the Gene Cluster and Elucidation of Compounds* **1** *and* **2**

To demonstrate the function of SlPKS4, a 9547 bp fragment containing the whole genomic sequence of SlPKS4, plus a downstream region of 526 bp containing the native terminator, were amplified via PCR and cloned into the expression vector pANU-*SlPKS4* by homologous recombination in *Saccharomyces cerevisiae*. The obtained construct, pANU-*SlPKS4*, was introduced into *A. nidulans* A1145 by polyethylene glycol (PEG)-mediated protoplast transformation. Integration transformants, including pANU-*SlPKS4*, were grown on liquid CD-Starch medium following selection by uridine and uracil autotrophy and subsequent confirmation by PCR amplification [17]. The cultures were extracted with ethyl acetate and analyzed by LC-MS for secondary metabolites. As shown in Figure 3, two additional peaks of compounds **1** and **2** were detected in the extract compared to the control strain containing the empty vector. The two compounds share similar UV spectra with absorption maxima at 240, 280 and 330 nm, together with $[M + H]^+$ ions at m/z = 207.1 and 223.2, respectively, indicating similar structures and differences coming from hydroxylation (Figure S2). Following large-scale fermentation, isolation and structural elucidation by 1D NMR analysis (Tables S3 and S4 and Figures S8–S11) confirmed compounds **1** and **2** to be similanpyrone B and pestapyrone A, respectively [17,18]. The literature review concludes that both compounds belong to the group of isocoumarins.

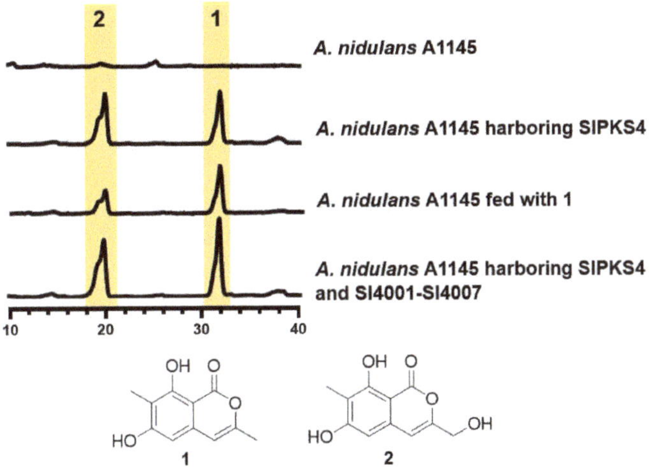

Figure 3. HPLC analysis of the secondary metabolites in *A. nidulans* strains. UV absorptions at 280 nm are illustrated. HPLC full chromatogram of the original *A. nidulans* and the strain harboring SlPKS4 were provided to prevent the presence of compound **2** in original *A.nidulans* (Figure S5). HPLC analysis method: 5:95 to100:0 MeOH-H_2O (with 0.1% trifluoroacetic acid), 40 min, 1 mL/min.

Similarly, the tailoring enzymes Sl4001-Sl4007 were separately cloned, constructed on expression plasmids and introduced into *A. nidulans* A1145 (Figure S4). Unexpectedly, after 4 days of culturing followed by extraction with ethyl acetate, no new compound, except **1** and **2**, was detected by LC-MS analysis (Figure 3). Double checking the gene transcription status by RT-PCR was performed, confirming that all seven tailoring enzymes, together with SlPKS4, were expressed properly (Figure S6), which exclude the possibility of unexpression.

Compound **1** has undergone investigation through chemical synthesis [19] and isotope labeling [20,21] since the 1980s, however, the enzyme responsible for the biosynthesis has not been reported until now. Our report about SlPKS4 is the first discovery of PKS responsible for compound **1**. Meanwhile, isocoumarins derivatives were generally discovered from the fungal genera *Penicillium, Ceratocystis, Fusarium, Artemisia, Aspergillus, Cladosporium, Oospora,* and *Hydrangea* [22]. To our best knowledge, there is no coumarins or isocoumarins reported from the fungal genus *Simplicillium,* so this is also the first time to prove that the fungal genus *Simplicillium* has the ability to produce isocoumarin derivatives.

2.3. *Origin Verification of Compound* **2** *by Biotransformation Assay*

Unexpected accumulation of compound **2** as a hydroxylated derivative of **1** raises the question on the origin of the hydroxylation activity, due to no corresponding enzymes for the related hydroxylation reaction. Inspired by a previous work by Li et al. [23], we postulated that endogenous enzymes from *A. nidulans* were also in charge of the conversion of **1** to **2**. Intrigued by this hypothesis, we conducted a feeding experiment of compound **1** in *A. nidulans* A1145. After four days feeding with compound **1**, compound **2** was clearly present in the culture extract after LC-MS analysis (Figure 3). This proved that *A. nidulans* can modify the initial polyketide product **1** by hydroxylation at the methyl (Figure 4). Unfortunately, no *A. nidulans* candidate enzymes could be anticipated for the process.

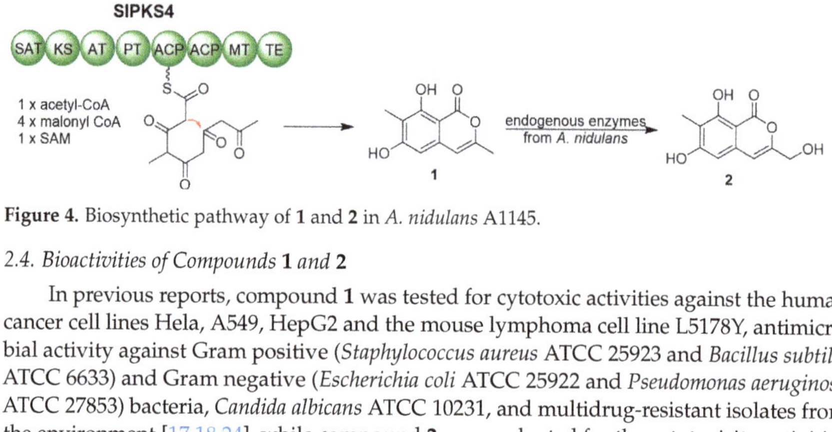

Figure 4. Biosynthetic pathway of **1** and **2** in *A. nidulans* A1145.

2.4. *Bioactivities of Compounds* **1** *and* **2**

In previous reports, compound **1** was tested for cytotoxic activities against the human cancer cell lines Hela, A549, HepG2 and the mouse lymphoma cell line L5178Y, antimicrobial activity against Gram positive (*Staphylococcus aureus* ATCC 25923 and *Bacillus subtilis* ATCC 6633) and Gram negative (*Escherichia coli* ATCC 25922 and *Pseudomonas aeruginosa* ATCC 27853) bacteria, *Candida albicans* ATCC 10231, and multidrug-resistant isolates from the environment [17,18,24], while compound **2** was evaluated for the cytotoxicity activities against a panel of cancer cell lines (A549, HL-60, K562 and L5178Y) [17,25]. All the above investigations proved to be inactive. In our research, the antimicrobial activities against MRCNS (Methicillin-Resistant Coagulase-Negative *Staphylococci*), MRSA (Methicillin-Resistant *Staphylococcus Aureus*), *S. aureus, Acinetobacter baumannii, B. cereus, B. subtilis, P. aeruginosa,* and *C. albicans* were evaluated, but none of them presented an antimicrobial effect under the concentration of 30 μM. Meanwhile, the radical scavenging assay based on 1,1-diphenyl-2-picrylhydrazyl radical 2,2-diphenyl-1-(2,4,6-trinitrophenyl) hydrazyl (DPPH) was used to test the radical scavenging activity of compounds **1** and **2**, and they showed similar effects with ED_{50} values of 67.4 μM and 104.2 μM (the value of ascorbic acid was 12.6 μM as positive control).

3. Discussion

Coumarins and isocoumarins are a large family of lactonic natural products, abundant in various organisms including bacteria, fungi, lichens, liverworts, sponges, plants and insects [22,26]. The widespread distribution and possession of a broad spectrum of pharmacological activities, including antifungal, anti-inflammatory, cytotoxic, and antimicrobial properties, have led to the continuous discovery of novel isocoumarin compounds [22,27,28]. Structurally, there are six chemical active positions (C_3 to C_8) of isocoumarin, which make a substantial contribution to the formation of diverse chemical derivatives containing alkyl, halogen, heterocyclic, aryl, etc. Isocoumarins have garnered significant attention in total synthesis due to their utility as key intermediates in the production of valuable compounds, such as isoquinolines and isochromenes [27,29,30]. These diverse chemical substitution patterns significantly augment the structural complexity and broaden the range of biological and pharmacological activities of isocoumarins [31–33].

Several biosynthetic gene and gene clusters responsible for coumarins and isocoumarins have been discovered [23,32–35]. In fungi, isocoumarin derivatives are primarily derived from the polyketide pathways, which are typically catalyzed by nonreducing polyketide synthase (nrPKS), containing domains such as starter unit ACP transacylase (SAT), β-ketoacyl synthase (KS), acyl transferase (AT), product template (PT), acyl carrier protein (ACP), methyltransferase (MT), and thioesterase (TE) [22,23,32]. However, despite recent significant advancements in the chemical synthesis of isocoumarins, there are relatively few PKSs responsible for their biosynthesis that have been documented, compared to the abundance of isocoumarin derivatives. In fungi, the elucidated biosynthetic PKSs could be divided into two categories. The first class, represented by Pcr9304 and cla3, have no MT domain, which corresponds with the absence of methyls on $C_{3,5,7}$, while only one report of the second class exists with $C_{3,5}$-dimethyl directed by AcreC. To date, there is no biosynthetic enzyme report about C_7-methyl isocoumarin derivatives.

Inspired by the uncommon carbon substituent at C_7-methyl of **1** and **2** with raising reported activities [22,26], we performed a comparison towards reported PKSs synthesizing isocoumarin derivatives by phylogenetic analysis (Figure 5). The results demonstrated that SlPKS4 is close to AcreC, which also contains an MT domain compared to others. NCBI BLASTP at the amino acid level between SlPKS4 and AcreC shows a sequence identity of 36.89%. However, the reason why the MT domain directed SAM on different sites remains unknown.

Despite being located in a cluster, the *SlPKS4* gene is sufficient for compound **1** production. The reasons why the tailoring enzymes were ineffective are still unknown. We proposed that it is possibly because of the inactivation of the genes encoding tailoring enzymes during the rearrangements and breakages, or as a result of being flanked by potential transposons [36–38]. Another hypothesis was that the post-translational modification rules may differ among different host strains, lead to wrong enzyme structures, and ultimately became inactive. Moreover, we did comparative analysis between the gene cluster containing SlPKS4 and other biosynthetic gene clusters producing isocoumarin using clinker (Figure S7). The results suggested that our gene cluster was low and/or similar as a whole, and the tailoring enzymes were least conversed.

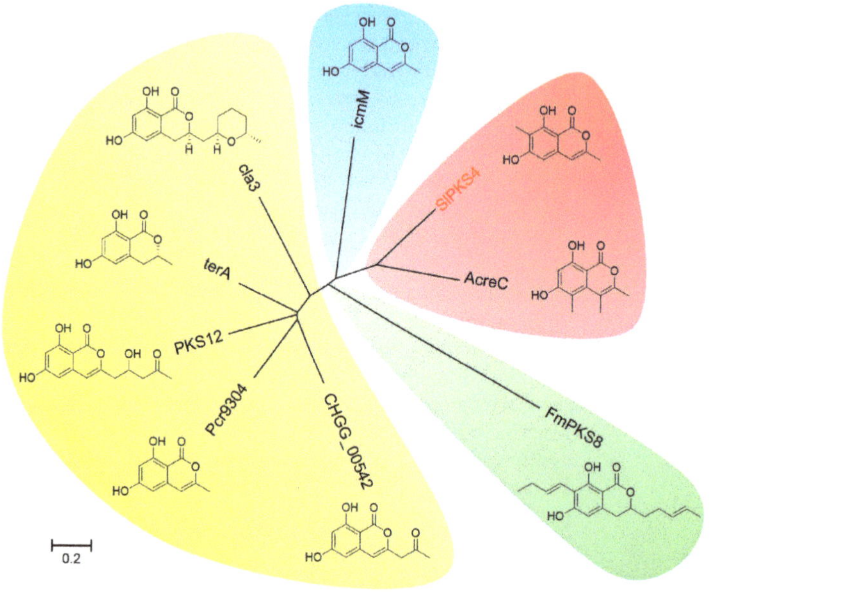

Figure 5. Phylogenetic tree analysis of SlPKS4 and reported PKSs synthesizing isocoumarin derivatives.

4. Materials and Methods

4.1. General Experimental Procedures

Genomic DNA samples were prepared using the CTAB isolation buffer at pH 8.0 (20 g/L cetyltrimethylammonium bromide, 1.4 M sodium chloride, and 20 mM EDTA). Polymerase chain reaction (PCR) was performed using Phusion® High-Fidelity DNA Polymerase (New England Biolabs, NEB, Beijing, China). PCR analyses were conducted using a 2×Hieff® PCR Master Mix (With Dye, Yeasen, Shanghai, China). DNA restriction enzymes were used as recommended by the manufacturer (New England Biolabs, NEB, Beijing, China). RT-PCR analysis was performed using Direct-zol™ RNA MiniPrep (Zymo Research, Irvine, CA, USA) and PrimeScript RT-PCR Kit (TaKaRa, Gunma, Japan). The gene-specific primers are listed in Table S1. Custom oligonucleotides synthesis and fragments sequencing were served by Shanghai Sangon DNA Technologies. LC-MS was performed using an Acquity UPLC H-Class coupled to a SQ Detector 2 mass spectrometer using a BEH C18 column (1.7 μm, 2.1 × 50 mm, 1 mL/min) (Waters Corporation, Milford, MA, USA). Semi-preparative HPLC (YMC Co., Ltd., Kyoto, Japan) was performed on an ODS column (YMC-Pack ODS-A, 10 × 250 mm, 5 μm, 3 mL/min). ^{1}H NMR and ^{13}C NMR spectra were recorded on an Agilent 500 MHz DD2 spectrometer (Agilent Technologies Inc., Santa Clara, CA, USA). The spectra were processed by the software MestReNova 6.1.0 (Metrelab, Coruña, Spain). NMR data are provided in Tables S3 and S4, and spectra in Figures S8–S11.

4.2. Materials and Culture Conditions

The fungal strain *S. lamellicola* HDN13-430 was isolated from marine sediment collected in Antarctic Pritz Bay. The strain was identified by an internal transcribed spacer (ITS) sequence and the sequence data were submitted to GenBank (GenBank accession number KY794926). The strains were deposited at the Ministry of Education of China, School of Medicine and Pharmacy, Ocean University of China, Qingdao, China.

For RNA isolation, *S. lamellicola* HDN13-430 was cultured at 28 °C on media potato dextrose agar (PDA, 20% potato, 2% dextrose, and 2% agar) plates for 5 days, and inoculated into Elenmeyer flasks containing 150 mL of 6 different liquid culture medium for 4 days.

For genomic DNA extraction, *S. lamellicola* HDN13-430 was cultured at 28 °C on PDA plates for 7 days.

Escherichia coli strain XL-1 was used for plasmids preservation and amplification. *Saccharomyces cerevisiae* Y31 was used for in vivo DNA recombination for plasmids construction.

A. nidulans A1145 was grown at 28 °C in CD (0.1% Glucose, 0.5 *v*/*v*% 20×Nitrate salts, 0.01 *v*/*v*% Trace elements, and 2% agar for solid media) media for sporulation, CDS (0.1% Glucose, 1.2 M D-sorbitol, 0.5 *v*/*v*% 20×Nitrate salts, 0.01 *v*/*v*% Trace elements, and 2% agar for solid media) to screen transformants or in CD-ST (2% starch, 2% Casamino acids, 5 *v*/*v*% 20×Nitrate salts, 0.1 *v*/*v*% Trace elements) media for heterologous expression and compound production. All medias were prepared with appropriate supplements, including 10 mM uridine, 5 mM uracil and/or 0.5 µg/mL pyridoxine HCl and/or 2.5 µg/mL riboflavin, depending on the plasmids being transformed.

4.3. Sequence Analysis of the SlPKS4 Gene

The whole genome sequencing data was analyzed by antiSMASH [39]. The phylogenetic analysis was conducted with the MEGA7 software [40], with the amino acid sequences of SlPKS4 and reported PKSs synthesizing isocoumarin derivatives retrieved from National Center for Biotechnology Information (NCBI). The conserved domain of the SlPKS4 protein was scanned by the InterProScan program [41]. Comparative analysis between the gene cluster and other isocoumarin BGCs (Figure S7) was conducted by clinker [42].

4.4. Heterologous Expression of SlPKS4 and Sl4001-Sl4007 in A. nidulans A1145

Genes *SlPKS4* and *Sl4001–Sl4007* were amplified from genomic DNA extract from *S. lamellicola* HDN13-430. Plasmids pANU, pANR, pANP with auxotrophic markers for uracil (pyrG), riboflavin (riboB) and pyridoxine (pyroA) were digested with PacI and NotI, PacI and BamHI, PacI and HindIII, respectively, and used as vectors to insert genes. The corresponding heterologous expression plasmids were obtained by in vivo yeast homologous recombination in *S. cerevisiae* Y31. The correct colonies checked by PCR were combined, and subjected to yeast miniprep to obtain the plasmids. The plasmids obtained from yeast miniprep using Plasmid Miniprep Kit (Zymo Research, Irvine, CA, USA) were introduced into the competent cells of *E. coli* XL-1. After plasmid extraction from *E. coli* to obtain transformants with a single plasmid, plasmids were sequenced to confirm identities.

For protoplast formation of *A. nidulans* A1145, the strain *A. nidulans* A1145 was first grown on CD plates at 37 °C for 3 days and fresh spores were collected and stored. Then, the spores were germinated in a 250 mL Erlenmeyer flask containing CD media at 37 °C and 180 rpm for about 8 h. Mycelia were gathered by centrifugation at 4000 rpm for 15 min, and washed by 25 mL osmotic buffer (1.2 M $MgSO_4$, 10 mM sodium phosphate, pH 5.8). Subsequently, the mycelia were suspended into 10 mL of osmotic buffer containing 30 mg lysing enzymes from Trichodema harzianum (Sigma) and 20 mg Yatalase (TaKaRa), transferred into an empty sterile bottle, and cultured in a shaker of 28 °C at 80 rpm overnight to form protoplast. After the whole night, the mixture was collected in a 50 mL centrifuge tube and covered gently with isopyknic protoplast trapping buffer (0.6 M sorbitol, 0.1 M pH 7.0 Tris-HCl). After centrifugation at 4000 rpm for 15 min at 4 °C, protoplasts were collected in the interface of the above two buffers. The protoplasts were then transferred to a sterile 50 mL centrifuge tube and washed by 20 mL STC buffer (1.2 M sorbitol, 10 mM $CaCl_2$, 10 mM pH 7.5 Tris-HCl). The protoplasts were resuspended in 2 mL STC buffer for transformation.

For protoplast transformation of *A. nidulans* A1145, the necessary plasmids and the corresponding empty plasmids (the desired strains were regarded as control in the following crude analysis) were added to 100 µL *A. nidulans* A1145 protoplast suspension prepared above and the mixture was incubated on ice for 60 min. Next, 600 µL of polyethylene glycol (PEG) solution (60% PEG, 50 mM calcium chloride and 50 mM pH 7.5 Tris-HCl) was added to the protoplast mixture, and the mixture was incubated at room temperature for an additional 25 min. The mixture was spread on the regeneration solid CDS medium with appropriate supplements, including 10 mM uridine, 5 mM uracil and/or 0.5 μg/mL pyridoxine HCl and/or 2.5 μg/mL riboflavin, depending on the plasmids being transformed and incubated at 37 °C for around 3 days.

4.5. Fermentation and Extraction

For small-scale analysis, the strains with expression plasmids and the control strains were grown on CD plates with appropriate supplements for 3 days at 37 °C. Shortly after sporing, they were inoculated into 150 mL of CD-ST medium with appropriate supplements and cultured at 28 °C, 180 rpm. Meanwhile, they were spread on solid CD-ST medium with appropriate supplements, respectively. Four days later, the cultures were extracted with ethyl acetate and the organic phase was evaporated and dissolved in MeOH, which was analyzed by HPLC.

For compound isolation, the selected strain was initially handled the same as above. Then, a large-scale fermentation was performed in 500 mL Erlenmeyer flasks (total 5 L) for further incubation. The broth was extracted three times with ethyl acetate to provide a total of 15 L of extract solution. The organic phase was evaporated under reduced pressure to afford a crude residue (6 g).

4.6. Compound Isolation

The extract was applied to MPLC (ODS) using a stepped gradient elution of MeOH-H_2O (40:60 to 100:0 for 60 min at 10 mL/min) to yield eight subfractions (Fr.1-Fr.8). Fr.4 was separated by semi-preparative HPLC (YMC-pack ODS, 10 × 250 mm, 3.0 mL/min) to afford **2** (65% MeOH in H_2O, 0.1% THF, 8.5 mg, t_R = 25 min) and Fr.5 was purified by semi-preparative HPLC to obtain **1** (72% MeOH in H_2O, 0.1% THF, 11.0 mg, t_R = 24 min). The purity of the compounds was checked by LC-MS and the structures were confirmed by ^1H and ^{13}C NMR spectra.

*4.7. Biotransformation Assay of **1** in A. nidulans*

For the biotransformation assay in *A. nidulans*, compound **1** was dissolved in a minimal amout of DMSO and then added into CD-ST liquid media at a final concentration of 100 μM. The strain *A. nidulans* A1145 was inoculated on the prepared medium and grown for 4 days at 28 °C 180 rpm. Meanwhile, a CD-ST medium with equal amount of compound **1** was prepared without any strain, and was cultivated under the same condition as the control. The cultures were then extracted by ethyl acetate and the organic phase was dried, dissolved in methanol and detected by LC-MS.

4.8. Assay of Antimicrobial and Antioxidant Activities

Antimicrobial activities of compounds **1** and **2** were tested by the micro broth dilution method, as mentioned in the previous study [43]. The microorganism suspension (198 μL, 106 cfu/mL) in MH medium (Casein Acid Hydrolysate 17.5 g/L, beef extract 2 g/L, soluble starch 1.5 g/L, pH 7.3) was added to each well of 96-well plates. Solutions (6 mM) of the compounds and positive drugs were made up in DMSO and dispensed into 96-well plates to provide 16 concentrations in the range of 30–0.02 μM. Incubated at 28 °C for 9 h (*Candida albicans* in 37 °C for 12 h), the growth inhibition was recorded.

In the DPPH scavenging assay, samples to be tested were dissolved in MeOH and the solution (160 μL) was dispensed into wells of a 96-well microtiter tray. Forty microliters of the DPPH solution in MeOH were added to each well. The mixture was shaken and left to stand for 30 min. After the reaction, absorbance was measured at 510 nm, and the percent inhibition was calculated. ED_{50} values denoted the concentration of sample required to scavenge 50% of the DPPH free radicals [44].

5. Conclusions

In summary, guided by the bioinformatic analysis of *S. lamellicola* HDN13-430, we successfully activated the silent nrPKS gene *SlPKS4* in *A. nidulans*, and proved its function as an isocoumarin synthase. Furthermore, we demonstrated by feeding experiment that **2** are modification products of **1** by uncharacterized endogenous host enzymes. Based on the DPPH scavenging assay, Compounds **1** and **2** showed moderate radical scavenging activities with ED_{50} values of 67.4 μM and 104.2 μM, respectively, which was the first time for this to be measured. Our results provide one additional example that the products of a heterologous expressed gene can be further converted by host enzymes. Furthermore, this is the first report of nrPKS responsible for the biosynthesis of compound **1**. This is also the first report that the fungal specie *S. lamellicola* could produce isocoumarin derivatives, which expands the production methods of isocoumarins compounds.

Supplementary Materials: The following supporting information can be downloaded at: https://www.mdpi.com/article/10.3390/md21090490/s1, Table S1: The primers used in this study; Table S2: Analysis of gene cluster; Table S3: NMR data of **1** in DMSO-d_6; Table S4: NMR data of **2** in CD_3OD; Figure S1: RT-PCR analysis for gene transcription status. The results showed that all seven tailoring enzymes together with SlPKS4 were totally silent under 6 regular laboratory conditions; Figure S2: AntiSMASH analysis results of the genome of the strain *S. lamellicola* HDN13-430; Figure S3: UV absorptions of compounds **1** and **2**; Figure S4: Maps of the vectors pANU-*SlPKS4*, pANU-*SlPKS4+Sl4004*, pANR-*Sl4001+5+7* and pANP-*Sl4002+3+6*; Figure S5: HPLC full chromatogram of the original *A. nidulans* and the strain harboring SlPKS4. The results prevent the presence of compound **2** in original *A.nidulans*; Figure S6: RT-PCR analysis for gene transcription status. The results showed that all seven tailoring enzymes together with SlPKS4 were expressed properly; Figure S7: Comparative analysis between the gene cluster and other isocoumarin BGCs by clinker; Figure S8: ^1H NMR (500 MHz, DMSO-d_6) spectrum of compound **1**; Figure S9: ^{13}C NMR (125 MHz, DMSO-d_6) spectrum of compound **1**; Figure S10: ^1H NMR (500 MHz, CD_3OD) spectrum of compound **2**; Figure S11: ^{13}C NMR (125 MHz, CD_3OD) spectrum of compound **2**. Gene Sequence of *SlPKS4*.

Author Contributions: The contributions of the respective authors are as follows: J.Y. drafted the work, constructed the plasmids and performed the fermentation, extraction, as well as the isolation. X.L., C.M. and Y.Z. were involved in the fermentation, extraction, elucidation of the constituents and bioinformatic analysis. J.Y., X.L. and C.L. performed the biological evaluations. Q.C. and G.Z. contributed to checking and confirming all the procedures of the isolation and identification. T.Z. and D.L. designed the study, supervised the laboratory work, and contributed to the critical reading of the manuscript, and was also involved in the structural determination and bioactivity elucidation. All authors have read and agreed to the published version of the manuscript.

Funding: This research was funded by the National Natural Science Foundation of China (U1906212, 81991522), Qingdao Marine Science and Technology Center (2022QNLM030003), the Fundamental Research Funds for the Central Universities (202172002 and 202262015), Hainan Provincial Joint Project of Sanya Yazhou Bay Science and Technology City (2021CXLH0012), Taishan Scholar Youth Expert Program in Shandong Province (tsqn 201812021, tsqn 202103153), Major Basic Research Programs of Natural Science Foundation of Shandong Province (ZR2021ZD28).

Data Availability Statement: The gene sequence of *SlPKS4* was uploaded to GeneBank (Accession Number: OR519897).

Acknowledgments: The authors recognize material contributions from Yi Tang (University of California, Los Angeles) for providing plasmids.

Conflicts of Interest: The authors declare no conflict of interest.

References

1. Butler, M.S.; Robertson, A.A.B.; Cooper, M.A. Natural Product and Natural Product Derived Drugs in Clinical Trials. *Nat. Prod. Rep.* **2014**, *31*, 1612–1661. [CrossRef] [PubMed]
2. Shin, H.J. Natural Products from Marine Fungi. *Mar. Drugs* **2020**, *18*, 230. [CrossRef] [PubMed]
3. Zhang, J.J.; Tang, X.; Moore, B.S. Genetic Platforms for Heterologous Expression of Microbial Natural Products. *Nat. Prod. Rep.* **2019**, *36*, 1313–1332. [CrossRef] [PubMed]

4. Wiemann, P.; Keller, N.P. Strategies for Mining Fungal Natural Products. *J. Ind. Microbiol. Biot.* **2014**, *41*, 301–313. [CrossRef]
5. Zou, R.; Chen, B.; Sun, J.; Guo, Y.-W.; Xu, B. Recent Advances of Activation Techniques-Based Discovery of New Compounds from Marine Fungi. *Fitoterapia* **2023**, *167*, 105503–105516. [CrossRef]
6. Yu, J.; Han, H.; Zhang, X.; Ma, C.; Sun, C.; Che, Q.; Gu, Q.; Zhu, T.; Zhang, G.; Li, D. Discovery of Two New Sorbicillinoids by Overexpression of the Global Regulator LaeA in a Marine-Derived Fungus *Penicillium dipodomyis* YJ-11. *Mar. Drugs* **2019**, *17*, 446. [CrossRef]
7. Yao, Y.; Wang, W.; Shi, W.; Yan, R.; Zhang, J.; Wei, G.; Liu, L.; Che, Y.; An, C.; Gao, S.-S. Overproduction of Medicinal Ergot Alkaloids Based on a Fungal Platform. *Metab. Eng.* **2022**, *69*, 198–208. [CrossRef]
8. Huo, L.; Hug, J.J.; Fu, C.; Bian, X.; Zhang, Y.; Müller, R. Heterologous Expression of Bacterial Natural Product Biosynthetic Pathways. *Nat. Prod. Rep.* **2019**, *36*, 1412–1436. [CrossRef]
9. Chiang, C.-Y.; Ohashi, M.; Tang, Y. Deciphering Chemical Logic of Fungal Natural Product Biosynthesis through Heterologous Expression and Genome Mining. *Nat. Prod. Rep.* **2023**, *40*, 89–127. [CrossRef]
10. Ma, C.; Zhang, K.; Zhang, X.; Liu, G.; Zhu, T.; Che, Q.; Li, D.; Zhang, G. Heterologous Expression and Metabolic Engineering Tools for Improving Terpenoids Production. *Curr. Opin. Biotechol.* **2021**, *69*, 281–289. [CrossRef]
11. Guo, W.; Wang, S.; Li, N.; Li, F.; Zhu, T.; Gu, Q.; Guo, P.; Li, D. Saroclides A and B, Cyclic Depsipeptides from the Mangrove-Derived Fungus *Sarocladium kiliense* HDN11-112. *J. Nat. Prod.* **2018**, *81*, 1050–1054. [CrossRef]
12. Shin, T.S.; Yu, N.H.; Lee, J.; Choi, G.J.; Kim, J.-C.; Shin, C.S. Development of a Biofungicide Using a Mycoparasitic Fungus *Simplicillium lamellicola* BCP and Its Control Efficacy against Gray Mold Diseases of Tomato and Ginseng. *Plant Pathol. J.* **2017**, *33*, 337–344. [CrossRef]
13. Le Dang, Q.; Shin, T.S.; Park, M.S.; Choi, Y.H.; Choi, G.J.; Jang, K.S.; Kim, I.S.; Kim, J.-C. Antimicrobial Activities of Novel Mannosyl Lipids Isolated from the Biocontrol Fungus *Simplicillium lamellicola* BCP against Phytopathogenic Bacteria. *J. Agric. Food Chem.* **2014**, *62*, 3363–3370. [CrossRef]
14. Abaya, A.; Serajazari, M.; Hsiang, T. Control of Fusarium Head Blight Using the Endophytic Fungus, *Simplicillium lamellicola*, and Its Effect on the Growth of *Triticum Aestivum*. *Biol. Control* **2021**, *160*, 104684–104693. [CrossRef]
15. Ahuja, M.; Chiang, Y.-M.; Chang, S.-L.; Praseuth, M.B.; Entwistle, R.; Sanchez, J.F.; Lo, H.-C.; Yeh, H.-H.; Oakley, B.R.; Wang, C.C.C. Illuminating the Diversity of Aromatic Polyketide Synthases in *Aspergillus nidulans*. *J. Am. Chem. Soc.* **2012**, *134*, 8212–8221. [CrossRef]
16. Matsuda, Y.; Wakimoto, T.; Mori, T.; Awakawa, T.; Abe, I. Complete Biosynthetic Pathway of Anditomin: Nature's Sophisticated Synthetic Route to a Complex Fungal Meroterpenoid. *J. Am. Chem. Soc.* **2014**, *136*, 15326–15336. [CrossRef] [PubMed]
17. Prompanya, C.; Dethoup, T.; Bessa, L.; Pinto, M.; Gales, L.; Costa, P.; Silva, A.; Kijjoa, A. New Isocoumarin Derivatives and Meroterpenoids from the Marine Sponge-Associated Fungus *Aspergillus similanensis* sp. Nov. KUFA 0013. *Mar. Drugs* **2014**, *12*, 5160–5173. [CrossRef]
18. Pérez Hemphill, C.F.; Daletos, G.; Liu, Z.; Lin, W.; Proksch, P. Polyketides from the Mangrove-Derived Fungal Endophyte *Pestalotiopsis clavispora*. *Tetrahedron Lett.* **2016**, *57*, 2078–2083. [CrossRef]
19. Carter, R.H.; Garson, M.J.; Hill, R.A.; Staunton, J.; Sunter, D.C. The Synthesis of Indan-1-ones and Isocoumarins. *J. Chem. Soc. Perkin Trans. 1* **1981**, *9*, 471–479. [CrossRef]
20. Lewis, C.N.; Staunton, J.; Sunter, D.C. Biosynthesis of Canescin, a Metabolite of *Aspergillus malignus*: Incorporation of Methionine, Acetate, Succinate, and Isocoumarin Precursors, Labelled with Deuterium and Carbon-13. *J. Chem. Soc. Perkin Trans. 1* **1988**, *8*, 747–754. [CrossRef]
21. Lewis, C.N.; Staunton, J.; Sunter, D.C. A ^{2}H and ^{13}C N.M.R. Study of the Biosynthesis of the Polyketide Isocoumarin Residue of Canescin in *Aspergillus malignus* from [1,2-$^{13}C_2$]-, and [1-^{13}C, 2-$^{2}H_3$]Acetates, [*Me*-^{13}C, $^{2}H_3$]Methionine, 6,8-Dihydroxy-3,7-Dimethylisocoumarin, and 6,8-Dihydroxy-7-Formyl-3-Methylisocoumarin. *J. Chem. Soc. Chem. Commun.* **1986**, *3*, 58–60. [CrossRef]
22. Saeed, A. Isocoumarins, Miraculous Natural Products Blessed with Diverse Pharmacological Activities. *Eur. J. Med. Chem.* **2016**, *116*, 290–317. [CrossRef] [PubMed]
23. Xiang, P.; Ludwig-Radtke, L.; Yin, W.-B.; Li, S.-M. Isocoumarin Formation by Heterologous Gene Expression and Modification by Host Enzymes. *Org. Biomol. Chem.* **2020**, *18*, 4946–4948. [CrossRef]
24. Zhou, J.; Li, G.; Deng, Q.; Zheng, D.; Yang, X.; Xu, J. Cytotoxic Constituents from the Mangrove Endophytic *Pestalotiopsis* Sp. Induce G_0/G_1 Cell Cycle Arrest and Apoptosis in Human Cancer Cells. *Nat. Prod. Res.* **2018**, *32*, 2968–2972. [CrossRef] [PubMed]
25. Ma, L.-Y.; Liu, D.-S.; Li, D.-G.; Huang, Y.-L.; Kang, H.-H.; Wang, C.-H.; Liu, W.-Z. Pyran Rings Containing Polyketides from *Penicillium raistrickii*. *Mar. Drugs* **2016**, *15*, 2. [CrossRef]
26. Shabir, G.; Saeed, A.; El-Seedi, H.R. Natural Isocoumarins: Structural Styles and Biological Activities, the Revelations Carry on.... *Phytochemistry* **2021**, *181*, 112568–112590. [CrossRef]
27. Pal, S.; Chatare, V.; Pal, M. Isocoumarin and Its Derivatives: An Overview on Their Synthesis and Applications. *Curr. Org. Chem.* **2011**, *15*, 782–800. [CrossRef]
28. Song, R.-Y.; Wang, X.-B.; Yin, G.-P.; Liu, R.-H.; Kong, L.-Y.; Yang, M.-H. Isocoumarin Derivatives from the Endophytic Fungus, *Pestalotiopsis* sp. *Fitoterapia* **2017**, *122*, 115–118. [CrossRef]
29. Saikia, P.; Gogoi, S. Isocoumarins: General Aspects and Recent Advances in Their Synthesis. *Adv. Synth. Catal.* **2018**, *360*, 2063–2075. [CrossRef]

30. Sarmah, M.; Chutia, K.; Dutta, D.; Gogoi, P. Overview of Coumarin-Fused-Coumarins: Synthesis, Photophysical Properties and Their Applications. *Org. Biomol. Chem.* **2022**, *20*, 55–72. [CrossRef]
31. Noor, A.O.; Almasri, D.M.; Bagalagel, A.A.; Abdallah, H.M.; Mohamed, S.G.A.; Mohamed, G.A.; Ibrahim, S.R.M. Naturally Occurring Isocoumarins Derivatives from Endophytic Fungi: Sources, Isolation, Structural Characterization, Biosynthesis, and Biological Activities. *Molecules* **2020**, *25*, 395. [CrossRef]
32. Cai, Y.; Rao, L.; Zou, Y. Genome Mining Discovery of a C_4 -Alkylated Dihydroisocoumarin Pathway in Fungi. *Org. Lett.* **2021**, *23*, 2337–2341. [CrossRef] [PubMed]
33. Tammam, M.A.; Gamal El-Din, M.I.; Abood, A.; El-Demerdash, A. Recent Advances in the Discovery, Biosynthesis, and Therapeutic Potential of Isocoumarins Derived from Fungi: A Comprehensive Update. *RSC Adv.* **2023**, *13*, 8049–8089. [CrossRef] [PubMed]
34. Atanasoff-Kardjalieff, A.K.; Seidl, B.; Steinert, K.; Daniliuc, C.G.; Schuhmacher, R.; Humpf, H.; Kalinina, S.; Studt-Reinhold, L. Biosynthesis of the Isocoumarin Derivatives Fusamarins Is Mediated by the PKS8 Gene Cluster in *Fusarium*. *ChemBioChem* **2023**, *24*, e202200342. [CrossRef] [PubMed]
35. Kahlert, L.; Villanueva, M.; Cox, R.J.; Skellam, E.J. Biosynthesis of 6-Hydroxymellein Requires a Collaborating Polyketide Synthase-like Enzyme. *Angew. Chem. Int. Ed.* **2021**, *60*, 11423–11429. [CrossRef]
36. Lind, A.L.; Wisecaver, J.H.; Lameiras, C.; Wiemann, P.; Palmer, J.M.; Keller, N.P.; Rodrigues, F.; Goldman, G.H.; Rokas, A. Drivers of Genetic Diversity in Secondary Metabolic Gene Clusters within a Fungal Species. *PLoS Biol.* **2017**, *15*, e2003583. [CrossRef]
37. Wiemann, P.; Guo, C.-J.; Palmer, J.M.; Sekonyela, R.; Wang, C.C.C.; Keller, N.P. Prototype of an Intertwined Secondary-Metabolite Supercluster. *Proc. Natl. Acad. Sci. USA* **2013**, *110*, 17065–17070. [CrossRef]
38. Stroe, M.C.; Netzker, T.; Scherlach, K.; Krüger, T.; Hertweck, C.; Valiante, V.; Brakhage, A.A. Targeted Induction of a Silent Fungal Gene Cluster Encoding the Bacteria-Specific Germination Inhibitor Fumigermin. *eLife* **2020**, *9*, e52541. [CrossRef]
39. Blin, K.; Shaw, S.; Augustijn, H.E.; Reitz, Z.L.; Biermann, F.; Alanjary, M.; Fetter, A.; Terlouw, B.R.; Metcalf, W.W.; Helfrich, E.J.N.; et al. AntiSMASH 7.0: New and Improved Predictions for Detection, Regulation, Chemical Structures and Visualisation. *NAR* **2023**, *51*, W46–W50. [CrossRef]
40. Kumar, S.; Stecher, G.; Tamura, K. MEGA7: Molecular Evolutionary Genetics Analysis Version 7.0 for Bigger Datasets. *Mol. Biol. Evol.* **2016**, *33*, 1870–1874. [CrossRef]
41. Mitchell, A.L.; Attwood, T.K.; Babbitt, P.C.; Blum, M.; Bork, P.; Bridge, A.; Brown, S.D.; Chang, H.-Y.; El-Gebali, S.; Fraser, M.I.; et al. InterPro in 2019: Improving Coverage, Classification and Access to Protein Sequence Annotations. *Nucleic Acids Res.* **2019**, *47*, D351–D360. [CrossRef]
42. Gilchrist, C.L.M.; Chooi, Y.-H. Clinker & Clustermap.Js: Automatic Generation of Gene Cluster Comparison Figures. *Bioinformatics* **2021**, *37*, 2473–2475. [CrossRef] [PubMed]
43. Yu, G.; Wu, G.; Sun, Z.; Zhang, X.; Che, Q.; Gu, Q.; Zhu, T.; Li, D.; Zhang, G. Cytotoxic Tetrahydroxanthone Dimers from the Mangrove-Associated Fungus *Aspergillus versicolor* HDN1009. *Mar. Drugs* **2018**, *16*, 335. [CrossRef] [PubMed]
44. Zhang, Z.; He, X.; Che, Q.; Zhang, G.; Zhu, T.; Gu, Q.; Li, D. Sorbicillasins A–B and Scirpyrone K from a Deep-Sea-Derived Fungus, *Phialocephala* sp. FL30r. *Mar. Drugs* **2018**, *16*, 245. [CrossRef] [PubMed]

MDPI
St. Alban-Anlage 66
4052 Basel
Switzerland
www.mdpi.com

Marine Drugs Editorial Office
E-mail: marinedrugs@mdpi.com
www.mdpi.com/journal/marinedrugs

www.ingramcontent.com/pod-product-compliance
Lightning Source LLC
LaVergne TN
LVHW070703100526
838202LV00013B/1022
* 9 7 8 3 0 3 6 5 9 1 6 6 7 *